THEORY OF MOLECULAR EXCITONS

THEORY OF MOLECULAR EXCITONS

A. S. Davydov
Kiev State University
Kiev, USSR

Translated from Russian by
Stephen B. Dresner

PLENUM PRESS • NEW YORK-LONDON • 1971

Aleksandr Sergeevich Davydov was graduated from Moscow State University in 1939 and defended his doctoral dissertation in 1943. A member of the Academy of Sciences of the Ukrainian SSR, he is director of the Department of Nuclear Theory of the Institute of Theoretical Physics, director of the Department of Theoretical Physics of the Institute of Physics (both of the Academy), and Professor of Theoretical Physics at Kiev State University. Working in the area of the theory of excitons in molecular crystals and the theory of the atomic nucleus, in 1948 he predicted the phenomenon that is known as "Davydov splitting." Davydov has investigated theoretically the collective excited states of nonspherical nuclei (Davydov-Filippov model and Davydov-Chaban model).

The original Russian text, published by Nauka Press in Moscow in 1968, has been corrected by the author for this edition. The English translation is published under an agreement with Mezhdunarodnaya Kniga, the Soviet book export agency.

Александр Сергеевич Давыдов

ТЕОРИЯ МОЛЕКУЛЯРНЫХ ЭКСИТОНОВ

TEORIYA MOLEKULYARNYKH EKSITONOV

THEORY OF MOLECULAR EXCITONS

Library of Congress Catalog Card Number 72-75767

SBN 306-30440-6

A Division of Plenum Publishing Corporation
227 West 17th Street, New York, N.Y. 10011

United Kingdom edition published by Plenum Press, London
A Division of Plenum Publishing Company, Ltd.
Donington House, 30 Norfolk Street, London W.C. 2, England

Printed in the United States of America

Preface to American Edition

My first book on the theory of excitons was published in Kiev in 1951 and in English translation (McGraw-Hill) in 1962. It reflected only the initial stage of development of the theory of excitons. The present monograph is entirely new.

It is my hope that this American edition will aid English-speaking scientists in becoming familiar with the latest stage of development of the theory. This edition also includes three recent original papers.

I take this opportunity to express my sincere thanks to the translator.

Kiev
September, 1969

A. S. Davydov

Foreword

In recent years, great advances have been made in the theoretical and experimental study of the optical properties of solids. These advances have to a considerable extent resulted from the wide use of the concept of the exciton, which was introduced as early as in 1931 by Ya. I. Frenkel.

Interest in the theory of exciton states in solids has increased significantly with the improvement of experimental techniques with polarized light at low temperatures and of methods of growing single crystals, with the extensive use of luminescent crystals, and with the establishment of the great role of energy-migration processes in solids and biological systems. Theoretical investigations of exciton states have made particularly great progress recently, because of the use of methods of quantum electrodynamics in the theory of solids.

Unfortunately, the progress in experimental and theoretical research on exciton states in solids has not yet been fully reflected in monographs. The author's monograph "Theory of Light Absorption in Molecular Crystals" was published in 1951, and an English translation, "Theory of Molecular Excitons," was published in 1962, but it represents only the initial stage of development of the theory. R. S. Knox, in his "Theory of Excitons," which was published in 1963 (the Russian edition appeared in 1966), succeeded in discussing in a small book the basic physical concepts underlying the theory of excitons, the limit models of excitons (Frenkel excitons and Wannier excitons), and a number of experimental studies whose interpretation required the concept of the exciton.

A recent book by Agranovich and Ginzburg, "Crystal Optics with Allowance for Spatial Dispersion, and the Theory of Excitons," presents a phenomenological investigation of the relationship between the theory of excitons and crystal optics. But this book does not discuss completely enough the methods for calculating the energy bands of excitons, nor does it examine the role of or methods for calculating the interactions of excitons with crystal-lattice vibrations. Such problems are discussed on the basis of microtheory in the present monograph.

Exciton states (currentless collective electronic excitations) appear in solids of various kinds: in molecular, ionic, and semiconductor crystals, and in complicated organic compounds. Excitons have begun to be mentioned even in analyses of the properties of normal and superconducting metals. The excitons in different solids have many features in common, but the methods for their experimental and theoretical study differ substantially from one type of solid to another. Certain problems of the theory of excitons in molecular crystals will be examined in this monograph, which is not a survey of the literature on the theory of excitons. The primary attention will be given to three problems in whose development the author participated. The references, which are given at the end of the book, are not meant to be complete.

The theory of excitons is progressing vigorously at the present time. This growth is occurring in a number of directions and through various methods. Some authors have expressed dissimilar and, in some cases, conflicting opinions about the nature of the observed phenomena. This monograph naturally reflects the point of view of its author. A great deal of attention has been given to the mathematical apparatus of the theory. Experimental research is discussed only to illustrate certain conclusions.

The formulas in each chapter are numbered by two figures separated by a dot. The first figure indicates the chapter section and the second indicates the formula number. References to formulas from other chapters contain a Roman numeral indicating the chapter number.

The author hopes that this monograph will further the development of the theory and the establishment of concepts that will best reflect the real phenomena, and that it will be useful to scientists and students specializing in solid state physics.

The author thanks V. M. Agranovich, A. F. Lubchenko, V. A. Onishchuk, É. N. Myasnikov, and B. M. Nitsovich for assistance in developing some of the topics in the monograph, and Yu. I. Sirotin, who read the manuscript and made some valuable comments.

A. S. Davydov

Contents

Chapter I

Fundamentals of the Phenomenological Theory of Electromagnetic Waves in Dielectric Media

1. The Macroscopic Body and the External Electromagnetic Field

When densities and internal energies are not very high (before photon–electron interconversions occur), any macroscopic body represents a universe of interacting electrons and atomic nuclei. The properties of ordinary macroscopic bodies, whose average number of particles per unit volume does not exceed 10^{30} cm^{-3}, are chiefly determined by the electromagnetic interactions of their component particles. On the cosmic scale (earth, sun, stars, etc.), gravitational interactions play a considerable role. In extremely dense media (for example, white dwarfs), the macroscopic properties are also very greatly dependent upon nuclear forces.

Since the speed of transmission of an interaction between particles is finite, the state of a system is not determined by the location of its particles at a given moment but is a function of the entire previous history of their motion. In order to take into account the time lag, the macroscopic body can be considered a system of electrons and nuclei that interact through an electromagnetic field. Such a field is an integral part of any body. In some cases, when the effective radius of interaction between charged particles is less than the wavelength of the electromagnetic field, it is sufficient to take into account the Coulomb interaction. The energy states of atoms, molecules, and certain solids are usually calculated with this approximation.

We shall be concerned with the properties of macroscopic bodies with respect to external electromagnetic fields (fields of charged particles, light waves). To study theoretically the interaction of a solid with an external electromagnetic field, we must examine a system consisting of an external field and a macroscopic body. The division of the complete system into an external field and a body is to a certain extent arbitrary. This division is well justified only when the interaction is weak and perturbation theory can be used. This occurs in practice when one studies the interaction of an electromagnetic field with a rarefied gas. In general, the interaction of electromagnetic radiation with solids and liquids cannot be considered weak. This is easy to see, if we recall that the phase velocity of electromagnetic-wave propagation in such media can differ by a factor of 1.5-2 from their velocity in a vacuum.

Although the interaction of an external electromagnetic field with a macroscopic body is in general not weak, division of the complete system into two subsystems (body and field) is often mathematically convenient, since it allows the penetration of radiation through matter to be studied in two stages.

To illustrate this, let us consider a simple quantum-mechanical example. Let determination of the steady states of a complete system consisting of two subsystems A and B reduce to determination of the eigenvalues and eigenfunctions of the Hamiltonian operator

$$H = H_A(r) + H_B(R) + V(r, R), \tag{1.1}$$

where r is the set of coordinates of subsystem A; R the set of coordinates of subsystem B; and V(r, R) the interaction-energy operator of both subsystems. Let us assume that we found beforehand the eigenfunctions and eigenvalues only of the energy operator $H_A(r)$ of subsystem A, i.e., solved the Schrödinger equation

$$[H_A(r) - \varepsilon_n]\,\varphi_n(r) = 0.$$

The functions $\varphi_n(r)$ form a complete orthonormal set (for simplicity, we shall assume that the spectrum of eigenvalues of operator H_A is discrete). Therefore, the solution of the general equation

$$\{H_A(r) + H_B(R) + V(r, R) - E\}\Psi = 0 \tag{1.2}$$

can be sought as

$$\Psi = \sum_n \Phi_n(R)\,\varphi_n(r). \tag{1.3}$$

If we substitute (1.3) into (1.2), we find the system of equations

$$\{H_B(R) - [E - \varepsilon_m]\}\,\Phi_m(R) = -\sum_n V_{nm}(R)\,\Phi_n(R), \tag{1.4}$$

where

$$V_{nm} = \int \varphi_m^*(r)\,V(r, R)\,\varphi_n(r)\,dr.$$

System (1.4) is exact. Inasmuch as this system of equations contains only the variables R of the second subsystem, it is sometimes a simpler task to find its solutions than to find the solutions of (1.2).

Equation (1.2) can be solved in two steps by another method. Let us examine a subsystem with the Hamiltonian operator

$$H'(r, R) = H_A(r) + V(r, R),$$

where the variables R of the second system are considered parameters. Calculation of the eigenfunctions and eigenvalues of $H'(r, R)$ boils down to solving the Schrödinger equation

$$[H'(r, R) - \varepsilon_n(R)]\,\varphi_n(r, R) = 0$$

for fixed R. After having found the set of functions φ_n, we can write the solution of Eq. (1.2) as

$$\Psi = \sum_n \Phi_n(R)\,\varphi_n(r, R).$$

If the interaction operator V(r, R) does not contain derivatives with respect to the coordinates R, then finding the functions $\Phi_n(R)$ comes down to solving the system of equations

$$[E - \varepsilon_m(R)]\,\Phi_m(R) = \sum_n \widetilde{H}_{nm}(R)\,\Phi_n(R),$$

where

$$\tilde{H}_{nm}(R) = \int \varphi_m^*(r, R)\, H_B(R)\, \varphi_n(r, R)\, dr$$

is a new operator.

A modification of the latter of the above methods is often used in theoretical investigation of the penetration of electromagnetic radiation through a homogeneous macroscopic body. A system consisting of a dielectric and an electromagnetic field of a certain frequency is considered. The wavelength, which corresponds to optical frequencies (visible and ultraviolet regions), considerably exceeds the interatomic distance of the solid. An electromagnetic field that corresponds to such waves we shall call a *longwave* or *macroscopic field*. A dielectric may be considered a continuous body with respect to a longwave field. The passage of optical waves through a dielectric is studied in two steps. First of all, a microscopic theory of dielectric polarization and magnetization is developed, i.e., the mean specific electric **P** and magnetic **M** polarizations by the electric **E** and magnetic **H** fields of the wave, which is considered an external perturbation, are calculated. In nonmagnetic dielectrics **M** = 0, and **P**, **E**, and **H** determine the electric-current density, dielectric displacement **D**, and magnetic induction **B** through the relations

$$\mathbf{j} = \frac{\partial \mathbf{P}}{\partial t}, \quad \mathbf{B} = \mathbf{H}, \quad \mathbf{D} = \mathbf{E} + 4\pi\mathbf{P}. \tag{1.5}$$

The second part of the problem reduces to solution of the macroscopic Maxwell equations

$$\left.\begin{aligned} &c\,\mathrm{rot}\,\mathbf{H} = \frac{\partial \mathbf{D}}{\partial t}, \quad c\,\mathrm{rot}\,\mathbf{E} = -\frac{\partial \mathbf{B}}{\partial t}, \\ &\mathrm{div}\,\mathbf{D} = \mathrm{div}\,\mathbf{B} = 0. \end{aligned}\right\} \tag{1.6}$$

Thus, the above method of theoretical investigation of electromagnetic-wave propagation in a nonmagnetic dielectric medium amounts to solution of a self-consistent problem: to determination of the polarization **P** under the influence of a field **E**, which itself is a function of **P**.

Usually, the field in the medium is weak; therefore, the resulting specific polarization is a linear function of the field. In this case, systems (1.5) and (1.6) are linear and homogeneous, so their solutions are linear functions of **E**. In some cases (for example, laser radiation), the radiation density is high enough to make systems (1.5) and (1.6) nonlinear.

In calculating specific polarization **P(E)**, the dielectric may be considered to be a system consisting of a large number of neutral subsystems, which are formed by molecules in molecular crystals or by groups of ions in ionic crystals. The field changes the state of electron motion in the subsystems when the position of the atomic nuclei in them is fixed (adiabatic approximation). The excited states that arise are unsteady. In due course, the electronic-excitation energy is redistributed with respect to all degrees of freedom: it converts to thermal motion of atomic nuclei. The rate of this transition depends upon the relationship between the motions of the electrons and atomic nuclei. This relationship usually increases as the crystal temperature increases. Besides energy redistribution, this relationship can cause the motion of atomic nuclei to interact in part directly with the electric field of the light wave. Both of these processes must be taken into account in calculating **P(E)**.

A feature of the above method is that the electromagnetic field is matched to the dielectric polarization it produces in the second stage of calculation, in solution of the Maxwell equations. Thus, the matching is accomplished after the effect of the motion of atomic nuclei has been taken into account.

There exists a second method for theoretical investigation of electromagnetic waves in dielectrics. In this method, the electric field of the light wave is first of all matched to the dielectric polarization it causes for rigidly fixed atomic nuclei. The electronic subsystem of the dielectric and the electromagnetic field are considered as a whole and the elementary excitations of such a system are studied. Interaction with the motion of the atomic nuclei is incorporated in the second stage of calculation.

Both methods have their advantages and disadvantages.

2. The Dielectric Constant of a Macroscopic Body

Let us consider a nonmagnetic dielectric, i.e., a medium whose magnetic permeability is unity and whose static conductivity is zero. The longwave electromagnetic radiation in such a medium ($\lambda \gg a$) is determined by the electric **E** and magnetic **H** field strengths and by the dielectric-displacement vector

$$\mathbf{D} = \mathbf{E} + 4\pi\mathbf{P}, \tag{2.1}$$

where **P** is the polarization vector for a unit volume. In the fields, which change with time, due to delay effects $\mathbf{P}(\mathbf{r}, t)$ is a function of **E** at point **r** and in its immediate vicinity during all of the preceding time. Therefore, the relationship between the vector **D** and field strength for an unbounded spatially homogeneous medium in linear approximation is expressed by the integral

$$\mathbf{D}(\mathbf{r}, t) = \mathbf{E}(\mathbf{r}, t) + \int_0^\infty d\tau \int_{c\tau \geqslant |\rho|} S(\rho, t)\, E(t-\tau, \mathbf{r}+\rho)\, d^3\rho, \tag{2.2}$$

where $S(\boldsymbol{\rho}, t)$ is a real tensor of rank two whose value is determined by the properties of the medium, particularly by the delay time of processes leading to the establishment of dielectric polarizability. In accordance with the principle of causality and the finite velocity of interaction propagation (the wave front is always propagated at the velocity of light c), integration in (2.2) extends to the region within the minus light cone. The displacement vector $\mathbf{D}(\mathbf{r}, t)$ must not be a function of **E** at very distant times and at very distant points in space; it is necessary, therefore, that the tensor $S(\boldsymbol{\rho}, \tau)$ approach zero fairly rapidly as τ and $\boldsymbol{\rho}$ approach infinity.

One of the problems of the quantum theory of solids is to establish a relationship such as (2.2) or its equivalent [see (2.7)]. The solution of this problem for certain particular cases will be considered in subsequent chapters. Now we shall consider the problem of the penetration of an electromagnetic wave through a dielectric medium for which the tensor $S(\boldsymbol{\rho}, \tau)$ is known.

The electromagnetic field in a nonmagnetic dielectric that is devoid of free electric charges is determined by Maxwell equations

(1.6). Let us assume that the dielectric occupies a large volume in the form of a cube with edge L. If L is sufficiently large, the choice of the boundary conditions on the surface of this volume is not of great importance for studying the spatial properties. For simplicity, let us use as the boundary conditions the conditions of periodicity with a long period L. As a basis, let us consider the set of functions orthonormal in this volume

$$\varphi_{\mathbf{Q}}(\mathbf{r}) = L^{-3/2} \exp\{i\mathbf{Qr}\},$$

where the components of the real wave vector **Q** are determined by

$$Q_i = L^{-1} 2\pi\nu_i, \qquad i = 1, 2, 3, \ \nu_i = 0, \pm 1, \pm 2, \ldots$$

The solutions of system (1.6) can be sought as

$$\left.\begin{aligned} \mathbf{D} &= L^{-3/2} \sum_{\mathbf{Q}} \mathbf{D}(\mathbf{Q}) \exp\{i(\mathbf{kr} - \omega t)\}, \\ \mathbf{E} &= L^{-3/2} \sum_{\mathbf{Q}} \mathbf{E}(\mathbf{Q}) \exp\{i(\mathbf{kr} - \omega t)\}, \\ \mathbf{H} &= L^{-3/2} \sum_{\mathbf{Q}} \mathbf{H}(\mathbf{Q}) \exp\{i(\mathbf{kr} - \omega t)\} \end{aligned}\right\} \tag{2.3}$$

for the two possible types of electromagnetic waves in crystals.

A. Spatially homogeneous, or normal, electromagnetic waves in crystals. In this case, in (2.3) the wave vectors are real, since $\mathbf{k} = \mathbf{Q}$, while the frequencies ω are generally complex functions of the real vector, i.e.,

$$\omega = \omega(\mathbf{Q}) - \frac{i}{2}\gamma(\mathbf{Q}), \tag{2.4}$$

where $\omega(\mathbf{Q})$ and $\gamma(\mathbf{Q})$ are real functions of the wave vector. The amplitudes of normal electromagnetic waves are independent of the space coordinates but vary with time as

$$\exp\left\{-\frac{1}{2}\gamma(\mathbf{Q})t\right\}.$$

In an infinite volume at fixed time, homogeneous waves can be normalized to delta functions.

B. Spatially inhomogeneous electromagnetic waves in crystals. In this case, in (2.3) frequency ω is real and is determined by the external sources, while the wave vector **k** is in general complex, i.e.,

$$\mathbf{k} = \mathbf{Q}(\omega) + i\mathbf{\Lambda}(\omega), \tag{2.5}$$

where $\mathbf{Q}(\omega)$ and $\Lambda(\omega)$ are real functions of frequency ω. Spatially inhomogeneous waves cannot be normalized in an infinite crystal.

When solving Eqs. (1.6) in the form of (spatially homogeneous) normal waves (case A), the problem reduces to calculation of the law of their dispersion, i.e., their frequency $\omega(\mathbf{Q})$ and attenuation $\gamma(\mathbf{Q})$ must be calculated for each fixed value of the real wave vector **Q**.

The frequency of spatially inhomogeneous electromagnetic waves (case B) is determined by the external sources. In this case, therefore, solution of Eqs. (1.6) comes down to calculation of the vectors

$$\mathbf{Q} = \mathbf{Q}(\omega) \quad \text{and} \quad \mathbf{\Lambda} = \mathbf{\Lambda}(\omega) \tag{2.6}$$

for each fixed real frequency ω.

Henceforth, we shall be chiefly concerned with the inhomogeneous electromagnetic waves that are generated in a crystal when light of a particular frequency penetrates it.

The interaction of longwave electromagnetic radiation with matter is described by the macroscopic dielectric constant $\varepsilon(\mathbf{k}, \omega)$, which defines the linear relationship

$$\mathbf{D}(\mathbf{Q}) = \varepsilon(\mathbf{k}, \omega)\,\mathbf{E}(\mathbf{Q}) \tag{2.7}$$

between the vectors $\mathbf{D}(\mathbf{Q})$ and $\mathbf{E}(\mathbf{Q})$ in (2.3) (providing $Qa \ll 1$). If we substitute expressions (2.3) and (2.7) into Eq. (2.2), we obtain an explicit expression of the dielectric-constant tensor in terms of the tensor $S(\boldsymbol{\rho}, \tau)$:

$$\varepsilon(\omega, \mathbf{k}) = 1 + \int_0^\infty d\tau \int_{c\tau \geqslant |\varrho|} d^3\rho\, S(\rho, \tau)\, e^{-\Lambda\rho} \exp\{i[\omega\tau + \mathbf{Q}\rho]\}. \tag{2.8}$$

The tensor $\varepsilon(\mathbf{k}, \omega)$ is an analytic function of the complex vector **k** and of the complex variable ω, which is situated in the upper half-plane (i.e., at $\operatorname{Im}\omega \geq 0$) [1, 2].

In this case, the function $\varepsilon(\mathbf{k}, \omega)$ has neither poles nor zeros in the upper half-plane of the complex variable ω. If we take advantage of this property, we can calculate the tensor $\varepsilon(\mathbf{k}, \omega)$ for real ω and $\mathbf{k} = \mathbf{Q}$ and continue it analytically to the region of complex ω (at Im $\omega \geq 0$) or of complex $\mathbf{k}$.

In a medium without free electric charges, which is described by Maxwell equations (1.6), a given real frequency ω corresponds to specific functions (2.6), i.e., the values ω, $\mathbf{Q}$, and $\mathbf{\Lambda}$ in (2.8) are related by (2.6). In the presence of free electric charges, ω and $\mathbf{Q}$ can take independent values. For example, the field of an electric charge moving in a medium with velocity $\mathbf{v}$ is characterized by frequency $\omega = \mathbf{vQ}$. For a given $\mathbf{Q}$, therefore, frequency ω can take various values proportional to velocity.

The dependence of the dielectric-constant tensor upon frequency ω is called the *time dispersion*, and its dependence upon $\mathbf{Q}$ is called the *spatial dispersion*.[1] Spatial dispersion is a result of the nonlocal nature of the relationship (2.2) between the displacement and electric vectors: the displacement at point $\mathbf{r}$ is determined by the electric-field strength not only at point $\mathbf{r}$ but also in its vicinity. The time (or frequency) dispersion is determined by the electric-polarization transition time.

In some cases, the tensor $S(\boldsymbol{\rho}, \tau)$ reduces to a product of the form

$$S(\rho, \tau) = S(\tau)\,\delta(\rho).$$

Then (2.8) is converted to the simple form

$$\varepsilon(\omega) = 1 + \int_0^\infty d\tau\, S(\tau) \exp(i\omega\tau). \tag{2.8a}$$

In this case, therefore, the dielectric-constant tensor is a function only of frequency (spatial dispersion is absent). In general, the tensor $\varepsilon(\omega, \mathbf{k})$ is complex, satisfying, because the tensor $S(\boldsymbol{\rho}, \tau)$ is real, the equation

$$\varepsilon(\omega, \mathbf{k}) = \varepsilon^*(-\omega, -\mathbf{k}^*).$$

Time dispersion plays a large role at frequencies close to the resonant frequencies (ω_i) of the medium, since the dependence of the tensor $\varepsilon(\mathbf{Q}, \omega)$ upon ω is determined by terms proportional to

[1] A systematic discussion of crystal optics taking into account spatial dispersion has been presented by Agranovich and Ginzburg [1].

the ratio $\omega/(\omega_i - \omega)$. The role of spatial dispersion in optically inactive (nongyrotropic) crystals is usually very small, since it is a function of the small parameter

$$p = \left(\frac{nR\omega}{c}\right)^2,$$

where R are the linear dimensions of the region encompassed by excitation and n is the refractive index of the medium at frequency ω. When absorption is present, the refractive index rarely exceeds 10; therefore, at optical frequencies, $p \approx 10^{-2}$-10^{-4}.

If we substitute expressions (2.3) into Eqs. (1.6), we find a system of equations for each wave vector **Q**:

$$\mathbf{kD}(\mathbf{Q}) = 0, \qquad \mathbf{kH}(\mathbf{Q}) = 0,$$
$$[\mathbf{k}\times\mathbf{E}(\mathbf{Q})] = \frac{\omega}{c}\mathbf{H}(\mathbf{Q}), \quad [\mathbf{k}\times\mathbf{H}(\mathbf{Q})] = -\frac{\omega}{c}\mathbf{D}(\mathbf{Q}).$$

If we eliminate the magnetic-field strength from the latter two equations, we obtain

$$\frac{\omega^2}{c^2}\mathbf{D}(\mathbf{Q}) = \mathbf{k}^2\mathbf{E}(\mathbf{Q}) - \mathbf{k}(\mathbf{kE}(\mathbf{Q})). \tag{2.9}$$

If we substitute expression (2.7) into Eq. (2.9), we obtain a homogeneous equation for the electric-field strength[2]:

$$\left\{\frac{\omega^2}{c^2}\varepsilon_{jl}(\omega, \mathbf{k}) + k_j k_l - \delta_{jl}\mathbf{k}^2\right\} E_l(\mathbf{Q}) = 0.$$

The condition for nontrivial solvability of this equation boils down to the dispersion equation

$$\det\left\{\frac{\omega^2}{c^2}\varepsilon_{jl}(\omega, \mathbf{k}) + k_j k_l - \delta_{jl}\mathbf{k}^2\right\} = 0. \tag{2.10}$$

If in the dispersion equation with known tensor $\varepsilon(\omega, \mathbf{k})$ we let $\mathbf{k} = \mathbf{Q}$, then when $\varepsilon(\omega, \mathbf{k})$ is complex the equation determines the dispersion and time attenuation of the corresponding normal electromagnetic waves with wave vector **Q** (case A). But for spatially inhomogeneous electromagnetic waves (case B), dispersion equation (2.10) enables us to determine $\mathbf{k} = \mathbf{Q}(\omega) + i\mathbf{\Lambda}(\omega)$ for each fixed real frequency. Here, the vector $\mathbf{\Lambda}(\omega)$ determines the decrease

[2]Here and throughout the book, summation is over the twice-encountered tensor indices.

in amplitude of the wave as it is propagated. Naturally, when the tensor $\varepsilon(\omega, \mathbf{k})$ is real and positive, both types of waves are identical, since $\mathbf{\Lambda}(\omega) = 0$ and $\gamma(\mathbf{Q}) = 0$. In this case, the dispersion equation determines either the function $\omega = \omega(\mathbf{Q})$ or the inverse function $\mathbf{Q} = \mathbf{Q}(\omega)$, depending upon whether the wave vector or the frequency is fixed.

If for some ω and $\mathbf{Q}$ the tensor $\varepsilon(\omega, \mathbf{k})$ is real and negative, then spatially homogeneous (normal) waves with corresponding ω and $\mathbf{Q}$ will be impossible. But solutions of type B, with the proper analytic continuation of $\varepsilon(\omega, \mathbf{k})$, are possible in this case. They will represent the spatially attenuated waves (in the absence of true absorption).

In some crystals in the presence of absorption (complex tensor $\varepsilon(\omega, \mathbf{k})$), the senses of vectors $\mathbf{Q}(\omega)$ and $\mathbf{\Lambda}(\omega)$ may not coincide. We shall, however, consider only the case of "homogeneous" plane waves, in which these vectors are parallel. In inhomogeneous plane waves, the planes of constant displacement are perpendicular to the vector $\mathbf{Q}$, and we can write

$$\mathbf{k} = \mathbf{Q} + i\mathbf{\Lambda} = \frac{\omega}{c}\mathbf{s}(n + i\varkappa), \qquad (2.11)$$

where $\mathbf{s}$ is a unit vector, and n and $\varkappa$ are the refractive index and the attenuation factor[3] of a homogeneous plane wave of frequency ω propagated along $\mathbf{s}$ and with a specific orientation of magnetic-field strength.

If in optically isotropic crystals at certain frequencies and senses of $\mathbf{s}$

$$\varepsilon(\omega, \mathbf{k}) = 0, \qquad (2.12)$$

then, according to (2.7) and (2.9), the displacement and electric-field strength must satisfy the equations

$$\mathbf{D}(\mathbf{Q}) = 0, \quad k^2\mathbf{E}(\mathbf{Q}) = \mathbf{k}(\mathbf{k}\mathbf{E}(\mathbf{Q})). \qquad (2.13)$$

Such electromagnetic waves are called longitudinal, since Eq. (2.13) is satisfied when

$$\mathbf{E} = \mathbf{E}^{\parallel} \equiv \mathbf{s}E(\mathbf{Q}).$$

---

[3] Unfortunately, there is no firmly established name for $\varkappa$ in the literature. For example, Landau and Lifshits [2] call $\varkappa$ the absorption coefficient, and Knox [3] and others call it the extinction coefficient.

In optically anisotropic crystals, the condition for the existence of longitudinal waves, i.e., waves for which $\mathbf{E} = \mathbf{E}^{\parallel} \mathbf{Q} \neq 0$, when $\mathbf{D}(\mathbf{Q}) = 0$ reduces, according to Eq. (2.9), to a system of three equations:

$$\varepsilon_{ij}(\mathbf{Q}, \omega) E_j^{\parallel}(\mathbf{Q}) = 0.$$

Generally speaking, these equations have solutions only for selected senses of $\mathbf{Q}$, providing that

$$\det\{\varepsilon_{ij}(\mathbf{Q}, \omega)\} = 0, \tag{2.12a}$$

which also determines the law of dispersion of longitudinal waves.

In general, $\det\{\varepsilon(\omega, \mathbf{Q})\} \neq 0$ and the electric-field strength of the wave has both transverse $\mathbf{E}^{\perp}(\mathbf{Q})$ and longitudinal $\mathbf{E}^{\parallel}(\mathbf{Q})$ components:

$$\mathbf{E}(\mathbf{Q}) = \mathbf{E}^{\perp}(\mathbf{Q}) + \mathbf{E}^{\parallel}(\mathbf{Q}),$$

where

$$\left.\begin{aligned} &\mathbf{s}\mathbf{E}^{\perp}(\mathbf{Q}) = 0, \quad \mathbf{E}^{\parallel}(\mathbf{Q}) = \mathbf{s}(\mathbf{s}\mathbf{E}(\mathbf{Q})), \\ &\mathbf{E}^{\perp}(\mathbf{Q}) = \mathbf{E}(\mathbf{Q}) - \mathbf{s}(\mathbf{s}\mathbf{E}(\mathbf{Q})). \end{aligned}\right\} \tag{2.14}$$

Using (2.11) and (2.14), we can transform Eq. (2.9) to

$$\mathbf{D}(\mathbf{Q}) = (n + i\varkappa)^2 \mathbf{E}^{\perp}(\mathbf{Q}). \tag{2.15}$$

where the refractive index and the absorption coefficient are functions of the real frequency ω.

3. The "Transverse" Dielectric Constant

In a number of cases, in dielectrics without free charges, it is convenient to use not the general dielectric-constant tensor, which determines the relationship (2.7) between the displacement vector and the total electric-field strength, but the auxiliary "transverse" dielectric-constant tensor $\varepsilon^{\perp}(\omega, \mathbf{k})$, which is introduced by

$$\mathbf{D}(\mathbf{Q}) = \varepsilon^{\perp}(\omega, \mathbf{k}) \mathbf{E}^{\perp}(\mathbf{Q}). \tag{3.1}$$

Equation (3.1) relates the displacement vector to the transverse part of the electric vector. The transverse dielectric-constant tensor was first introduced by Pekar [4], and then by Agranovich and Ginzburg [5].

The transverse dielectric-constant tensor can be introduced only for the three directions in which homogeneous plane waves can be propagated in crystals.

In dielectrics without free electric charges, the displacement vector is always transverse, $\mathbf{sD(Q)} = 0$, i.e., it lies in a plane perpendicular to the vector **s**. To eliminate indeterminacy in the choice of $\varepsilon^{\perp}(\omega, \mathbf{k})$, we shall require that

$$\mathbf{s}\,\varepsilon^{\perp}(\omega, \mathbf{k}) = \varepsilon^{\perp}(\omega, \mathbf{k})\,\mathbf{s} = 0 \tag{3.1a}$$

or

$$\varepsilon^{\perp}(\omega, \mathbf{k})\,\mathbf{E}(\mathbf{Q}) = \varepsilon^{\perp}(\omega, \mathbf{k})\,\mathbf{E}^{\perp}(\mathbf{Q}).$$

be satisfied. Considering this, we can rewrite (3.1) as

$$\mathbf{D}(\mathbf{Q}) = \varepsilon^{\perp}(\omega, \mathbf{k})\,\mathbf{E}(\mathbf{Q}). \tag{3.2}$$

Since vectors $\mathbf{D(Q)}$ and $\mathbf{E}^{\perp}\mathbf{(Q)}$ lie in a plane perpendicular to the wave vector **Q**, then, by virtue of (3.1), the tensor $\varepsilon^{\perp}(\omega, \mathbf{k})$ can be considered a tensor that acts upon the vectors lying in this plane. Let us introduce the coordinate system ξ, η, ζ, which is related to the sense of the wave vector such that the ζ axis is directed along **Q**, while the ξ and η axes are along the principal directions of the tensor $\varepsilon^{\perp}$. We shall call this the system of wave-vector coordinates. In this coordinate system

$$\varepsilon^{\perp}(\omega, \mathbf{k}) = \begin{pmatrix} \varepsilon^{\perp}_{\xi\xi} & 0 & 0 \\ 0 & \varepsilon^{\perp}_{\eta\eta} & 0 \\ 0 & 0 & 0 \end{pmatrix}. \tag{3.3}$$

The tensor $\varepsilon^{\perp}(\omega, \mathbf{k})$ completely determines the properties of the dielectric with respect to plane waves propagated in a given direction **Q**. In particular, this tensor is comparatively simply related to the refractive index and the attenuation factor of these waves. In fact, comparison of (2.15) and (3.1) gives

$$\varepsilon^{\perp}(\omega, \mathbf{k})\,\mathbf{E}^{\perp}(\mathbf{Q}) = (n + i\varkappa)^2\,\mathbf{E}^{\perp}(\mathbf{Q}). \tag{3.4}$$

In the wave-vector coordinate system, this equation comes down to two equations

$$\varepsilon^{\perp}_{\xi\xi}(\omega, \mathbf{k})\,E^{\perp}_{\xi} = (n_{\xi} + i\varkappa_{\xi})^2 E^{\perp}_{\xi},$$

$$\varepsilon^{\perp}_{\eta\eta}(\omega, \mathbf{k})\,E^{\perp}_{\eta} = (n_{\eta} + i\varkappa_{\eta})^2\,E^{\perp}_{\eta},$$

from which follow the simple relations

$$\varepsilon^{\perp}_{\xi\xi} = (n_\xi + i\varkappa_\xi)^2, \qquad \varepsilon^{\perp}_{\eta\eta} = (n_\eta + i\varkappa_\eta)^2. \tag{3.5}$$

Since the tensor $\varepsilon^{\perp}$ is itself a function of the wave vector **Q**, which, in accordance with (2.11), is a function of n, system (3.5) is, in general, nonlinear relative to n and $\varkappa$, and its solution is very complicated.

According to Eqs. (2.1) and (3.1), in a dielectric without free charges, an external spatially homogeneous transverse electric field

$$\mathbf{E} = \mathbf{E}^{\perp}(\mathbf{Q}) \exp\{i(\mathbf{Qr} - \omega t)\}$$

gives rise to a specific electric dipole moment **P**, whose transverse component is determined with the aid of the transverse dielectric-constant tensor[4]:

$$\mathbf{P}^{\perp}(\mathbf{Q}; \mathbf{r}, t) = \frac{\varepsilon^{\perp}(\mathbf{Q}, \omega) - 1}{4\pi} E^{\perp}(\mathbf{Q}) \exp\{i(\mathbf{Qr} - \omega t)\}. \tag{3.6}$$

Relation (3.6) is often used in quantum-mechanical calculation of the dielectric-constant tensor of dielectrics.

For dielectrics ($\rho_f = 0$), we can find a simple relationship between the total and transverse dielectric-constant tensors. Let us represent in (2.7) the total field as the sum of the transverse $\mathbf{E}^{\perp}(\mathbf{Q})$ and longitudinal $\mathbf{s}E^{\parallel}(\mathbf{Q})$ fields. Then we obtain

$$\mathbf{D}(\mathbf{Q}) = \varepsilon(\mathbf{k}, \omega)\,[\mathbf{E}^{\perp}(\mathbf{Q}) + \mathbf{s}E^{\parallel}(\mathbf{Q})]. \tag{3.7}$$

From the condition $\mathbf{sD}(\mathbf{Q}) = 0$ for transversality of the vector $\mathbf{D}(\mathbf{Q})$, with the aid of (3.7), let us express the longitudinal field in terms of the transverse:

$$E^{\parallel} = -(s_i \varepsilon_{ij} E_j^{\perp})(s_i \varepsilon_{ij} s_j)^{-1}.$$

Substituting the obtained value into Eq. (3.7), we have

$$D_i(\mathbf{Q}) = \left[\varepsilon_{il} - \frac{\varepsilon_{ij} s_j s_m \varepsilon_{ml}}{s_i \varepsilon_{ij} s_j}\right] E_l^{\perp}(\mathbf{Q}). \tag{3.8}$$

---

[4] Here and below, the expression $\varepsilon^{\perp}(\mathbf{Q}, \omega)$ is used to denote the transverse dielectric constant, which characterizes the response of the system to an external wave with a fixed real wave vector.

By comparing (3.8) and (3.1), we find the desired relationship between the tensor components:

$$\varepsilon_{il}^{\perp} = \varepsilon_{il} - \frac{\varepsilon_{ij} s_j s_m \varepsilon_{ml}}{s_i \varepsilon_{ij} s_j}. \tag{3.9}$$

From (3.9), it follows that

$$s_i \varepsilon_{ij}^{\perp} = 0, \tag{3.10}$$

which can be used to eliminate ambiguity when $\varepsilon^{\perp}(\mathbf{k}, \omega)$ is determined by relation (3.1).

4. The Dielectric Constant that Determines the Response of the System to an External Influence

A crystal is an inhomogeneous medium whose properties remain unchanged when its lattice is displaced by any vector

$$\mathbf{n} = \sum_{i=1}^{3} \mathbf{a}_i n_i \qquad (n_i = 0, \pm 1, \pm 2, \ldots),$$

where $\mathbf{a}_i$ are the basis vectors of a unit cell. The possibility of using the dielectric-constant tensor as a macroscopic characteristic of a crystal that determines its interaction with a macroscopic (longwave, $\lambda \gg a$) field has already been discussed several times in the literature [1, 5-8]. We shall consider this problem in detail in Section 7 of Chapter IV. Here, we shall proceed from the assumption that when investigating an external longwave influence, the crystal can be considered a continuous medium. In this case, the dielectric constant characterizes the macroscopic response of such a system to an external influence.

Let us calculate the electromagnetic field created in a dielectric by given extrinsic currents. We shall assume, first of all, that the dielectric has no free charges and that the extrinsic current density is determined by the expression

$$\mathbf{j}^{\mathrm{ex}}(\mathbf{r}, t) = \{\mathbf{j}^{\perp}(\mathbf{Q}) \exp[i(\mathbf{Qr} - \omega t)] + \text{complex conj.}\} e^{\eta t}, \tag{4.1}$$

where

$$\mathbf{Q}\mathbf{j}^{\perp}(\mathbf{Q}) = 0, \qquad \mathbf{j}^{\perp}(\mathbf{Q}) = \mathbf{j}^{*\perp}(-\mathbf{Q}),$$

and η is a small positive value, indicating that the extrinsic current was switched on in the infinite past. In the final expressions, we must proceed to the limit $\eta \to 0$.

In the absence of free electric charges and in the presence of extrinsic currents, the Maxwell equations have the form

$$\left.\begin{aligned} &\operatorname{rot}\mathbf{B} - \frac{1}{c}\frac{\partial \mathbf{D}}{\partial t} = \frac{4\pi}{c}\mathbf{j}^{ex},\\ &\operatorname{rot}\mathbf{E} + \frac{1}{c}\frac{\partial \mathbf{B}}{\partial t} = 0,\\ &\operatorname{div}\mathbf{D} = \operatorname{div}\mathbf{B} = 0. \end{aligned}\right\} \tag{4.2}$$

Using the relations

$$\mathbf{E} = -\frac{1}{c}\frac{\partial \mathbf{A}}{\partial t} - \operatorname{grad}\varphi, \quad \mathbf{B} = \operatorname{rot}\mathbf{A} \tag{4.3}$$

let us express the fields **E** and **B** in terms of the vector potential **A**, which satisfies the Coulomb calibration (see Sections 2 and 6 in [9]):

$$\operatorname{div}\mathbf{A} = 0, \quad \nabla^2\varphi = 0. \tag{4.4}$$

Let

$$\mathbf{A}(\mathbf{r}, t) = \{\mathbf{A}^{\perp}(\mathbf{Q})\exp[i(\mathbf{Qr} - \omega t)] + \text{complex. conj.}\}. \tag{4.5}$$

It follows from (4.4) that vector potential (4.5) is transverse, i.e., $\mathbf{QA}^{\perp}(\mathbf{Q}) = 0$. Using (4.3), we have

$$\mathbf{E}(\mathbf{r}, t) = \{\mathbf{E}^{\perp}(\mathbf{Q})\exp[i(\mathbf{Qr} - \omega t)] + \text{complex. conj.}\} - \operatorname{grad}\varphi, \tag{4.6}$$

where

$$\mathbf{E}^{\perp}(\mathbf{Q}) = \frac{i\omega}{c}\mathbf{A}^{\perp}(\mathbf{Q}).$$

Now, using (3.1), we can write

$$\mathbf{D}(\mathbf{r}, t) = \{\varepsilon^{\perp}(\omega, \mathbf{Q})\mathbf{E}^{\perp}(\mathbf{Q})\exp[i(\mathbf{Qr} - \omega t)] + \text{complex. conj.}\}. \tag{4.7}$$

Substitution of (4.1), (4.6), and (4.7) into Eq. (4.2) gives

$$\left\{-\frac{\omega^2}{c^2}\varepsilon^{\perp}(\omega, \mathbf{Q}) + \hat{I}\mathbf{Q}^2\right\}\mathbf{A}^{\perp}(\mathbf{Q}) = \frac{4\pi}{c}\mathbf{j}^{\perp}(\mathbf{Q}), \tag{4.8}$$

where $\hat{I}$ is a unit tensor with components δ_{jl}. From Eq. (4.8) it follows that extrinsic currents cause in a dielectric a field the components of whose vector potential are

$$A_l^{\perp}(\mathbf{Q}) = \frac{4\pi}{c}\left\{-\frac{\omega^2}{c^2}\varepsilon^{\perp}(\omega, \mathbf{Q}) + \hat{I}\mathbf{Q}^2\right\}^{-1}_{lm} j_m^{\perp}(\mathbf{Q}). \tag{4.9}$$

In this case, according to relation (4.6), vector potential (4.9) determines only the transverse component of the electric-field strength. The dielectric-constant tensor $\varepsilon^{\perp}$ takes into account the entire Coulomb interaction between the charges that are a part of the neutral molecules of the crystal (this problem is discussed in greater detail in Section 7 of Chapter IV).

In the more general case when free charges are present in the dielectric, the system of Maxwell equations takes the form

$$\left.\begin{array}{ll} \operatorname{rot}\mathbf{B}-\frac{1}{c}\frac{\partial\mathbf{D}}{\partial t}=\frac{4\pi}{c}\mathbf{j}, & \operatorname{rot}\mathbf{E}+\frac{1}{c}\frac{\partial\mathbf{B}}{\partial t}=0, \\ \operatorname{div}\mathbf{B}=0, & \operatorname{div}\mathbf{D}=4\pi\rho\ (\mathbf{r},t). \end{array}\right\} \quad (4.10)$$

We shall assume, further, that

$$\mathbf{j}^{\mathrm{ex}}(\mathbf{r},t)=\{\mathbf{j}(\mathbf{Q})\exp[i(\mathbf{Q}\mathbf{r}-\omega t)]+\text{compl. conj.}\},$$
$$\operatorname{div}\mathbf{j}^{\mathrm{ex}}+\frac{\partial\rho^{\mathrm{ex}}}{\partial t}=0.$$

We introduce the vector potential **A** by means of the relations

$$\mathbf{E}=-\frac{1}{c}\frac{\partial\mathbf{A}}{\partial t},\quad \mathbf{B}=\operatorname{rot}\mathbf{A},\quad \operatorname{div}\mathbf{A}\neq 0.$$

Let

$$\mathbf{A}(\mathbf{r},t)=\{\mathbf{A}(\mathbf{Q})\exp[i(\mathbf{Q}\mathbf{r}-\omega t)]+\text{complex conj.}\},$$
$$\mathbf{E}(\mathbf{r},t)=\{\mathbf{E}(\mathbf{Q})\exp[i(\mathbf{Q}\mathbf{r}-\omega t)]+\text{complex conj.}\},$$
$$\mathbf{D}(\mathbf{r},t)=\{\varepsilon(\omega,\mathbf{Q})\mathbf{E}(\mathbf{Q})\exp[i(\mathbf{Q}\mathbf{r}-\omega t)]+\text{complex conj.}\},$$

where the vectors **A(Q)** and **E(Q)** have nonzero longitudinal components. Then, we obtain from (4.10) an equation for the amplitudes of the vector potential:

$$\left\{-\frac{\omega^2}{c^2}\varepsilon_{lm}(\omega,\mathbf{Q})+\delta_{lm}\mathbf{Q}^2-Q_lQ_m\right\}A_m(\mathbf{Q})=\frac{4\pi}{c}j_l^{\mathrm{ex}}(\mathbf{Q}). \quad (4.11)$$

Equation (4.11) expresses in terms of the total-dielectric-constant tensor the relationship between the vector-potential components and the density of the extrinsic currents, which also contain longitudinal components.

5. Phenomenological Theory of Excitons

In the macroscopic Maxwell equations (1.6), the properties of the medium are characterized by the tensors of the total $\varepsilon(\mathbf{Q},\omega)$

and transverse $\varepsilon^{\perp}(\mathbf{Q}, \omega)$ dielectric constants. With the aid of (2.7) and (3.1), the dielectric constants link the displacement vector of the electromagnetic field

$$\mathbf{D}(\mathbf{Q}, \omega) = \mathbf{E}(\mathbf{Q}, \omega) + 4\pi\mathbf{P}(\mathbf{Q}, \omega) \tag{5.1}$$

to the vectors of the total $\mathbf{E}(\mathbf{Q}, \omega)$ and transverse $\mathbf{E}^{\perp}(\mathbf{Q}, \omega)$ electric fields. Here, in accordance with Eq. (2.9), the relationship between the real wave vector $\mathbf{k} = \mathbf{Q}$ and frequency for various types of elementary excitations within it and the field of the crystal is determined by the system of equations

$$\frac{\omega^2}{c^2}\mathbf{D}(\mathbf{Q}, \omega) = Q^2\mathbf{E}^{\perp}(\mathbf{Q}, \omega), \tag{5.2}$$

where

$$\mathbf{E}^{\perp}(\mathbf{Q}, \omega) = \mathbf{E}(\mathbf{Q}, \omega) - \mathbf{s}(\mathbf{s}\cdot\mathbf{E}(\mathbf{Q}, \omega)), \quad \mathbf{s} = \frac{\mathbf{Q}}{|\mathbf{Q}|}. \tag{5.3}$$

In the absence of absorption, Eq. (5.2) determines the steady states of the system. In these states, the frequency (energy) is a real function of the real wave vector. In the presence of slight absorption, the elementary excitations become quasi-steady, and a complex frequency is associated with them.

According to Frenkel's definition [10], excitons correspond to elementary excitations in crystals in which only the total Coulomb interaction between the charges in the molecules is taken into account and which are characterized by a particular value of the real wave vector. According to the quantum theory of radiation [9], in order to take into account only the Coulomb interaction between the charges of the system, we must eliminate from the total interaction carried by the electromagnetic field the transverse component of the electric field. Therefore, exciton states with wave vector $\mathbf{Q}$ are elementary excitations in which there is no transverse electric field with the same wave vector $\mathbf{Q}$, i.e., exciton states must satisfy the equation

$$\mathbf{E}^{\perp}(\mathbf{Q}, \omega) = 0. \tag{5.4}$$

According to Eqs. (2.7), (3.1), and (5.2), two types of exciton states are possible.

1) Longitudinal excitons, for which $\mathbf{D}(\mathbf{Q}, \omega) = 0$. In this case, according to Eqs. (5.1) and (5.4), the specific polarization

of the crystal is longitudinal, i.e.,

$$\mathbf{P}(\mathbf{Q}, \omega) = -\frac{1}{4\pi}\mathbf{E}^{\parallel}(\mathbf{Q}, \omega) \neq 0. \tag{5.5}$$

If $\mathbf{D}(\mathbf{Q}, \omega) = 0$, from (2.7) and (5.4) follows the system of equations

$$\varepsilon(\mathbf{Q}, \omega)\mathbf{E}^{\parallel}(\mathbf{Q}, \omega) = 0. \tag{5.6}$$

System (5.6) has nontrivial solutions if

$$\det\{\varepsilon(\mathbf{Q}, \omega)\} = 0, \tag{5.7}$$

which is the dispersion equation for longitudinal excitons, i.e., it determines the relationship between their frequency and wave vector. In anisotropic crystals, longitudinal excitons are possible only for particular senses of $\mathbf{Q}$. In isotropic crystals, the tensor $\varepsilon(\mathbf{Q}, \omega)$ decomposes to a scalar, and Eq. (5.7) reduces to the simple equation

$$\varepsilon(\mathbf{Q}, \omega) = 0. \tag{5.8}$$

Thus, according to Eqs. (5.7) and (5.8), the zeroes of the total dielectric-constant tensor correspond to longitudinal excitons. However, not all of the zeros of the tensor $\varepsilon(\mathbf{Q}, \omega)$ correspond to longitudinal excitons, since, besides Eq. (5.7), Eq. (5.6) at $\mathbf{E}^{\parallel} \neq 0$ must be satisfied.

2) Nonlongitudinal excitons. Elementary excitations of a crystal in which $\mathbf{E}^{\perp} = 0$ and the specific polarization $\mathbf{P}$ contains a nonzero transverse component we shall call simply excitons. It follows from Eq. (5.1) that when $\mathbf{P}^{\perp} \neq 0$ and $\mathbf{E}^{\perp} = 0$, the displacement vector cannot be equal to zero,

$$\mathbf{D}(\mathbf{Q}, \omega) \neq 0. \tag{5.9}$$

From (5.2) we find that if $\mathbf{E}^{\perp} = 0$, (5.9) can be satisfied only for very long waves ($\omega a/c \to 0$), i.e., when the delay effect is ignored. Excitons, therefore, are idealized elementary excitations in whose consideration we ignore the delay effect and take into account only Coulomb interaction. Such an idealization is very useful in the first stage of theoretical calculation. In a number of cases, in the strong interaction of electronic excitations with phonons, for example, the delay effect plays a minor role. Let us rewrite (3.1) as

$$\mathbf{E}^{\perp}(\mathbf{Q}, \omega) = \{\varepsilon^{\perp}(\mathbf{Q}, \omega)\}^{-1}\mathbf{D}(\mathbf{Q}, \omega). \tag{5.10}$$

Then, it follows from conditions (5.4) and (5.9) that the exciton states of a crystal are determined by the system of equations

$$\{\varepsilon^{\perp}(\mathbf{Q}, \omega)\}^{-1} \mathbf{D}(\mathbf{Q}, \omega) = 0. \tag{5.11}$$

The condition

$$\det \{[\varepsilon^{\perp}(\mathbf{Q}, \omega)]^{-1}\} = 0 \tag{5.12}$$

of nontrivial solvability of system (5.11) is the exciton dispersion equation.

Thus, the exciton frequencies (energies) correspond to the zeros of the tensor $[\varepsilon^{\perp}(\mathbf{Q}, \omega)]^{-1}$ or to the poles of the tensor $\varepsilon^{\perp}(\mathbf{Q}, \omega)$. The reverse is also true: if the exciton energy is determined by $E(\mathbf{Q})$, then in the vicinity of one pole the dielectric constant can be represented as

$$\varepsilon^{\perp}(\mathbf{Q}, \omega) = \varepsilon_0 + A\{E(\mathbf{Q}) - \hbar\omega\}^{-1}. \tag{5.13}$$

In the presence of slight absorption, the dielectric constant is a complex value, and in a single-pole approximation for an isotropic crystal it is represented by the function

$$\varepsilon^{\perp}(\mathbf{Q}, \omega) = \varepsilon_0 + A\left\{E(\mathbf{Q}) - \hbar\omega - \frac{i\hbar}{2}\Gamma(\mathbf{Q})\right\}^{-1}. \tag{5.14}$$

The pole of the function $\varepsilon^{\perp}(\mathbf{Q}, \omega)$ is located in the lower half-plane of the complex exciton energy

$$E(\mathbf{Q}) - \frac{i}{2}\hbar\Gamma(\mathbf{Q}). \tag{5.15}$$

The imaginary part of the exciton energy (5.15) determines the law of time attenuation ($\sim \exp[-\Gamma(\mathbf{Q})t]$) of exciton states with energy $E(\mathbf{Q})$.

Photoexcitons. Elementary excitations and fields that also contain a transverse electric component we shall call photoexcitons. To find elementary excitations in which $\mathbf{E}^{\perp}(\mathbf{Q}, \omega) \neq 0$, let us transform Eqs. (5.2) and (3.1) to

$$\left\{\mathbf{Q}^2 - \frac{\omega^2}{c^2}\varepsilon^{\perp}(\mathbf{Q}, \omega)\right\}\mathbf{E}^{\perp}(\mathbf{Q}, \omega) = 0. \tag{5.16}$$

Then the condition

$$\det\left\{\mathbf{Q}^2 - \frac{\omega^2}{c^2}\varepsilon^{\perp}(\mathbf{Q}, \omega)\right\} = 0 \tag{5.17}$$

of nontrivial solvability of the system of homogeneous equations will be the dispersion equation for photoexcitons.[5]

In isotropic crystals, in particular, Eq. (5.17) reduces to

$$\Phi(\mathbf{Q}, \omega) \equiv \left[\frac{c^2Q^2}{\omega^2} - \varepsilon^{\perp}(\mathbf{Q}, \omega)\right] = 0. \tag{5.18}$$

In other words, the frequencies of photoexcitons $\omega_l(\mathbf{Q})$ as functions of the real wave vectors $\mathbf{Q}$ are the zeros of the function $\Phi(\mathbf{Q}, \omega)$, which is defined by Eq. (5.18).

In the frequency range ω in which the dielectric constant is real and is represented in a single-pole approximation by expression (5.13), Eq. (5.18) converts to a second-degree equation in ω^2. In this case, each value of $E(\mathbf{Q})$ will correspond to two elementary excitations, i.e., the two photoexciton branches $\omega_1^2(\mathbf{Q})$ and $\omega_2^2(\mathbf{Q})$ with energies

$$\mathscr{E}_l(\mathbf{Q}) = \hbar\omega_l(\mathbf{Q}), \qquad l = 1, 2. \tag{5.19}$$

In the presence of absorption, the dielectric constant of an isotropic crystal is complex and can be represented as

$$\varepsilon^{\perp}(\mathbf{Q}, \omega) = \varepsilon'(\mathbf{Q}, \omega) + i\varepsilon''(\mathbf{Q}, \omega), \tag{5.20}$$

where ε' and ε'' are real functions. If absorption is slight, $\varepsilon' \gg \varepsilon''$ and the solutions of Eq. (5.18) can be represented as

$$\omega_l = \omega_l(\mathbf{Q}) - i\frac{1}{2}\gamma_l(\mathbf{Q}), \qquad l = 1, 2. \tag{5.21}$$

In the first approximation, the photoexciton frequencies, i.e., the functions $\omega_l(\mathbf{Q})$, are the solutions of the equation

$$\frac{c^2Q^2}{\omega^2} - \varepsilon'(\mathbf{Q}, \omega) = 0, \tag{5.22}$$

while the imaginary part of expression (5.21) characterizes the time attenuation ($\sim \exp\{-\gamma(\mathbf{Q})t\}$) of photoexcitons and is deter-

[5] According to (4.8), the dispersion equation for photoexcitons coincides with the condition of nontrivial solvability of the Maxwell equations for a transverse field in the absence of external transverse currents. Therefore, the problem of finding elementary excitations in the form of photoexcitons is equivalent (in the absence of absorptions) to solving the self-consistent problem in Section 1.

mined by the approximation equation

$$\gamma_l(\mathbf{Q}) = \frac{\omega\varepsilon''(\mathbf{Q}, \omega_l(\mathbf{Q}))}{\varepsilon'(\mathbf{Q}, \omega_l(\mathbf{Q}))}. \tag{5.23}$$

In particular, in a single-pole approximation, when the dielectric constant is determined by expression (5.14), each exciton state with energy (5.17) corresponds to two branches of photoexcitons, whose frequencies $\omega(\mathbf{Q})$ are calculated from

$$\frac{c^2 Q^2}{\omega^2(\mathbf{Q})} = \varepsilon_0 + \frac{A\,[E(\mathbf{Q}) - \hbar\omega(\mathbf{Q})]}{[E(\mathbf{Q}) - \hbar\omega(\mathbf{Q})]^2 + \frac{1}{4}\hbar^2\Gamma^2(\mathbf{Q})}.$$

The time attenuation of photoexcitons in this approximation is given by

$$\gamma_l(\mathbf{Q}) = \frac{\hbar\omega_l(\mathbf{Q})\,\Gamma(\mathbf{Q})/2}{\varepsilon_0\left\{[E(\mathbf{Q}) - \hbar\omega_l(\mathbf{Q})]^2 + \frac{1}{4}\hbar^2\Gamma^2(\mathbf{Q})\right\} + A\,[E(\mathbf{Q}) - \hbar\omega_l(\mathbf{Q})]}.$$

Chapter II

Elementary Theory of Excitons in Coordinate Representation

1. Frenkel Excitons and Wannier Excitons

The steady internal states of atoms and molecules are characterized by discrete sets of physical quantities (energy, angular momentum, parity, etc.), which have particular values in these states. In solids, due to the presence of translational symmetry, one of the quantities that characterize steady states is the real wave vector **k** (or quasi-momentum $\hbar$**k**).

The translational symmetry of a crystal of infinite dimensions manifests itself as invariance of the energy operator H of the crystal in translation of any lattice vector **n**, which is defined by

$$\mathbf{n} = \sum_{i=1}^{3} n_i \mathbf{a}_i, \qquad n_i = 0 \ \pm 1, \pm 2, \ldots \tag{1.1}$$

in terms of the three noncoplanar basis vectors $\mathbf{a}_i$ of a unit cell. A unit cell is the smallest part of a crystal by translation of whose basis vectors the entire crystal can be formed. Everywhere in this book, with the exception of Section 6 of Chapter II, a unit cell will be in the form of a parallelepiped with sides $\mathbf{a}_1$, $\mathbf{a}_2$, $\mathbf{a}_3$ and volume $v = \mathbf{a}_1[\mathbf{a}_2 \times \mathbf{a}_3]$. This is one of the simplest unit cells. The disadvantage of this choice is that such a unit cell does not reflect all of the properties of crystal symmetry. When one wishes to take into account other properties of crystal symmetry, it is convenient to use the Wigner – Seitz unit cell. If from a center point in the crystal we draw lines to all of the nearest equivalent (i.e., those obtained by translation) lattice points, the Wigner–Seitz cell will be bounded by planes that divide these lines in half and are perpendicular to them.

Let $T_\mathbf{n}$ be the translation operator for lattice vector **n**. The effect of the translation operator on the wave function, which is a function of the radius vector **r**, is determined by

$$T_\mathbf{n}\psi(\mathbf{r}) = \psi(\mathbf{r}+\mathbf{n}). \tag{1.2}$$

Owing to translational symmetry, the translation operator $T_\mathbf{n}$ commutes with the energy operator of the crystal. Therefore, the eigenfunctions of all of the steady states are simultaneously eigenfunctions of the translation operator. The eigenfunctions $\psi_\mathbf{k}$ and eigenvalues $t_\mathbf{n}$ of the translation operator are determined by

$$T_\mathbf{n}\psi_\mathbf{k}(\mathbf{r}) = t_\mathbf{n}\psi_\mathbf{k}(\mathbf{r}).$$

If we compare this equation with (1.2), we find

$$t_\mathbf{n} = \exp(i\mathbf{k}\mathbf{n}), \qquad \psi_\mathbf{k}(\mathbf{r}) = A\exp(i\mathbf{k}\mathbf{r}), \tag{1.3}$$

where A is a normalization factor and **k** is an arbitrary real vector, which is called the *wave vector*. The set of all values of the wave vectors forms a three-dimensional vector space, which we shall call the ***k*-space**.

The wave vector in (1.3) is determined within the accuracy of the transformation

$$\mathbf{k} \to \mathbf{k} + \mathbf{g}_m,$$

where

$$\mathbf{g}_m = 2\pi\sum_{i=1}^{3} m_i\mathbf{b}_i; \tag{1.4}$$

the subscript m in $\mathbf{g}_m$ indicates the set of three integers m_i; and $\mathbf{b}_i$ are the basis vectors of the *reciprocal lattice*, which are defined in terms of the basis vectors of the lattice by

$$\mathbf{b}_1 = v^{-1}[\mathbf{a}_2\times\mathbf{a}_3], \qquad \mathbf{b}_2 = v^{-1}[\mathbf{a}_3\times\mathbf{a}_1], \qquad \mathbf{b}_3 = v^{-1}[\mathbf{a}_1\times\mathbf{a}_2],$$
$$v = \mathbf{a}_1[\mathbf{a}_2\times\mathbf{a}_3].$$

They satisfy the equations

$$\mathbf{a}_i\mathbf{b}_j = \delta_{ij}. \tag{1.5}$$

The equivalence of vectors $\mathbf{k}$ and $\mathbf{k} + \mathbf{g}_m$ is easily seen with the aid of definitions (1.1) and (1.4) and Eqs. (1.5).

The set of all vectors $\mathbf{g}_m$ (1.4) forms in the $\mathbf{k}$-space a lattice with the basis vectors $2\pi\mathbf{b}_1$, $2\pi\mathbf{b}_2$, and $2\pi\mathbf{b}_3$. Since the vectors $\mathbf{k}$ and $\mathbf{k} + \mathbf{g}_m$ are equivalent, for a single-valued classification of the eigenvalues (1.3) of the translation operator we need consider only the vectors $\mathbf{k}$ that lie in the first Brillouin zone. This zone is the part of the $\mathbf{k}$-space bounded by the planes that divide in half and are perpendicular to the lines connecting the point $\mathbf{k} = 0$ and the nearest points determined by the vectors $\mathbf{g}_m$. The choice of the main region of nonequivalent (reduced) wave vectors is not a unique one. Instead of the first Brillouin zone, which in general is represented by a rather complicated polyhedron, the main region of the $\mathbf{k}$-space can be the parallelepiped formed by the vectors $2\pi\mathbf{b}_1$, $2\pi\mathbf{b}_2$, and $2\pi\mathbf{b}_3$ whose center is at $\mathbf{k} = 0$. With this as the main region in the $\mathbf{k}$-space, the reduced wave vectors take all possible values that satisfy the inequalities

$$-\pi < \mathbf{k}\mathbf{a}_i \leqslant \pi, \qquad i = 1, 2, 3. \tag{1.6}$$

Everywhere in this book, with the exception of Section 6 of Chapter II, we shall use the second method to determine the reduced wave vectors. The region of the $\mathbf{k}$-space defined by inequalities (1.6) we shall call the *basic cell of the k-space*. In a simple cubic lattice, the basic cell in the $\mathbf{k}$-space coincides with the first Brillouin zone.

Translational symmetry is lost at the boundaries of a crystal of finite dimensions. If we are interested only in volume effects, then in crystals whose linear dimensions are great in comparison with the lattice constant we can preserve translational symmetry artificially by introducing cyclic boundary conditions. Let the crystal have the shape of a parallelepiped with edges $N_i a_i$, where N_i are large numbers. Then the cyclic boundary conditions reduce to the requirements that the functions $\psi_\mathbf{k}(\mathbf{r})$ satisfy the conditions

$$\psi_\mathbf{k}(\mathbf{r}) = \psi_\mathbf{k}(\mathbf{r} + N_1\mathbf{a}_1) = \psi_\mathbf{k}(\mathbf{r} + N_2\mathbf{a}_2) = \psi_\mathbf{k}(\mathbf{r} + N_3\mathbf{a}_3).$$

If we substitute into these equations the explicit form of eigenfunctions (1.3), we see that the cyclic boundary conditions are satisfied if the wave vector takes discrete values defined by

$$\mathbf{k} = \sum_{i=1}^{3} \frac{2\pi}{N_i} \nu_i \mathbf{b}_i, \tag{1.7}$$

where ν_i are integers that satisfy the inequalities

$$-\frac{1}{2}N_i < \nu_i \leqslant \frac{1}{2}N_i, \qquad i = 1, 2, 3.$$

The number of possible values of the wave vectors defined by Eq. (1.7) is equal to the number of unit cells in the crystal $N = N_1N_2N_3$.

In the ground state, the wave vector $\mathbf{k} = 0$. The energies of elementary excited states of a crystal $E_f(\mathbf{k})$ are usually read from the energy of the ground state. The energy of each elementary excitation is a function of the wave factor and of the set of quantum numbers f, which, together with $\mathbf{k}$, completely determine the excitation state.

Thus, the wave functions of all steady states of an ideal crystal can be characterized by the values of the wave vectors $\mathbf{k}$ (lying in the basic cell of the k-space). In this case,

$$T_{\mathbf{n}}\psi_{\mathbf{k}f}(\mathbf{r}) = \exp(i\mathbf{kn})\,\psi_{\mathbf{k}f}(\mathbf{r}),$$
$$H\psi_{\mathbf{k}f}(\mathbf{r}) = F_f(\mathbf{k})\,\psi_{\mathbf{k}f}(\mathbf{r}),$$

where f indicates the other quantum numbers that determine the excitation state.

In dielectric crystals, the first electronic excitation states do not give rise to electrical conductivity in a constant electric field. Such "currentless" excited states in dielectric crystals may be divided into three types.

1) Phonon excitations, which are elementary collective excitations that correspond to lattice vibrations, i.e., to vibrations about the equilibrium positions of ions in ionic lattices or to vibrations and rotational oscillations about the equilibrium positions of molecules in molecular crystals. These excitations are associated with comparatively low frequencies: from zero to a few hundreds of inverse centimeters.[1]

2) Excitons, which are currentless collective excitations that correspond to a change in the state of motion of loosely bound electrons in ions and molecules of crystals. These excitations are associated with frequencies on the order of tens of thousands of

[1] Frequencies in inverse seconds are obtained by multiplying frequencies in inverse centimeters by $3 \cdot 10^{10}$ cm/sec.

inverse centimeters. Excitations in molecular crystals that correspond to intramolecular vibrations with frequencies exceeding 100 cm^{-1} are often classified as excitons.

3) High-frequency excitons, which are associated with a change in the state of motion of "internal," tightly bound electrons in atoms or ions and with frequencies on the order of hundreds of thousands of inverse centimeters.

The concept of excitons was first introduced into physics by Frenkel [1, 2] in 1931. Later, a number of papers appeared in which this idea was developed [3, 4] and the existence of excitons was studied theoretically [5, 6]. In [4], Wannier showed that in ionic crystals the exciton state could be considered a hydrogen-like state of an electron and a hole that interact according to Coulomb's law.

In this book, we shall investigate the elementary excited states of crystals that correspond to excitons. Exciton states are usually studied theoretically in an adiabatic approximation, which is based on the assumption that exciton frequencies considerably exceed phonon frequencies. Then, in the zeroth approximation it can be assumed that the molecules (atoms, ions) are rigidly fixed in equilibrium positions that determine the space lattice. In this approximation, the energies $E_f(\mathbf{k})$ of the elementary excitations (excitons) are the eigenvalues of the Hamiltonian operator H_{ex}, which contains the electron-kinetic-energy operators and the Coulomb-interaction operators of the electrons of the crystal and the fixed (in equilibrium positions) atomic nuclei.

The values of $E_f(\mathbf{k})$, which pertain to various $\mathbf{k}$ at fixed f, form the f-th exciton band of elementary excited states. Each elementary excitation with a particular $\mathbf{k}$ value is distributed throughout the entire crystal. In crystals with finite dimensions, the number of different possible $\mathbf{k}$ values, and, therefore, the number of sublevels in each band, is equal to the number of unit cells in the crystal. In general, for fixed f the energy $E_f(\mathbf{k})$ is a function of the sense and absolute value of $\mathbf{k}$. For crystals with a center of symmetry, for a fixed sense of $\mathbf{s} = \mathbf{k}/|\mathbf{k}|$ at small $\mathbf{k}$ ($\mathbf{k}a_i \ll 1$) the exciton energy can be written as

$$F_f(\mathbf{s}k) = E_f(\mathbf{s}) + \frac{\hbar^2\mathbf{k}^2}{2m^*(\mathbf{s})} + \dots \qquad (1.8)$$

According to expression (1.8), the exciton energy at small **k** coincides with the energy of a quasi-particle with "effective mass" $m^*(\mathbf{s})$ and "internal energy" $E_f(\mathbf{s})$, which is, in general, a function of the sense of **s**.

The effective mass of an exciton can be either positive or negative. In the first case, as the quasi-momentum $\hbar\mathbf{k}$ increases, the exciton energy increases; in the second case, it decreases. The quasi-momentum of an exciton $\hbar\mathbf{k}$ is in a certain respect analogous to the momentum vector of a free particle. It should be borne in mind, however, that this analogy is far from complete. The quasi-momentum $\hbar\mathbf{k}$ is not unique, since the vectors **k** and **k** + **g** are equivalent. Moreover, the eigenfunctions ψ_{kf} are not eigenfunctions of the momentum operator $i\hbar\nabla$.

In accordance with the selection of the basis functions for constructing the electronic "currentless" states of a crystal, it is customary to distinguish two limiting types of excitons: Frenkel excitons and Wannier excitons.

Frenkel excitons (excitons of "small radius") are used to describe the elementary collective electronic excitations of molecular crystals that consist of weakly interacting molecules. In this case, in a first approximation, an intramolecular electronic excitation of a free molecule may be compared to each elementary collective excitation of the crystal (exciton). The wave functions of the steady states of the molecules are the principal basis functions. It is assumed, here, that the wave functions of adjacent molecules overlap slightly. Therefore, the region in which the basis wave functions differ from zero coincides approximately with a unit cell. The basis molecular functions form the many-electron wave functions, which determine the ground and excited states of the entire crystal as a whole. Frenkel excitons are often called *molecular excitons*. The basic problem of the theory of molecular excitons is determination of the nature of change of intramolecular excitations when a crystal is formed of molecules.

As we shall see below, the characteristics of molecular excitons are greatly dependent upon the structure of the crystal. The study of molecular excitons, therefore, can serve as a method for studying the structure of molecular crystals and its variation under different external mechanical influences. On the other hand, if we know how the spectrum of excited states of a molecule varies in a crystal, we can, by studying the excited states of the crystal, determine certain properties of the excited states of complex molecules.

The Frenkel exciton model has been used extensively by the author [7-9] in constructing a theory of light absorption in complex molecular crystals. The first convincing experimental studies of Frenkel excitons were made by Prikhot'ko [10, 11], and McClure and Craig et al. [12, 13]. The first reviews of experimental work on the spectra of exciton absorption of light were written by McClure [14] and Wolf [15].

Wannier excitons [4, 16-18] are usually used to describe the collective electronic excited states of semiconductors and dielectrics with a high dielectric constant. In the zeroth approximation, the currentless electronic excitation of ionic crystals corresponds to the transfer of an electron from one ion to another, which, in the language of band theory, corresponds to the transfer of an electron from a filled valence band to an empty conduction band. Owing to the high dielectric constants ε of such crystals, the interaction between an electron that has been transferred to the conduction band and the hole (which corresponds to a positive charge) that remains in the valence band is considerably weakened. The excited states of such a "system" correspond to the excitations of a hydrogen-like atom with charge $Z = 1/\varepsilon$ and reduced mass $1/\mu = (1/m_e^*) + (1/m_h^*)$, where m_e^* is the effective mass of an electron in the conduction band of the corresponding crystal and m_h^* is the effective mass of a hole in the valence band of the same crystal. Thus, the "internal energy," i.e., the energy of a Wannier exciton at rest, is determined in the zeroth approximation by

$$E_n = -\frac{\mu e^4}{2\hbar^2\varepsilon^2 n^2}, \qquad n = 1, 2, \ldots \tag{1.9}$$

If translational symmetry is also taken into account, then to internal energy (1.9) we must add the "kinetic energy" of the exciton. Therefore, the exciton energy

$$E_n(\mathbf{k}) = -\frac{\mu e^4}{2\hbar^2\varepsilon^2 n^2} + \frac{\hbar^2\mathbf{k}^2}{2(m_e^* + m_h^*)}. \tag{1.10}$$

The Bohr radius of the "internal" state of an exciton with quantum number n is expressed by the formula

$$R_n = \frac{n^2\varepsilon m_e}{\mu} a_B, \tag{1.11}$$

where m_e is the mass of a free electron, and

$$a_B = \hbar^2 (m_e e^2)^{-1} \approx 0.5 \cdot 10^{-8}\,\text{cm}$$

is the Bohr radius of a hydrogen atom. In a CdS crystal, $R_1 \approx 30 a_B$, and in a Cu_2O crystal, $R_1 \approx 17 a_B$. In these crystals, therefore, the two-particle wave function of the internal exciton states is actually distributed over a region whose radius considerably exceeds the lattice constant.

In more exact calculations, it should be remembered that the effective masses m_e^* and m_h^* are tensors and that interaction between an electron and a hole reduces to the operator $e^2/\varepsilon r$ only when r considerably exceeds the lattice constant.

The main efforts of theoreticians in research on Wannier excitons are directed at calculating the "internal" excited states of excitons with allowance for the complicated structure of the valence and conduction bands (the presence of degeneracy, anisotropy, the presence of band extremes at $\mathbf{k} \neq 0$, etc.) [17, 19]. The principal methods for calculating these states are reviewed in Knox' book [20], which gives the original references.

An experimental proof of the existence of Wannier excitons has evidently been obtained by Gross et al. [21]. The concept of Wannier excitons has been used by Zhuze and Ryvkin [22] to explain the photoconductive effect in cuprous oxide, by Apker and Taft [23] to explain experiments on the photoemissive effect, and by a number of others. A direct proof of the movement of excitons in crystals has been obtained by Hopfield and Thomas [24].

Frenkel excitons and Wannier excitons are two limiting models of elementary collective currentless excitations of dielectrics in which the excitation is distributed throughout a large region of the crystal. In a number of cases, excited states of an intermediate type are realized. For example, in molecular crystals composed of atoms of inert gases [20, 25, 26], the transition of atoms to an excited state is associated with a considerable increase in the effective radius of distribution of the electron wave function. Because of this, an electron in an excited state moves in part in a region occupied by other atoms.

In spite of the differences between the models of the two limiting types of excitons, they have some properties in common, as follows.

1) The energy of exciton states in crystals is a function of the wave vector, i.e., the energy states form quasi-continuous bands.

2) In steady states, an excitation (exciton) is distributed throughout the entire crystal.

3) The properties of excitons depend upon the structure of the crystal as a whole.

4) A quasi-momentum can be associated with each steady state.

These characteristics of exciton states determine their common properties in interaction with light, with lattice vibrations, and other perturbations.

In this monograph, we shall study exciton states chiefly by the example of molecular crystals, using the Frenkel model of molecular excitons.

2. Molecular Excitons in Crystals with One Molecule in a Unit Cell

Molecular crystals are solids composed of molecules (or atoms of inert gases) with van der Waals interaction forces between them. The energy of such molecular interactions is very small in comparison with the electron bonding energy in the molecules. Typical molecular crystals are those formed from anisotropic molecules of aromatic compounds: anthracene, naphthalene, benzene, etc.

Theory must explain how the energy states of free molecules change when a crystal is formed and what the properties of the crystal will be with respect to interaction with electromagnetic radiation. Unlike a gas, in a crystal the molecules are arranged in a definite order close to one another. If the forces of intermolecular interaction could be brought to zero while preserving the order of arrangement of the molecules, the result would be a model of an oriented gas, which is sometimes used to explain certain optical properties of molecular crystals in a zeroth approximation. As we shall see below, however, even the presence of small intermolecular interactions will lead, in a number of cases, to very unusual characteristics in a real crystal, which will distinguish it from the oriented-gas model. These characteristics show up considerably when the excited states of the crystal are studied. Exciton or collective excited states can occur in a crystal that are not the simple sum of the excited states of the free molecules.

Exciton states in a solid are not steady. The excitation energy is emitted or converted to heat energy. In some cases, a portion of the elementary-excitation energy is emitted while the remaining part is converted to heat energy, i.e., is redistributed in many degrees of freedom at the low frequencies that are associated with vibrations of the molecules about the equilibrium positions. In order to simplify the calculations, we shall first of all ignore the interaction of intramolecular excitations with molecular vibrations. For this, we shall assume that the molecules are rigidly fixed in their equilibrium positions. In this section, we shall consider crystals that contain one molecule in each unit cell.

A unit cell is a parallelepiped with the three noncoplanar basis vectors $\mathbf{a}_1$, $\mathbf{a}_2$, $\mathbf{a}_3$. The space lattice of the crystal is formed by periodic repetition of the unit cell along the senses of the three basis vectors. The position of the space-lattice points is determined by the lattice vector

$$\mathbf{n} = \sum_{i=1}^{3} n_i \mathbf{a}_i, \tag{2.1}$$

where n_i are integers.

In crystals with one molecule in each unit cell, all of the molecules are the same and have identical orientations relative to the basis vectors. The molecules are situated at the space-lattice points; therefore, the position of their centers of inertia can be characterized by the lattice vectors. Let us assume that the crystal is a large parallelepiped with edges $N_1\mathbf{a}_1$, $N_2\mathbf{a}_2$, $N_3\mathbf{a}_3$. Then the total number of molecules in the crystal $N = N_1N_2N_3$. In large crystals, the numbers N_i can be considered even, since $N \approx N + 1$. The positions of the molecules (they are assumed to be rigidly fixed) in such a lattice are determined by lattice vectors (2.1) at n_i values that satisfy the inequalities

$$-\frac{1}{2} N_i < n_i \leqslant \frac{N_i}{2}, \qquad i = 1, 2, 3. \tag{2.2}$$

When studying the volume properties of large crystals, the choice of the boundary conditions at the surface of the crystal does not play an important role. Periodicity conditions with large periods $N_i\mathbf{a}_i$ are usually used as boundary conditions.

Let H_n be the energy operator of the molecule that occupies point **n**, and let V_{nm} be the interaction-energy operator of the two molecules at points **n** and **m**. Then the total energy operator of the crystal

$$H = \sum_{\mathbf{n}} H_{\mathbf{n}} + \frac{1}{2} \sum_{\mathbf{n},\mathbf{m}}{}' V_{\mathbf{nm}}. \tag{2.3}$$

Summation in (2.3) is over all values of the lattice vectors **n** and **m** defined by Eq. (2.1) under conditions (2.2). Here and below, the prime on the second summation indicates that terms with $\mathbf{n} = \mathbf{m}$ are absent in the sum.

First, let us consider the part of operator (2.3) without intermolecular interaction. This approximation corresponds to an oriented-gas model with properly situated molecules. Let the energy operator H_n have a set of eigenfunctions $\varphi_n^f(\xi_n)$ that correspond to the eigenvalues of the energy ε_f. The energy ε_f characterizes the internal electronic excitation of a molecule. The symbol ξ_n in parentheses indicates the set of coordinates corresponding to the internal degrees of freedom of the molecule occupying point **n**. The sign f indicates the set of quantum numbers of each steady state. The ground state of a molecule corresponds to $f = 0$. First of all, for simplicity we shall assume that the steady states of free molecules do not have degeneracy.

In the ground state and in excited states when the excitation energy is less than the ionization energy, the wave functions of adjacent molecules in molecular crystals overlap slightly. We shall be concerned with only small electronic excitations. In order to avoid considerable complications, therefore, we shall ignore the overlapping of the wave functions of adjacent molecules. This approximation works out particularly well in crystals consisting of molecules of aromatic compounds (benzene, naphthalene, anthracene, etc.). In such molecules, transitions to the first electronic excited states involve changes in the motion of the π electrons of the carbon atoms of the molecule. The hydrogen atoms, which are held by σ bonds, are situated at the periphery of the molecule. The first electron transitions in such molecules occur between states in which the spatial distributions of the π-electron densities differ little. The π electrons of aromatic molecules are in a sense similar to the electrons of the $4f$ shells in atoms of the rare-earth elements.

The ground state of a crystal with the energy operator $H_0 = \sum_n H_n$ has the energy $N\varepsilon_0$ and is described by the wave function

$$\psi^0 = \prod_n \varphi_n^0. \tag{2.4}$$

The excited states corresponding to the f-th excitation of one molecule of the crystal have the energy $(N-1)\varepsilon_0 + \varepsilon_f$. These states are N-fold degenerate, since any of the N molecules of the crystal can be in the f-th excited state. If the molecule at **n** is excited, the corresponding wave function will have the form

$$\psi_n^f = \varphi_n^f \prod_m \varphi_m^0, \qquad m \neq n. \tag{2.5}$$

Functions (2.4) and (2.5) must be antisymmetrized with respect to all electrons. This antisymmetrization results in additional energy terms that contain the overlap integrals of the wave functions of adjacent molecules [8]. In molecular crystals of the aromatic type, these terms are small, and we shall ignore them below.

Since all of the molecules in the crystal are the same, instead of functions (2.5), which indicate the location of the excited molecule in the crystal, we can introduce N new orthonormal functions

$$|\mathbf{k}f\rangle = N^{-1/2} \sum_n \psi_n^f \exp(i\mathbf{kn}), \tag{2.6}$$

which differ in the values of the wave vector **k**, which is defined by

$$\mathbf{k} = \sum_{i=1}^{3} \frac{2\pi}{N_i} \nu_i \mathbf{b}_i, \qquad -\frac{N_i}{2} < \nu_i \leqslant \frac{N_i}{2}, \tag{2.7}$$

where $\mathbf{b}_i$ are the basis vectors of the reciprocal lattice, which are related to the basis vectors of the lattice by $\mathbf{b}_i \mathbf{a}_j = \delta_{ij}$. The set of all vectors (2.7) forms in the **k**-space the first unit cell. The functions $\Omega(\mathbf{k}, \mathbf{n}) = (1/\sqrt{N}) \exp(i\mathbf{kn})$ make up a complete set of functions that satisfy the relations

$$\sum_k \Omega(\mathbf{k}, \mathbf{n})\, \Omega^*(\mathbf{k}, \mathbf{m}) = \delta_{n,m}; \qquad \sum_n \Omega(\mathbf{k}, \mathbf{n})\, \Omega^*(\mathbf{k}', \mathbf{n}) = \delta_{k,k'}.$$

Transformation (2.6) may be considered a unitary transformation to a new representation of the wave functions.

In a system with operator H_0, all of the states (2.6), which differ in the values of the wave vector **k**, have the same energy. This degeneracy is removed if we take into account the interactions between molecules. In molecular crystals, the intermolecular interaction is small, so in the first approximation of perturbation theory the energy of the crystal is equal to the mean value of operator (2.3) in states corresponding to wave functions (2.4) and (2.6) of the zeroth approximation. Thus, we obtain the energy of the ground state and the energy of the excited states that correspond to functions (2.6). If we take their difference, we find the excitation energy for transition from the ground state (2.4) to excited state (2.6):

$$E_f(\mathbf{k}) = \Delta\varepsilon_f + D_f + L_f(\mathbf{k}), \tag{2.8}$$

where $\Delta\varepsilon_f$ is the excitation energy of a free molecule;

$$D_f = \sum_{\mathbf{m}}{}' \left\{ \int |\varphi_{\mathbf{n}}^f|^2 V_{\mathbf{nm}} |\varphi_{\mathbf{m}}^0|^2 d\tau - \int |\varphi_{\mathbf{n}}^0|^2 V_{\mathbf{nm}} |\varphi_{\mathbf{m}}^0|^2 d\tau \right\} \tag{2.9}$$

is the change in the interaction energy of one molecule with all of the surrounding molecules in its transition to the f-th excited state; and

$$L_f(\mathbf{k}) = \sum_{\mathbf{m}}{}' M_{\mathbf{nm}}^f \exp\{i\mathbf{k}(\mathbf{n}-\mathbf{m})\} \tag{2.10}$$

is an addition to the excitation energy, which is a function of the wave vector of the excited state. In this case, the matrix element

$$M_{\mathbf{nm}}^f = \int \varphi_{\mathbf{n}}^{*0} \varphi_{\mathbf{m}}^{*f} V_{\mathbf{nm}} \varphi_{\mathbf{m}}^0 \varphi_{\mathbf{n}}^f d\tau \tag{2.11}$$

determines the transition of excitation f from molecule **n** to molecule **m**. In the initial state $\varphi_{\mathbf{m}}^0\varphi_{\mathbf{n}}^f$, molecule **m** is not excited, while molecule **n** is in the f-th excited state. In the final state $\varphi_{\mathbf{n}}^0\varphi_{\mathbf{m}}^f$, the excitation has moved from molecule **n** to molecule **m**. With this interpretation in mind, we shall call $M_{\mathbf{nm}}^f$ the matrix element or integral of the excitation transfer between molecules **n** and **m**.

Usually, an excited molecule interacts with surrounding molecules more strongly than does an unexcited molecule. Therefore, correction (2.9) to the excitation energy of the crystal has a negative sign and brings about a decrease in this excitation energy. The third term in (2.8) is a function of the N values of the wave vector **k**.

Thus, N different excited states correspond to a nondegenerate excited state of a free molecule in the crystal. In large crystals, adjacent values of **k** differ little from one another, so the N values of the excitation energy (2.8) form a quasi-continuous band of excited states that consists of N sublevels. Each of these excited states, which pertains to a specific value of the wave vector **k**, is a collective excited state of the entire crystal. Such elementary excitations were first examined by Frenkel [1] and are called e x c i t o n s.

The exciton states of a crystal are characterized by the quasi-momentum $\hbar\mathbf{k}$ and energy (2.8). The wave function of an exciton, with allowance for its time dependence, can be represented as

$$\Phi_{\mathbf{k}f}(\ldots\xi_{\mathbf{n}}\ldots) = N^{-1/2}\sum_{\mathbf{n}}\psi_{\mathbf{n}}^{f}(\ldots\xi_{\mathbf{n}}\ldots)\exp\{i\,[\mathbf{kn} - \omega_f(\mathbf{k})\,t]\}, \quad (2.12)$$

where $\omega_f(\mathbf{k}) = \hbar^{-1}E_f(\mathbf{k})$. It follows from the form of function (2.12) that all of the molecules of the crystal play an identical role in the formation of an exciton state. In other words, the excitation that is associated with exciton state (2.12) is distributed throughout the entire crystal and is not concentrated on one molecule.

The states associated with excitation of a region of the crystal are represented by wave packets, i.e., linear combinations of functions (2.12), for example

$$\Psi = \sum_{\mathbf{k}} w(\mathbf{k})\,\Phi_{\mathbf{k}f}, \quad (2.13)$$

where the functions $w(\mathbf{k})$ differ from zero only for **k** values that satisfy the conditions

$$k_x^0 - \Delta k \leqslant k_x \leqslant k_x^0 + \Delta k, \quad k_y = k_y^0, \quad k_z = k_z^0.$$

The excitations described by wave packets (2.13) do not have a definite energy or a definite wave-vector value. The smaller the region of excitation Δx, the greater the indeterminacy of Δk_x,

$$\Delta x\,\Delta k_x \approx 2\pi. \quad (2.14)$$

According to (2.7), in a cubic crystal the total range of variation of k_x in one band is $2\pi/a$. Let us assume that $\Delta k_x = \xi \cdot 2\pi/a$, where $\xi \leq 1$. Then, it follows from (2.14) that $\Delta x = a/\xi$. Thus, if $\xi = 0.1$, i.e., the indeterminacy of k is 0.1 of the possible k values in one band, then the effective region covered by the excitation in the crystal is equal to 10 lattice constants.

When electromagnetic radiation acts on the crystal, the excited region has linear dimensions on the order of the radiation wavelength. For visible and ultraviolet light, the wavelength is considerably greater than the lattice constant; therefore, $\Delta x \gg a$ and $\Delta k \ll 2\pi/a$. The excited states produced by light, therefore, are described by wave packets with a very small spread of wave-vector values. Such excitations can, with good approximation, be classified as idealized exciton states (2.12), which correspond to a definite **k** value.

The explicit dependence of the energy of exciton states upon the wave vector is defined by the sum (2.10). Methods for calculating these sums will be indicated in Section 5. In crystals with a center of symmetry, at small **k** for each given sense, which we shall denote by the unit vector **s**, excitation energy (2.8) is a quadratic function of the absolute value of **k**:

$$E(\mathrm{s}k) = E(\mathrm{s}) + \frac{\hbar^2\mathbf{k}^2}{2m^*(\mathrm{s})}. \tag{2.15}$$

The effective mass of an exciton m^* **(s)** can be either positive or negative.

An excitation that covers a certain region of a crystal and is described by wave packets (2.13), in which w(k) differs from zero only for **k** directed along **s** and having its absolute value in a small range about k_0, moves through the crystal along the sense of **s** with the group velocity

$$\mathrm{v}(\mathrm{s}k_0) = \left\{\mathrm{s}\frac{d\omega_f(k)}{dk}\right\}_{k=k_0} = \mathrm{s}\frac{\hbar k_0}{m^*(\mathrm{s})}.$$

For a rough estimate of the effective mass and group velocity of an exciton, we can consider a crystal whose unit cell is formed by three mutually perpendicular basis vectors $\mathbf{a}_i$. In this case, expression (2.10) can be represented (see Section 5) as

$$L(\mathbf{k}) \approx \frac{4\pi(\mathbf{dk})^2}{v\mathbf{k}^2} + \sum_{i=1}^{3} M_i \cos(\mathbf{ka}_i), \tag{2.16}$$

where v is the volume of a unit cell, and **d** is the effective dipole moment of a quantum transition in a molecule. Therefore, the effective mass of an exciton moving along $\mathbf{a}_i$ is

$$m_i^* = -\hbar^2 (M_i \mathbf{a}_i^2)^{-1} \text{ at } \mathbf{k}\mathbf{a}_i \ll 1, \tag{2.17}$$

and its group velocity

$$|\mathbf{v}_i| = \hbar^{-1} |M_i| k\mathbf{a}_i^2. \tag{2.18}$$

The time for movement of an excitation one lattice constant in the sense of $\mathbf{a}_i$ is

$$\tau = \frac{|\mathbf{a}_i|}{|\mathbf{v}_i|} = \frac{\hbar}{|M_i| ka_i}. \tag{2.19}$$

In the approximation employed, the widths of the exciton bands for excitons whose wave vectors are parallel to $\mathbf{a}_i$ are

$$\Delta L_i \equiv |L(0) - L(\pi/a)| = 2|M_i|.$$

The absolute values of the exciton effective masses (2.17) can be expressed in terms of the widths of the exciton bands:

$$|m_i^*| = 2\hbar^2 (\mathbf{a}_i^2 \Delta L_i)^{-1}. \tag{2.20}$$

In three-dimensional optically isotropic crystals, the ground state of the molecules has a totally symmetric representation, and the first electronic excited state is triply degenerate (such as, for example, the 1s and 2p states of Wannier excitons in cubic crystals). The electron dipole moment produced by light in such molecules is always perpendicular to the wave vector **k**, regardless of the sense of the latter. In optically isotropic crystals, therefore, the excitons produced by light are transverse, and in formulas (2.16)-(2.18), $M_1 = M_2 = M_3 > 0$. Longitudinal excitons, for which the dipole moment of the transition is parallel to the vector **k**, can be formed by other means of crystal excitation. The energy of longitudinal excitons in cubic crystals is usually higher than the energy of transverse excitons (see Section 5). Near $k \approx 0$, the energy of longitudinal excitons exceeds the energy of transverse excitons (see Section 5) by $4\pi\mathbf{d}^2/v\varepsilon_0$, where v is the volume of a unit cell; **d** is the electric dipole moment of a quantum transition of a molecule to the f-th electronic state; and ε_0 the dielectric constant that is determined by all of the electronic states of the molecules except the f-th.

The difference between the energies of longitudinal and transverse excitons is due to long-range dipole–dipole resonance interactions among the molecules of the crystal. In calculations of exciton energy, this effect has been studied by Heller and Marcus [27] and been discussed repeatedly [28-31] without taking into account ε_0. The effect of long-range dipole–dipole interaction on exciton energy has been investigated by Agranovich and Ginzburg [31] on the basis of the phenomenological Maxwell equations. In noncubic crystals, excitons can be separated into longitudinal and transverse only for certain senses of the wave vector. In these crystals, however, the exciton energy is not an analytic function of the wave vector at $\mathbf{k} \approx 0$ (see [32-34] and Section 5).

3. Molecular Excitons in Crystals with Several Molecules in a Unit Cell

Let a crystal contain σ molecules in a unit cell and let the positions of the molecules in the crystal be determined by the vectors

$$\mathbf{R}_{\mathbf{n}\alpha} = \mathbf{n} + \boldsymbol{\rho}_\alpha, \qquad \alpha = 1, 2, \ldots, \sigma, \tag{3.1}$$

where the lattice vector $\mathbf{n}$ indicates the location of the unit cell and the vector $\boldsymbol{\rho}_\alpha$ determines the position of molecule number α in the unit cell. The subscript α will also indicate the orientation of the molecule. Molecules that have the same indices α coincide in lattice-constant translations, so they are called *translational-equivalent*.

The energy operator of a crystal with fixed molecules can be written as

$$H = \sum_{\mathbf{n},\,\alpha} H_{\mathbf{n}\alpha} + \frac{1}{2} \sum_{\mathbf{n},\,\alpha;\,\mathbf{m},\,\beta}{}' V_{\mathbf{n}\alpha,\,\mathbf{m}\beta}\,. \tag{3.2}$$

Summation in (3.2) is over all σN molecules of the crystal. The prime on the second summation indicates that terms for which $\mathbf{n} = \mathbf{m}$ and $\alpha = \beta$ simultaneously are absent in the sum. The operators $H_{\mathbf{n}\alpha}$ determine the internal states of the molecules that occupy positions $\mathbf{n}\alpha$ in the crystal. For all practical purposes, we do not know the operators of free molecules and their eigenvalues and eigenfunctions. Therefore, it is convenient to consider the operators $H_{n\alpha}$ as molecule operators that take into account the possible deformations of molecules in the crystal due to their nonresonance interaction with their closest surroundings. In particular, the operators $H_{n\alpha}$ must reflect not the symmetry of free molecules but the local symmetry of the molecules located in positions $\mathbf{n}\alpha$ of the lattice. In this case, the operators $V_{\mathbf{n}\alpha,\mathbf{m}\beta}$ must take into account only the remaining part of the interaction between molecules $\mathbf{n}\alpha$ and $\mathbf{m}\beta$, which cannot be included in the operators $H_{\mathbf{n}\alpha}$. With this treatment, the eigenvalues ε_f of the operators $H_{\mathbf{n}\alpha}$ and certain combinations of the matrix elements of the operators $V_{\mathbf{n}\alpha,\mathbf{m}\beta}$, which would have been calculated with the aid of the eigenfunctions $\varphi^f_{\mathbf{n}\alpha}$ of the operators $H_{\mathbf{n}\alpha}$, may be considered as phenomenological parameters of the theory.

If all of the molecules of the crystal are the same, the wave function of the ground state of the crystal in the zeroth approximation (i.e., ignoring $V_{\mathbf{n}\alpha,\mathbf{m}\beta}$) will have the form

$$\psi^0 = \prod_{\mathbf{n}\alpha} \varphi^0_{\mathbf{n}\alpha}. \tag{3.3}$$

The energy of this state in the first approximation is

$$\mathscr{E}_0 = N\sigma\varepsilon_0 + \frac{1}{2}\sum{}' \int |\varphi^0_{\mathbf{n}\alpha}|^2 V_{\mathbf{n}\alpha,\,\mathbf{m}\beta} |\varphi^0_{\mathbf{m}\beta}|^2 \, d\tau.$$

The states in which only one molecule $\mathbf{n}\alpha$ is excited are represented in the zeroth approximation by the wave function

$$\psi^f_{\mathbf{n}\alpha} = \varphi^f_{\mathbf{n}\alpha} \prod_{\mathbf{m},\,\beta} \varphi^0_{\mathbf{m}\beta}\,, \qquad \mathbf{m},\,\beta \neq \mathbf{n},\,\alpha. \tag{3.4}$$

Instead of (3.4), it is more convenient to consider another set of orthonormal functions

$$\psi^f_{\alpha}(\mathbf{k}) = N^{-1/2} \sum_{\mathbf{n}} \psi^f_{\mathbf{n}\alpha} \exp(i\mathbf{k}\mathbf{n}), \tag{3.5}$$

where the wave vector is defined by Eqs. (2.7).

The excited states of the crystal are determined by the energy operator

$$\Delta H = H - \mathscr{E}_0,$$

where H is given by expression (3.2). The operator ΔH in the representation of wave functions (3.5) has the form of a matrix with matrix elements

$$\mathscr{L}_{\alpha\beta}(\mathbf{k}) \equiv \langle\alpha|\Delta H|\beta\rangle = (\Delta\varepsilon_f + D_f)\,\delta_{\alpha\beta} + L^f_{\alpha\beta}(\mathbf{k}), \tag{3.6}$$

where

$$D_f = \sum_{\mathbf{n},\,\alpha}' \left\{\int |\varphi^f_{\mathbf{m}\beta}|^2 V_{\mathbf{n}\alpha,\,\mathbf{m}\beta} |\psi^0_{\mathbf{n}\alpha}|^2 d\tau - \int |\varphi^0_{\mathbf{m}\beta}|^2 V_{\mathbf{n}\alpha,\,\mathbf{m}\beta} |\varphi^0_{\mathbf{n}\alpha}|^2 d\tau\right\}, \tag{3.6a}$$

$$L^f_{\alpha\beta}(\mathbf{k}) = \sum_{\mathbf{n}}' M^f_{0\beta,\,\mathbf{n}\alpha} \exp(i\mathbf{kn}). \tag{3.7}$$

In formula (3.7)

$$M^f_{0\beta,\,\mathbf{n}\alpha} = \int \varphi^{*0}_{\mathbf{n}\alpha} \varphi^{*f}_{0\beta} V_{\mathbf{n}\alpha,\,0\beta} \varphi^0_{0\beta} \varphi^f_{\mathbf{n}\alpha}\, d\tau \tag{3.8}$$

are the matrix elements of the transfer of excitation f from molecule $\mathbf{n}\alpha$ to molecule $\mathbf{m}\beta$. These elements are real and symmetric about the indices $\mathbf{n}\alpha$ and 0β.

Matrix elements (3.8) form a Hermitian matrix. Considering that matrix elements (3.8) are real and symmetric, it follows from (3.7) that

$$L^f_{\alpha\beta}(\mathbf{k}) = L^{*f}_{\beta\alpha}(\mathbf{k}) = L^f_{\alpha\beta}(-\mathbf{k}).$$

The Hermitian matrix formed by matrix elements (3.7) will be called the r e s o n a n c e - i n t e r a c t i o n m a t r i x .

The eigenvalues of the operator ΔH with matrix elements (3.6) determine the energies of the exciton states in the crystal. To find these eigenvalues, we must convert from basis functions (3.5) to the new basis functions $\Phi^f_\mu(\mathbf{k})$, relative to which the operator ΔH is diagonal. This conversion is accomplished by the unitary transformation

$$\psi^f_\alpha(\mathbf{k}) = \sum_{\mu=1}^{\sigma} u_{\alpha\mu} \Phi^f_\mu(\mathbf{k}), \tag{3.9}$$

$$\sum_{\alpha=1}^{\sigma} u^*_{\alpha\mu} u_{\alpha\nu} = \delta_{\mu\nu}. \tag{3.9a}$$

In general, the coefficients of unitary transformation (3.9) are complex functions of the wave vector. They are determined from

$$\sum_{\alpha,\beta} u^*_{\alpha\nu}\mathscr{L}^f_{\alpha\beta}(\mathbf{k})\,u_{\beta\mu} = E^f_\mu(\mathbf{k})\,\delta_{\nu\mu}$$

or the equivalent equations

$$\sum_{\beta=1}^{\sigma} \mathscr{L}^f_{\alpha\beta}(\mathbf{k})\,u_{\beta\mu} = E^f_\mu(\mathbf{k})\,u_{\alpha\mu}.$$

Thus, the σ coefficients $u_{\alpha\mu}$ at fixed μ form the eigenfunctions of the Hermitian matrix operator $\mathscr{L}_{\alpha\beta}$ for the eigenvalue $E_\mu(\mathbf{k})$. To determine these eigenvalues and eigenfunctions, we must solve the system of homogeneous equations

$$\sum_{\beta=1}^{\sigma} \mathscr{L}_{\alpha\beta}(\mathbf{k})\,u_\beta = E^f(\mathbf{k})\,u_\alpha. \tag{3.10}$$

The condition of nontrivial solvability of this system reduces to an equation of degree σ in $E^f(\mathbf{k})$. Since $\mathscr{L}_{\alpha\beta}$ is Hermitian, all of the roots of this equation $\mathbf{E}^f_\mu(\mathbf{k})$, $\mu = 1, 2,..., \sigma$, are real functions of $\mathbf{k}$. Therefore, each excited state of a molecule in the crystal will be associated with σ bands of quasi-continuous excited states of the crystal. According to (3.9), the band $\mathbf{E}^f_\mu(\mathbf{k})$ corresponds to the wave function

$$\Phi^f_\mu(\mathbf{k}) = \sum_\alpha \psi^f_\alpha(\mathbf{k})\,u^*_{\alpha\mu}. \tag{3.11}$$

When $\mathbf{k} \neq 0$, matrix elements (3.7) are, generally speaking, complex functions of the wave vector. For crystals with centers of symmetry, it is often convenient to consider the real and symmetric about the indices α and β matrix elements $\widetilde{L}_{\alpha\beta}(\mathbf{k})$, which are introduced by the unitary transformation

$$\widetilde{L}_{\alpha\beta}(\mathbf{k}) = L_{\alpha\beta}(\mathbf{k})\,e^{i\mathbf{k}(\rho_\alpha-\rho_\beta)} = \sum_{\mathbf{n}}{}' M^f_{0\beta,\,\mathbf{n}\alpha}e^{i\mathbf{k}\mathbf{r}_\mathbf{n}}, \tag{3.12}$$

where

$$\mathbf{r}_\mathbf{n} = \mathbf{n} + \rho_\alpha - \rho_\beta$$

is the radius vector from molecule 0β to molecule $\mathbf{n}\alpha$. Simultaneously with unitary transformation (3.12), we must make the transformation of functions

$$\tilde{u}_\alpha = u_\alpha \exp(i\mathbf{k}\boldsymbol{\rho}_\alpha). \tag{3.13}$$

Then system (3.10) is transformed to

$$\sum_{\beta=1}^{\sigma} \{(\Delta\varepsilon_f + D_f)\,\delta_{\alpha\beta} + \tilde{L}^f_{\alpha\beta}(\mathbf{k})\}\,\tilde{u}_\beta = E^f(\mathbf{k})\,\tilde{u}_\alpha, \tag{3.14}$$

where

$$\tilde{L}_{\alpha\alpha}(\mathbf{k}) = L_{\alpha\alpha}(\mathbf{k}).$$

The solutions $\tilde{u}_\alpha(\mathbf{k})$ of Eqs. (3.14) will be real functions of **k**. Here, the wave functions of the exciton states (3.11) are written as

$$\Phi^{\overline{f}}_{\mu}(\mathbf{k}) = \sum_{\alpha} \psi^f_\alpha(\mathbf{k})\,\tilde{u}_{\alpha\mu} \exp(-i\mathbf{k}\boldsymbol{\rho}_\alpha). \tag{3.15}$$

As an illustration of the above relationships, let us consider monoclinic crystals such as anthracene, naphthalene, etc., which have a center of symmetry and contain two identical molecules in a unit cell. After unitary transformation (3.12), the matrix elements (3.6) of the operator of the excited states of the crystal form a square matrix of order two

$$\begin{pmatrix} \Delta\varepsilon_f + D_f + \tilde{L}_{11}(\mathbf{k}), & \tilde{L}_{12}(\mathbf{k}) \\ \tilde{L}_{12}(\mathbf{k}), & \Delta\varepsilon_f + D_f + \tilde{L}_{22}(\mathbf{k}) \end{pmatrix}. \tag{3.16}$$

As will be shown in Section 5, for the senses of the wave vector that are perpendicular or parallel to the plane of symmetry of the crystal, we have

$$\tilde{L}_{11}(\mathbf{k}) = \tilde{L}_{22}(\mathbf{k}).$$

In this case, matrix (3.16) of the energy operator of the excited states of the crystal is diagonalized by a transformation matrix that is independent of the vector **k**:

$$(\tilde{u}_{\alpha\mu}) = \frac{1}{\sqrt{2}} \begin{pmatrix} 1, & 1 \\ 1, & -1 \end{pmatrix}. \tag{3.17}$$

Here, according to (3.14) and (3.15), a molecular excitation $\Delta\varepsilon_f$ is associated with two bands of excited states.

1) The exciton band

$$E_1^f(\mathbf{k}) = \Delta\varepsilon_f + D_f + \tilde{L}_{11}(\mathbf{k}) + \tilde{L}_{12}(\mathbf{k}) \quad (3.18)$$

with the wave functions (with accuracy to the common phase factor)

$$\Phi_1^f(\mathbf{k}) = \frac{1}{\sqrt{2}}\{\psi_1^f(\mathbf{k}) + \psi_2^f(\mathbf{k})\,e^{i\varphi(\mathbf{k})}\}, \quad (3.18a)$$

where

$$\exp\{i\varphi(\mathbf{k})\} = \exp\{i\mathbf{k}(\rho_1 - \rho_2)\} = \frac{L_{12}(\mathbf{k})}{|L_{12}(\mathbf{k})|}. \quad (3.19)$$

2) The exciton band

$$E_2^f(\mathbf{k}) = \Delta\varepsilon_f + D_f + L_{11}(\mathbf{k}) - L_{12}(\mathbf{k}) \quad (3.20)$$

with the wave functions

$$\Phi_2^f(\mathbf{k}) = \frac{1}{\sqrt{2}}\{\psi_1^f(\mathbf{k}) - \psi_2^f(\mathbf{k})\,e^{i\varphi(\mathbf{k})}\}. \quad (3.20a)$$

In general (see Section 5), matrix elements $\tilde{L}_{11}$ and $\tilde{L}_{12}$ are not analytic functions of k in the area of k ≈ 0, and they cannot be expanded into a series in the vicinity of this point. When the sense of k is fixed, however, this expansion is possible. Let **s** be a unit vector that determines one of the senses for which (3.18) and (3.20) are valid. Then k = **s**k, and at small k we can write

$$E_1(\mathrm{s}k) = E_1(\mathrm{s}) + \frac{\hbar^2k^2}{2m_1^*(\mathrm{s})}, \qquad E_2(\mathrm{s}k) = E_2(\mathrm{s}) + \frac{\hbar^2k^2}{2m_2^*(\mathrm{s})},$$

where m_1^* (**s**) and m_2^* (**s**) are the effective masses of the exciton states in the direction of **s**. They are determined by

$$[m_1^*(\mathrm{s})]^{-1} = \hbar^{-2}\left\{\frac{\partial^2}{\partial k^2}[\tilde{L}_{11}(\mathrm{s}k) + \tilde{L}_{12}(\mathrm{s}k)]\right\}_{k=0},$$

$$[m_2^*(\mathrm{s})]^{-1} = \hbar^{-2}\left\{\frac{\partial^2}{\partial k^2}[\tilde{L}_{11}(\mathrm{s}k) - \tilde{L}_{12}(\mathrm{s}k)]\right\}_{k=0}.$$

Generally speaking, the effective mass of an exciton is a function of the direction of its motion and of the number of the exciton band. If

$$\left|\left\{\frac{\partial^2}{\partial k^2}\tilde{L}_{11}(\mathrm{s}k)\right\}_{k=0}\right| < \left|\left\{\frac{\partial^2}{\partial k^2}\tilde{L}_{12}(\mathrm{s}k)\right\}_{k=0}\right|,$$

then the effective masses of excitons in the two bands at k ≈ 0 will have opposite signs. If the inequality is reversed, both effective masses will have the same sign.

If a unit cell contains two different molecules or the sense of the wave vector is not perpendicular to the plane of symmetry or does not lie in that plane, then $\tilde{L}_{11}(\mathbf{k})$ cannot be equal to $\tilde{L}_{22}(\mathbf{k})$ and system (3.14) is transformed to

$$\sum_{\beta=1} \mathscr{L}^{f}_{\alpha\beta}(\mathbf{k})\,\tilde{u}_{\beta} = E(\mathbf{k})\,\tilde{u}_{\alpha}, \tag{3.21}$$

where

$$\mathscr{L}^{f}_{\alpha\beta}(\mathbf{k}) \equiv (\Delta\varepsilon_{f} + D_{f})\,\delta_{\alpha\beta} + \tilde{L}_{\alpha\beta}(\mathbf{k}).$$

By solving system (3.21), we find the energies of the exciton states

$$E_{\mu}(\mathbf{k}) = \frac{1}{2}\{\mathscr{L}_{11}(\mathbf{k}) + \mathscr{L}_{22}(\mathbf{k}) - (-1)^{\mu}[(\mathscr{L}_{11} - \mathscr{L}_{22})^{2} + 4\tilde{L}^{2}_{12}]^{1/2}\} \tag{3.22}$$

and the transformation matrix

$$(\tilde{u}_{\alpha\mu}) = \begin{pmatrix} \cos\frac{\gamma}{2} & \sin\frac{\gamma}{2} \\ \sin\frac{\gamma}{2} & -\cos\frac{\gamma}{2} \end{pmatrix}, \tag{3.23}$$

where $\alpha, \mu = 1, 2$, and the angle γ is determined from

$$\operatorname{tg}\gamma = 2\tilde{L}_{12}(\mathbf{k})\,[\mathscr{L}_{11}(\mathbf{k}) - \mathscr{L}_{22}(\mathbf{k})]^{-1}.$$

According to (3.15), the wave functions of the exciton states associated with energies (3.22) have the form

$$\left.\begin{aligned} \Phi_{1}(\mathbf{k}) &= \psi_{1}(\mathbf{k})\cos\frac{\gamma}{2} + \psi_{2}(\mathbf{k})\,e^{i\varphi(\mathbf{k})}\sin\frac{\gamma}{2}, \\ \Phi_{2}(\mathbf{k}) &= \psi_{1}(\mathbf{k})\sin\frac{\gamma}{2} - \psi_{2}(\mathbf{k})\,e^{i\varphi(\mathbf{k})}\cos\frac{\gamma}{2}, \end{aligned}\right\} \tag{3.24}$$

where $\varphi(\mathbf{k})$ is determined by Eq. (3.19).

If the inequality $|\mathscr{L}_{11} - \mathscr{L}_{22}| \ll |\tilde{L}_{12}|$ is satisfied, then $\gamma \approx \pi/2$, and wave functions (3.24) and excitation energies (3.22) will differ little from the corresponding values given by expressions (3.18) and (3.20). But if

$$|\mathscr{L}_{11} - \mathscr{L}_{22}| \gg |\tilde{L}_{12}|,$$

then $\gamma \approx 0$, and the excited states of the crystal are determined by the expressions

$$E_1(\mathbf{k}) = \Delta\varepsilon_1 + D_1 + L_{11}(\mathbf{k}) + \frac{\widetilde{L}_{12}^2(\mathbf{k})}{|\mathscr{L}_{11} - \mathscr{L}_{22}|},$$
$$E_2(\mathbf{k}) = \Delta\varepsilon_2 + D_2 + L_{22}(\mathbf{k}) - \frac{\widetilde{L}_{12}^2(\mathbf{k})}{|\mathscr{L}_{11} - \mathscr{L}_{22}|}$$

and correspond to the functions

$$\Phi_1(\mathbf{k}) \approx \psi_1(\mathbf{k}), \quad \Phi_2(\mathbf{k}) = \psi_2(\mathbf{k}).$$

In this case, the molecules of type (or orientation) 1 participate in the formation of the excited states of the crystal practically independently of the molecules of type (or orientation) 2.

Thus, one nondegenerate excited state of a free molecule in crystals containing two of the same molecules in each unit cell corresponds not to one but to two bands of exciton states. This splitting was first examined by the author [7, 8, 35, 36], and it is usually called Davydov splitting [14, 37-41] to distinguish it from Bethe splitting [42] of degenerate energy levels of atoms and molecules in crystals by their internal electric fields, which remove the degeneracy.

In the particular case when the energy of the exciton states is determined by expressions (3.18) and (3.20), the absolute value of the splitting when the wave vector is fixed is

$$|\Delta E(\mathbf{k})| = 2|\widetilde{L}_{12}(\mathbf{k})|. \tag{3.25}$$

If a unit cell contains σ identical molecules, then the excitation energy of the crystal will split up into σ bands of excited states, whose energies and wave functions (3.11) are determined by solving a system of σ homogeneous equations (3.10) for each value of the wave vector **k**. In general, such calculations are complicated. In particular cases, however, they can be simplified considerably. In crystals that have a center of symmetry and consist of identical molecules, for example, system (3.10) can be replaced by system (3.14), which contains only real, symmetric matrix elements $\widetilde{L}_{\alpha\beta}(\mathbf{k})$. In this case, the solutions of (3.14) are expressed in terms of the real matrices $(\widetilde{u}_{\alpha\mu})$. For senses of the wave vectors that are perpendicular to or lie in the plane of symmetry of the crystal, all of the diagonal matrix elements $L_{\alpha\alpha}(\mathbf{k})$ are equal to one another. In these cases, all of the matrix elements of the unitary matrix $(\widetilde{u}_{\alpha\mu})$ are independent of **k** and have the same absolute values, which, along with the unitarity condition, leads to the system of equations

$$\sum_{\alpha} \tilde{u}_{\alpha\mu} \tilde{u}_{\alpha\nu} = \delta_{\nu\mu}, \qquad \tilde{u}^2_{\alpha\mu} = \text{const},$$

which uniquely determine all of the values $\tilde{u}_{\alpha\mu}$. For example, for crystals with four identical molecules in a unit cell

$$(\tilde{u}_{\alpha\mu}) = \frac{1}{2} \begin{pmatrix} 1 & 1 & 1 & 1 \\ 1 & -1 & 1 & -1 \\ 1 & -1 & -1 & 1 \\ 1 & 1 & -1 & -1 \end{pmatrix}.$$

4. Exciton States and the Dielectric Constant

Exciton states occur when a crystal absorbs light and by other means of crystal excitation. A monochromatic light wave produces in the crystal a variable specific polarization **P** (or a variable current density $j = \partial \mathbf{P}/\partial t$), which is proportional to the electric-field strength of the light wave. The proportionality factor is a function of the dielectric constant of the crystal (see Sections 2 and 3 of Chapter I), which, in turn, is determined by the spectrum and properties of the excited states of the crystal. Let us examine the contribution of the exciton states to the dielectric constant of the crystal. For this, we must study the changes in the states of the crystal produced by an external light wave of frequency ω. We shall assume that frequency ω corresponds to the transparency range of the crystal. In this case, we may ignore the interaction of excitons with phonons, which is the cause of light absorption by the crystal (see Chapter IV).

If we ignore the interaction of excitons with phonons, the electronic state of the crystal will be described by operator (3.2), which does not take into account motion of the molecules. Let this operator be H_{ex}. We shall assume that we know the eigenvalues $\hbar\omega_l$ and eigenfunctions of the operator H_{ex}. The field of a light wave with frequency ω and wave vector **Q** is determined by the transverse vector potential

$$\mathbf{A}(\mathbf{r}, t) = \mathbf{A}(\mathbf{Q}) \exp\{i(\mathbf{Q}\mathbf{r} - \omega t)\} + \text{complex conj.}, \tag{4.1}$$

where

$$\operatorname{div} \mathbf{A} = 0, \quad \mathbf{A}(\mathbf{Q}) = \frac{c}{i\omega} \mathbf{E}^{\perp}(\mathbf{Q}); \tag{4.2}$$

c is the velocity of light; and $\mathbf{E}^{\perp}(\mathbf{Q})$ is the transverse part of the longwave electric field of the light wave, which is considered an external (perturbing) field. All of the remaining electromagnetic field (including the total Coulomb field) in the crystal characterizes the interaction between the electrons and atomic nuclei. It is assumed to be taken into account in the operator H_{ex}.

The interaction operator of the light wave with the crystal can be written as

$$H_{\text{int}} = -\frac{1}{c} \mathbf{A}(\mathbf{Q}) \mathbf{J}_{\perp}(\mathbf{Q}) e^{-i\omega t} + \text{Hermitian conj.}, \tag{4.3}$$

where

$$\mathbf{J}_{\perp}(\mathbf{Q}) \equiv \frac{e}{m} \sum_{\mathbf{n}, j} \mathbf{p}_{\mathbf{n}j}^{\perp} \exp\{i\mathbf{Q}(\mathbf{n} + \mathbf{x}_{\mathbf{n}j})\}; \tag{4.4}$$

e, m are the electron charge and mass; $\mathbf{x}_{\mathbf{n}j}$ is the radius vector (at lattice point $\mathbf{n}$) of the j-th electron, which is a part of the molecules of the $\mathbf{n}$-th unit cell; and $\mathbf{p}_{\mathbf{n}j}^{\perp}$ is the transverse component of the momentum operator of the same electron. We shall assume that interaction operator (4.3) is adiabatically inserted in the infinite past. For this, it is sufficient to let frequency ω in (4.3) contain a small positive imaginary part $i\eta$. In all of the final results, we should pass to the limit $\eta \to +\infty$.

Let the crystal be in the state Φ_0 at time $t = -\infty$. Then, by time t, under the influence of perturbation operator (4.3), the crystal will adiabatically change to a state that in the first order of perturbation theory is described by the function

$$|t\rangle = \{\Phi_0 + \varphi e^{-i\omega t} + \chi e^{i\omega^* t}\} e^{-i\omega_0 t}, \tag{4.5}$$

where

$$\left.\begin{aligned} \varphi &= -\frac{1}{c} [\hbar(\omega_0 + \omega) - H_{ex}]^{-1} \mathbf{A}(\mathbf{Q}) \mathbf{J}_{\perp}(\mathbf{Q}) \Phi_0, \\ \chi &= -\frac{1}{c} [\hbar(\omega_0 - \omega^*) - H_{ex}]^{-1} \mathbf{A}^{+}(\mathbf{Q}) \mathbf{J}_{\perp}^{+}(\mathbf{Q}) \Phi_0. \end{aligned}\right\} \tag{4.6}$$

In order to calculate the current density in the state represented by function (4.5), let us examine the current-density operator $\hat{\mathbf{j}}_{\perp}$. In the presence of the field determined by the transverse potential A, we can write

$$\hat{\mathbf{j}}_\perp(\mathbf{r}) = \frac{e}{m}\sum_{\mathbf{n},j}\delta(\mathbf{r}-\mathbf{n}-\mathbf{x}_{\mathbf{n}j})\left[\mathbf{p}_{\mathbf{n}j}^\perp - \frac{e}{c}\mathbf{A}(\mathbf{n}+\mathbf{x}_{\mathbf{n}j};\,t)\right].$$

Using the delta-function representation

$$\delta(r) = \frac{1}{V}\sum_{\mathbf{k}}\exp(i\mathbf{kr}),$$

where V is the volume of the crystal and k is wave vector (2.7), let us transform $\hat{\mathbf{j}}_\perp(\mathbf{r})$ to

$$\hat{\mathbf{j}}_\perp(\mathbf{r}) = \frac{1}{V}\sum_{\mathbf{k}} e^{i\mathbf{kr}}\mathbf{J}_\perp(-\mathbf{k}) - \frac{e^2 n_e}{mc}\{\mathbf{A}(\mathbf{Q})\,e^{i(\mathbf{Qr}-\omega t)} + \text{Hermitian conj.}$$

where n_e is the number of electrons per unit volume of the crystal, and the operator $\mathbf{J}_\perp(\mathbf{k})$ is determined by Eq. (4.4)

In a linear approximation, the mean current density in state (4.5) is

$$\mathbf{j}_\perp(\mathbf{r},t) \equiv \langle t|\hat{\mathbf{j}}_\perp(\mathbf{r})|t\rangle = \sum_{\mathbf{k}}\langle t|\hat{\mathbf{j}}_{\mathbf{k}}^\perp(\mathbf{r})|t\rangle - $$

$$-\frac{e^2 n_e}{mc}\{\mathbf{A}(\mathbf{Q})\exp[i(\mathbf{Qr}-\omega t)] + \text{Hermitian conj.}\},\tag{4.7}$$

where

$$\langle t|\hat{\mathbf{j}}_{\mathbf{k}}^\perp(\mathbf{r})|t\rangle = \frac{1}{cV}\sum_l{}' \left[\left\{\frac{\langle\Phi_0|\mathbf{J}_\perp(-\mathbf{k})|\Phi_l\rangle\langle\Phi_l|\mathbf{A}(\mathbf{Q})\,\mathbf{J}_\perp(\mathbf{Q})|\Phi_0\rangle}{\hbar(\omega_{l0}-\omega)} + \right.\right.$$

$$\left.\left. + \frac{\langle\Phi_0|\mathbf{A}(\mathbf{Q})\,\mathbf{J}_\perp(\mathbf{Q})|\Phi_l\rangle\langle\Phi_l|\mathbf{J}_\perp(-\mathbf{k})|\Phi_0\rangle}{\hbar(\omega_{l0}+\omega)}\right\} e^{-i\omega t + i\mathbf{kr}} + \text{Hermitian conj.}\right];\tag{4.8}$$

$\omega_{l0} \equiv \omega_l - \omega_0$; and the prime on the summation indicates that summation is over all states l that do not coincide with the electronic ground state 0.

Let us use expression (4.7) in the case of the exciton states of a crystal that contains σ identical molecules in a unit cell. In our approximation (the molecules are rigidly fixed in the lattice), the excited states are characterized by a set of three values: f, μ, $\mathbf{k}$, where f denotes the corresponding molecular excitation, μ the number of the exciton band, and k the wave vector of the exciton state. The wave functions of the exciton states corresponding to the excitation energy

$$\hbar\omega_{l0} = E^f_\mu(\mathbf{k}) \tag{4.9}$$

have, according to formulas (3.5) and (3.11), the form

$$\Phi_{f\mu}(\mathbf{k}) = \frac{1}{\sqrt{N}} \sum_{\mathbf{n},\alpha} e^{i\mathbf{kn}} \psi^f_{\mathbf{n}\alpha} u^*_{\alpha\mu}(\mathbf{k}). \tag{4.10}$$

In the dipole approximation (see [43], Section 43, p. 333),

$$\langle \psi^f_{\mathbf{n}\alpha} | e^{i\mathbf{Qx}^\perp_{\mathbf{n}\alpha}} \frac{e}{m} \mathbf{p}^\perp_{\mathbf{n}\alpha} | \psi^0_{\mathbf{n}\alpha} \rangle = i\Omega_f \mathbf{d}_{f\alpha}, \tag{4.11}$$

where $\mathbf{x}^\perp_{\mathbf{n}\alpha}$ and $\mathbf{p}^\perp_{\mathbf{n}\alpha}$ are the total transverse (to the vector **k**) components of the radius vector and the momentum operator of all of the electrons of molecule $\mathbf{n}\alpha$; Ω_f is the frequency of the intramolecular transition $0 \to f$; and

$$\mathbf{d}^\perp_{f\alpha} \equiv \langle \psi^f_{\mathbf{n}\alpha} | e\mathbf{x}^\perp_{\mathbf{n}\alpha} | \psi^0_{\mathbf{n}\alpha} \rangle \tag{4.12}$$

is the transverse (to the vector **k**) component of the electric dipole moment of the transition $0 \to f$ in molecule α. Therefore, if we substitute (4.9) and (4.10) into expression (4.8), we obtain

$$\langle t | \mathbf{j}_\mathbf{k}(\mathbf{r}) | t \rangle = \frac{1}{cv} \sum_{f,\mu} \left[\left\{ \frac{\mathbf{p}^{*\perp}_{f\mu} (\mathbf{A}(\mathbf{Q})\, \mathbf{p}^\perp_{f\mu})}{E^f_\mu(\mathbf{k}) - \hbar\omega} e^{i\mathbf{kr}} \delta(\mathbf{k} - \mathbf{Q} + 2\pi\mathbf{b}\nu) + \right.\right.$$
$$\left.\left. + \frac{(\mathbf{A}(\mathbf{Q})\, \mathbf{p}^{*\perp}_{f\mu}) \mathbf{p}^\perp_{f\mu}}{E^f_\mu(\mathbf{k}) + \hbar\omega} e^{-i\mathbf{kr}} \delta(\mathbf{k} + \mathbf{Q} + 2\pi\mathbf{b}\nu) \right\} \Omega^2_f e^{-i\omega t} + \text{Hermitian conj.} \right], \tag{4.13}$$

where v is the volume of a unit cell;

$$\mathbf{p}^\perp_{f\mu} \equiv \sum_\alpha \mathbf{d}^\perp_{f\alpha} u^*_{\alpha\mu}(\mathbf{k}) \tag{4.14}$$

is the transverse component of the electric dipole moment of a unit cell, which is due to transition of the crystal from the ground state to exciton state μ, **k** which corresponds to intramolecular excitation f; **b** is the reciprocal-lattice vector; and $\nu = 0, \pm 1, \pm 2, \ldots$.

It follows from (4.13) that excitons are produced by monochromatic electromagnetic wave (4.1) only when the following three conditions are fulfilled simultaneously:

1) when the scalar product of the transverse (to the vector **Q**) component of the electric-dipole-moment vector (4.14) of one

unit cell and the vector $\mathbf{A}(\mathbf{Q})$ differs from zero (polarization condition);

2) when the law of conservation of quasi-momentum is satisfied: $\mathbf{k} = \mathbf{Q}$. For visible and ultraviolet light, $\mathbf{Q}\mathbf{a}_i \ll 1$. The values of $\mathbf{k} = \mathbf{Q} \pm 2\pi\mathbf{b}\nu$ at $\nu \neq 0$ correspond to shortwave fields. When studying the optical properties of crystals in the visible and ultraviolet regions, only $\nu = 0$ should be taken into account in expression (4.13), since the contribution of shortwave fields is very small (see Section 7 of Chapter (IV);

3) when the law of conservation of energy is satisfied:

$$E^f_\mu(\mathbf{Q}) = \hbar\omega.$$

The electric dipole moment and its transverse component (4.14) of a quantum transition in one unit cell are easily calculated if we know the coefficients $u_{\alpha\mu}$ and the directions of the dipole moments $\mathbf{d}_{f\alpha}$ of the molecules that form the unit cell. For example, in an anthracene crystal, which belongs to the monoclinic system [44] with two molecules in a unit cell, the lattice vector is determined by the expression

$$\mathbf{n} = \mathbf{a}n_1 + \mathbf{b}n_2 + \mathbf{c}n_3, \quad n_i = 0, \pm 1, \pm 2, \ldots$$

where $i = 1, 2, 3$; and $\mathbf{a}, \mathbf{b}, \mathbf{c}$ are the three basis vectors of the base crystal, of which $\mathbf{b}$ coincides with the monoclinic axis; $\mathbf{ab} = \mathbf{bc} = 0$; and $\mathbf{ac} = ac \cos 125°$. The centers of mass of the molecules are determined by the vector

$$\mathbf{n} + \boldsymbol{\rho}_\alpha, \quad \alpha = 1, 2, \tag{4.15}$$

where

$$\boldsymbol{\rho}_\alpha = \begin{cases} 0, & \text{if } \alpha = 1; \\ \frac{1}{2}(\mathbf{a} + \mathbf{b}), & \text{if } \alpha = 2. \end{cases}$$

In this case, according to formulas (3.13) and (3.17), the transformation matrix has the form

$$u_{\alpha\mu} = \frac{1}{\sqrt{2}} \begin{pmatrix} 1 & \exp\left[-\frac{i\mathbf{k}}{2}(\mathbf{a} + \mathbf{b})\right] \\ 1 & -\exp\left[-\frac{i\mathbf{k}}{2}(\mathbf{a} + \mathbf{b})\right] \end{pmatrix}.$$

Therefore, transitions to exciton states (3.18) and (3.20) are associated with the dipole moments

$$\left.\begin{aligned} \mathbf{p}_{f1} &= \frac{1}{\sqrt{2}}\left(\mathbf{d}_{f1} + \mathbf{d}_{f2}\exp\left\{i\mathbf{k}\frac{\mathbf{a}+\mathbf{b}}{2}\right\}\right) && \text{in band } E_1^f(\mathbf{k}), \\ \mathbf{p}_{f2} &= \frac{1}{\sqrt{2}}\left(\mathbf{d}_{f1} - \mathbf{d}_{f2}\exp\left\{i\mathbf{k}\frac{\mathbf{a}+\mathbf{b}}{2}\right\}\right) && \text{in band } E_2^f(\mathbf{k}). \end{aligned}\right\} \quad (4.15\text{a})$$

At fixed **k**, the scalar product of these vectors is

$$(\mathbf{p}_{f1}\mathbf{p}_{f2}) = 2\,(\mathbf{d}_{f1}\mathbf{d}_{f2})\sin\left[\frac{\mathbf{k}}{2}(\mathbf{a}+\mathbf{b})\right]. \quad (4.16)$$

For all of the senses of **k** that are perpendicular to the vector **a** + **b**, in particular, those perpendicular to the cleavage plane **ab** of the crystal, the dipole moments of the transitions to the two bands [$E_1^f(\mathbf{k})$ and $E_2^f(\mathbf{k})$] are mutually orthogonal. This orthogonality is also realized for all **k** that satisfy the inequality $\mathbf{k}(\mathbf{a}+\mathbf{b}) \ll 1$, particularly at $\mathbf{k} = \mathbf{Q}$.

Thus, excitations that are associated with different bands of exciton states, although they are associated with a single nondegenerate energy state in a free molecule, have not only different energies but also different polarizations. The polarization is characterized by the symmetry properties of the crystal and emphasizes the collective nature of the exciton states, which is a result of interactions between molecules. If the molecular crystal were a simple combination of oriented anisotropic molecules (oriented-gas model), the above-mentioned polarization would be absent.

This assertion is illustrated schematically in Fig. 1, where the case of the oriented-gas model is shown on the left. In this case, the directions of the dipole moments of the transition in individual molecules are independent and are related to a single frequency ω_0. The rate of absorption and emission of light of frequency ω_0 varies little when the polarization changes. The more real case in which both molecules participate simultaneously in light absorption and emission is shown on the right in Fig. 1. Frequency ω_1 corresponds to a collective excitation in which the dipole moment of a unit cell (solid arrow) is formed by addition of the dipole moments of the transitions in the molecules (broken arrows). Frequency ω_2 corresponds to a collective excitation in

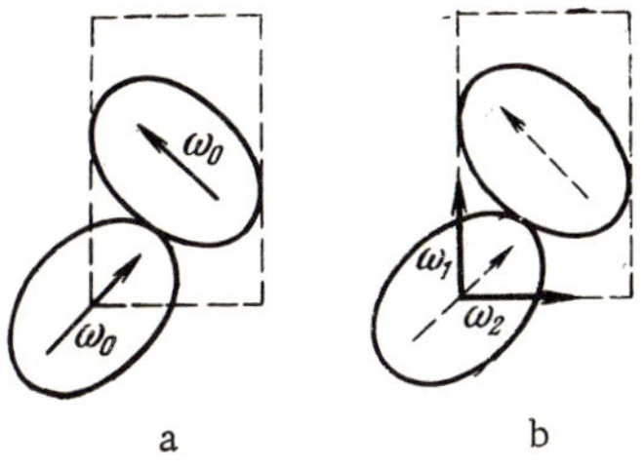

Fig. 1. Dipole transitions in molecular crystal: a) for oriented-gas model; b) with allowance for exciton states.

which the dipole moment of a unit cell is equal to the difference between the dipole moments of the transitions in the molecules. Therefore, frequencies ω_1 and ω_2 will appear in different components of the polarized light.

Now let us calculate the optical properties of the crystal. If we substitute matrix element (4.13) into expression (4.7) and take (4.2) into account, we find

$$\langle t|\hat{\mathbf{j}}_\perp(\mathbf{r})|t\rangle = \frac{\partial \mathbf{P}^\perp}{\partial t} = -i\omega\beta\,\mathbf{E}^\perp(\mathbf{Q})\,e^{i(\mathbf{Qr}-\omega t)} + \text{Hermitian conj.}, \qquad (4.17)$$

where

$$\mathbf{P}^\perp = \mathbf{P}^\perp(\mathbf{Q})\exp\{i[\mathbf{Qr}-\omega t]\} + \text{Hermitian conj.}; \qquad (4.18)$$
$$\mathbf{P}^\perp(\mathbf{Q}) = \beta\mathbf{E}^\perp(\mathbf{Q})$$

is the vector of specific electric polarization of the crystal; and β is a tensor whose components are determined by

$$\beta_{xy} = -\frac{e^2 n_e}{m\omega^2}\delta_{xy} + \frac{1}{v}\sum_{f,\mu}\frac{\Omega_f^2}{\omega^2}\left\{\frac{\overset{*}{\mathbf{p}}{}^{\perp x}_{f\mu}\,\mathbf{p}^{\perp y}_{f\mu}}{E^f_\mu(\mathbf{Q})-\hbar\omega} + \frac{\mathbf{p}^{\perp x}_{f\mu}\,\overset{*}{\mathbf{p}}{}^{\perp y}_{f\mu}}{E^f_\mu(\mathbf{Q})+\hbar\omega}\right\}. \qquad (4.19)$$

The obtained expression is valid only in the frequency range outside of the poles, i.e., at $\omega \neq \hbar^{-1}E^f_\mu(\mathbf{Q})$.

From the transversality condition

$$\mathbf{s}\cdot\mathbf{D}(\mathbf{Q}) = 0, \qquad \mathbf{s} = \frac{\mathbf{Q}}{Q}$$

of the displacement vector

$$\mathbf{D}(\mathbf{Q}) = \mathbf{E}(\mathbf{Q}) + 4\pi\mathbf{P}(\mathbf{Q}) \qquad (4.20)$$

it follows that the longitudinal components of the electric field and of the specific electric dipole moment are related by

$$\mathbf{E}^{\parallel}(\mathbf{Q}) = -4\pi \mathbf{s}(\mathbf{s}\cdot\mathbf{P}(\mathbf{Q})). \tag{4.21}$$

From (4.20) and (4.21), we find

$$\mathbf{D}(\mathbf{Q}) = \mathbf{E}^{\perp}(\mathbf{Q}) + 4\pi \mathbf{P}^{\perp}(\mathbf{Q}). \tag{4.22}$$

Now, using Eq. (4.18), we obtain a relationship between the displacement vector and the transverse electromagnetic field

$$\mathbf{D}(\mathbf{Q}) = (1 + 4\pi\beta)\mathbf{E}^{\perp}(\mathbf{Q}). \tag{4.23}$$

Finally, comparison of (4.23) and (I.3.1) allows us to write

$$\varepsilon^{\perp}_{xy}(\mathbf{Q}, \omega) = 1 + 4\pi\beta^{\perp}_{xy} + \gamma\cdot s_x s_y, \tag{4.24}$$

where γ is an arbitrary constant. Therefore, Eqs. (4.23) and (I.3.1) do not uniquely determine the tensor $\varepsilon^{\perp}$. If we choose the constant γ such that relation (I.3.1a) is satisfied, then

$$\varepsilon^{\perp}_{xy} = \left(1 - \frac{\omega_p^2}{\omega^2}\right)\eta_{xy} + \frac{4\pi}{v}\sum_{f,\mu}\frac{\Omega_f^2}{\omega^2}\left\{\frac{p^{\perp x}_{f\mu}(\mathbf{Q})\, p^{\perp y}_{f\mu}(\mathbf{Q})}{E^f_\mu(\mathbf{Q}) - \hbar\omega} + \frac{p^{\perp x}_{f\mu}(\mathbf{Q})\, p^{\perp y}_{f\mu}(\mathbf{Q})}{E^f_\mu(\mathbf{Q}) + \hbar\omega}\right\}, \tag{4.25}$$

where $\omega_p^2 = 4\pi e^2/mv$ is the square of the plasma frequency, and η is the "transverse" projection tensor, whose components are determined by

$$\eta_{xy} = \delta_{xy} - s_x s_y. \tag{4.26}$$

It is easy to see that the effect of the tensor η on the vector $\mathbf{E}(\mathbf{Q})$ comes down to isolation of its transverse part

$$\mathbf{E}^{\perp}(\mathbf{Q}) = \eta\mathbf{E}(\mathbf{Q}).$$

In the coordinate system of the wave vector $\mathbf{Q}$ (when $\mathbf{s} = (0, 0, 1)$, see Section 3 of Chapter I), the components of the transverse-projection tensor are determined by the matrix

$$(\eta_{xy}) = \begin{pmatrix} 1 & 0 & 0 \\ 0 & 1 & 0 \\ 0 & 0 & 0 \end{pmatrix}.$$

In the coordinate system of the wave vector, therefore, tensor equation (4.26) amounts to two equations between the diagonal elements of the tensors

$$\varepsilon^{\perp}_{xx} = 1 + 4\pi\beta_{xx}, \qquad \varepsilon^{\perp}_{yy} = 1 + 4\pi\beta_{yy},$$

which, along with Eqs. (I.3.5), allows us to express the refractive index and the absorption coefficient for plane waves in terms of auxiliary tensor (4.22):

$$(n_x + i\varkappa_x)^2 = 1 + 4\pi\beta_{xx}, \quad (n_y + i\varkappa_y)^2 = 1 + 4\pi\beta_{yy}. \tag{4.27}$$

Using the approximate equation $E_\mu^f \approx \hbar\omega_f$, the identity

$$\left\{\frac{E_\mu^f}{\hbar\omega\sqrt{(E_\mu^f)^2 - \hbar^2\omega^2}}\right\}^2 = \{[E_\mu^f]^2 - \hbar^2\omega^2\}^{-1} + (\hbar\omega)^{-2}$$

and the sum rule

$$\frac{2m}{e^2\hbar}\sum_{\mu, f} E_\mu^f(\mathbf{Q})\, p_{f\mu}^x\, p_{f\mu}^y = \delta_{xy},$$

we can convert expression (4.25) to a similar form:

$$\varepsilon_{xy}^{\perp} = \eta_{xy} + \sum_{\mu, f}\frac{8\pi\hbar\omega_f p_{f\mu}^{\perp x}(\mathbf{Q})\, p_{f\mu}^{\perp y}(\mathbf{Q})}{v\,[(E_\mu^f(\mathbf{Q}))^2 - \hbar^2\omega^2]}.$$

5. Calculation of the Resonance-Interaction Matrix

It was shown in Section 3 that in exciton states with the wave vector **k**, the effect of intermolecular interaction is expressed in terms of D_f and the elements of the resonance-interaction matrix $L_{\alpha\beta}(\mathbf{k})$, which are determined by sums (3.7). In calculating these values, the interaction-energy operator $V_{n\alpha, m\beta}$ of the two molecules, which characterizes the Coulomb interaction of the electrons and nuclei of both molecules, is usually expanded into a series in inverse powers of the distance between the centers of the molecules. The individual terms of the series characterize multipole–multipole interactions of various orders. If the molecules are neutral, the first term of the series in the interaction operator corresponds to dipole–dipole interaction. According to (3.6a), D_f is a function of the mean values of the electric multipole moments in the ground and in the excited states of the interacting molecules. In molecules with a center of symmetry, in steady states the mean values of the dipole moments (and of all other odd moments) are equal to zero. Therefore, only the terms beginning with quadrupole–quadrupole interaction make a contribution to D_f.

In this case, the interaction energies between two molecules and their variation when one of the molecules is excited decrease with an increase in molecular distance R as R^{-5}. In calculating D_f, therefore, the interaction of a given molecule only with its nearest neighbors may be taken into account. Unfortunately, the wave functions of the steady states of the molecules in a crystal are poorly known, and the quadrupole moments of the corresponding states are not known. Therefore, D_f must be considered a phenomenological parameter that makes a contribution to the energy displacement of the steady states of the crystal relative to the energy of the corresponding states of the molecules.

With the above expansion of the operator $V_{n\alpha,m\beta}$, the elements of the resonance-interaction matrix

$$L^f_{\alpha\beta}(\mathbf{k}) = \sum_{\mathbf{n}}{}' M^f_{0\beta,\,\mathbf{n}\alpha} \exp(i\mathbf{kn}) \tag{5.1}$$

are also represented in the form of a series. If the electric dipole moment of the transition (4.12) in a molecule in a crystal differs from zero, usually only the first term of the series, which corresponds to the dipole–dipole interaction of the dipole moments of the transition (4.12), is retained. In this approximation

$$M^f_{0\beta,\,\mathbf{n}\alpha} = \frac{1}{\mathbf{r}_\mathbf{n}^5}\{(\mathbf{d}_{\alpha f}\mathbf{d}_{\beta f})\,\mathbf{r}_\mathbf{n}^2 - 3(\mathbf{d}_{\alpha f}\mathbf{r}_\mathbf{n})(\mathbf{d}_{\beta f}\mathbf{r}_\mathbf{n})\}, \tag{5.2}$$

where

$$\mathbf{r}_\mathbf{n} = \mathbf{n} + \rho_\alpha - \rho_\beta$$

is the radius vector from molecule 0β to molecule $\mathbf{n}\alpha$. Since only one excited state of a molecule is considered here and below in this section, the subscript f is omitted.

In particular, for crystals with one molecule in a unit cell, $\mathbf{r}_\mathbf{n} = \mathbf{n}$, $\mathbf{d}_\alpha = \mathbf{d}$, and matrix element (5.2) has the form

$$M_{0\mathbf{n}} = \mathbf{d}^2 |\mathbf{n}|^{-3}(1 - 3\cos^2\varphi), \tag{5.2a}$$

where φ is the angle between vectors $\mathbf{d}$ and $\mathbf{n}$. In this case, resonance-interaction matrix (5.1) reduces to a single element

$$L^f(\mathbf{k}) = 2\sum_{\mathbf{n}(\neq 0)} M_{0\mathbf{n}} \cos(\mathbf{kn}). \tag{5.3}$$

TABLE 1. Frequencies and Oscillator Strengths of Intramolecular Transitions.

Molecule	Transition frequency, cm^{-1}	Oscillator strength	10^8 R, cm	10^8 l, cm	$\eta = l/R$
Anthracene	~40,000	2.3	~9	1.3	0.14
Anthracene	~26,000	0.23	~9	1.0	0.11
Naphthalene	~32,000	0.001	~7	0.03	0.004

It is sometimes convenient to write matrix elements (5.2) in another form. For this, we introduce the rectangular coordinate system x, y, z with the z axis directed along the vector $\mathbf{r}_n$, and let $\cos\theta^x_{0\beta},\ldots,\cos\theta^z_{n\alpha}$ be the cosines of the angles between the x, y, and z axes and the electric moments of the transitions in the corresponding molecules. Then,

$$M_{0\beta,\mathbf{n}\alpha} = |\mathbf{r_n}|^{-3}\,\mathbf{d}^2_\alpha \{\cos\theta^x_{0\beta}\cos\theta^x_{\mathbf{n}\alpha} + \cos\theta^y_{0\beta}\cos\theta^y_{\mathbf{n}\alpha} - 2\cos\theta^z_{0\beta}\cos\theta^z_{\mathbf{n}\alpha}\}. \quad (5.4)$$

If all of the molecules of a unit cell are the same, then $\mathbf{d}^2_\alpha$ is not a function of the subscript α and can be expressed in terms of the oscillator strength F of the corresponding transition in the molecule, the cyclic transition frequency Ω, and the electron mass m and charge e:

$$\mathbf{d}^2_\alpha = \frac{e^2\hbar F}{2m\Omega}. \quad (5.5)$$

It follows directly from expression (5.4) that the matrix elements are real, symmetric, and, in the dipole approximation, proportional to the oscillator strength of the molecular transition and to the geometry factor, which is a function of the distance between tho molecules and their mutual orientation.

In a molecular crystal, the molecules are in immediate contact with one another. This causes doubt in the validity of expanding the Coulomb interaction operator $V_{0\beta,n\alpha}$ into a series in inverse powers of the distance between the centers of the molecules. It should be remembered that in reality the series expansion parameter in this case is not the ratio of the dimensions of the molecules to their mutual distance (this ratio is close to unity for adjacent molecules) but the ratio η of the length $l = d_\alpha/e$ of the dipole moment of the transition (4.12) to the intermolecular distance R. The values of some typical electron transitions in molecules of aromatic compounds are given in Table 1.

As can be seen from the table, even in the case of the very strong transitions that occur in the far ultraviolet region in anthracene, the parameter η is small for adjacent molecules. Smallness of the ratios η is characteristic of crystals consisting of aromatic molecules, in which, as was noted in Section 2, the π electrons responsible for the first excited states of the molecules are inner electrons.

The elements of resonance-interaction matrix (5.1) form a Hermitian matrix for each wave-vector value. As was pointed out in Section 3, in crystals with a center of symmetry, instead of matrix elements (5.1), it is convenient to consider the real and symmetric about α and β matrix elements

$$\widetilde{L}_{\alpha\beta}(\mathbf{k}) = \sum_{\mathbf{n}}{}' M_{0\beta,\mathbf{n}\alpha} \exp(i\mathbf{k}\mathbf{r}_{\mathbf{n}}). \tag{5.6}$$

Summation in (5.6) at $\alpha \neq \beta$ is over all lattice vectors, including $\mathbf{n} = 0$; at $\alpha = \beta$, summation is over all lattice vectors $\mathbf{n} \neq 0$.

Let us calculate the elements (5.6) of the resonance-interaction matrix for a one-dimensional crystal of infinite dimensions that contains two molecules in each unit cell and has a center of symmetry. The centers of gravity of the molecules of the n-th cell (Fig. 2) are situated at the points $\mathbf{n}_1 = n\mathbf{a}$; $\mathbf{n}_2 = (n + 1/2)\mathbf{a}$; $n = 0, \pm 1, \ldots$. The positions of the electric dipole moments in the molecules are determined by the unit vectors $\mathbf{e}_1$ and $\mathbf{e}_2$, wherein

$$\mathbf{a}\mathbf{e}_1 = -\mathbf{a}\mathbf{e}_2 = a \cos\delta.$$

Using formulas (5.6), (5.2), and (5.5), we obtain

$$\widetilde{L}_{11} = \sum_{n}{}' M_{01,\,n2}\, e^{i\mathbf{k}\mathbf{a}n} = M_1 \sum_{n=1}^{\infty} n^{-3} \cos(\mathbf{k}\mathbf{a}n), \tag{5.7}$$

where

$$M_1 = \frac{e^2\hbar F}{m\omega a^3}(1 - 3\cos^2\delta); \tag{5.8}$$

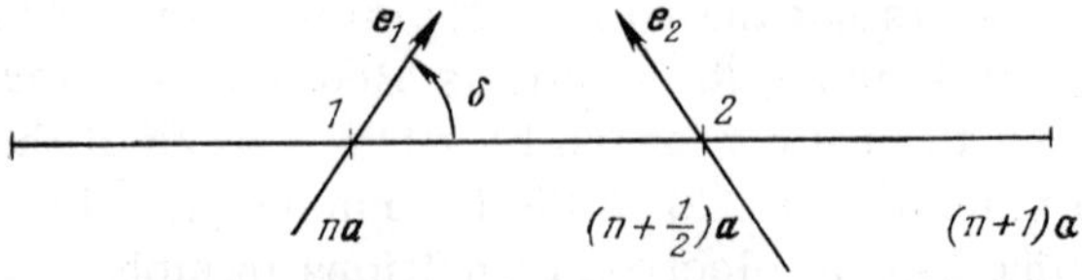

Fig. 2. Dipole moment of quantum transitions in one-dimensional crystal containing two molecules in a unit cell.

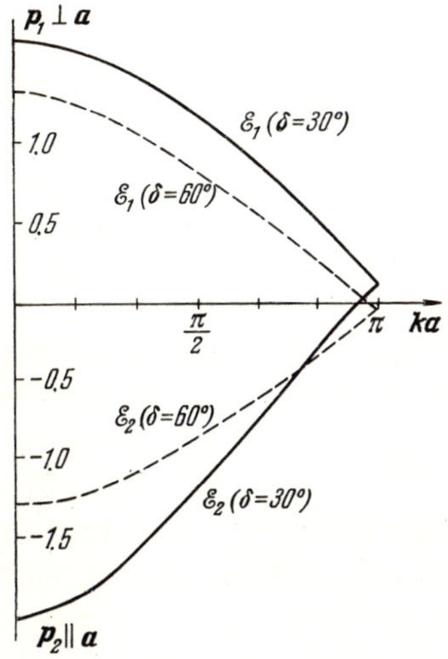

Fig. 3. Exciton energies in one-dimensional crystal as functions of **ka**.

and F is the oscillator strength of a quantum transition in a molecule.

The off-diagonal matrix element is determined by

$$\widetilde{L}_{12} = \sum_n M_{01,n2}\, e^{i\mathbf{ka}(n+1/2)} = M_2 \sum_{n=0}^{\infty} (2n+1)^{-3} \cos\left\{\mathbf{ka}\left(n+\frac{1}{2}\right)\right\}, \qquad (5.9)$$

where

$$M_2 = \frac{8e^2\hbar F}{m\omega a^3}(1+\cos^2\delta). \qquad (5.10)$$

The wave vector in (5.7) and (5.9) is directed along basis vector **a** of the crystal and takes continuous values in the range

$$-\pi \leqslant \mathbf{ka} \leqslant \pi. \qquad (5.11)$$

In accordance with (3.18), (3.20), (5.7), and (5.9), we find the energies of the two exciton bands associated with one intramolecular excitation:

$$\mathscr{E}_1(\mathbf{k}) \equiv \frac{E_1(\mathbf{k}) - \Delta\varepsilon - D}{M_2} = \chi(\delta) \sum_{n=1}^{\infty} \frac{\cos(\mathbf{ka}n)}{n^3} + \sum_{n=0}^{\infty} \frac{\cos[\mathbf{ka}(n+1/2)]}{(2n+1)^3}, \qquad (5.12a)$$

$$\mathscr{E}_2(\mathbf{k}) \equiv \frac{E_2(\mathbf{k}) - \Delta\varepsilon - D}{M_2} = \chi(\delta) \sum_{n=1}^{\infty} \frac{\cos(\mathbf{ka}n)}{n^3} - \sum_{n=0}^{\infty} \frac{\cos[\mathbf{ka}(n+1/2)]}{(2n+1)^3}, \qquad (5.12b)$$

where

$$\chi(\delta) = \frac{M_1}{M_2} = \frac{1 - 3\cos^2\delta}{8(1 + \cos^2\delta)}$$

is a monotone increasing function of a value from $-1/8$ to $1/8$ as δ varies from 0 to 90°. When $\delta = 54°44'$, this function vanishes. The exciton energies (in units $8e^2\hbar F/m\omega a^3$) in the two bands (5.12) as functions of **ka** in the range $0 \leq \mathbf{ka} \leq \pi$ are shown in Fig. 3. The solid curves are for $\delta = 30°$ and the broken curves are for $\delta = 60°$. The energies for negative wave vectors are determined by

$$\mathscr{E}_i(-\mathbf{k}) = \mathscr{E}_i(\mathbf{k}).$$

The electric dipole moments of the quantum transitions to states (5.12) are determined, according to (4.15a), in each unit cell by the expressions

$$\mathbf{p}_1 = \frac{1}{\sqrt{2}}\left(\mathbf{d}_1 + \mathbf{d}_2 e^{i\frac{\mathbf{ka}}{2}}\right),$$

$$\mathbf{p}_2 = \frac{1}{\sqrt{2}}\left(\mathbf{d}_1 - \mathbf{d}_2 e^{i\frac{\mathbf{ka}}{2}}\right).$$

When $\mathbf{ka} \ll 1$, the vector $\mathbf{p}_1$ is perpendicular to and the vector $\mathbf{p}_2$ parallel to basis vector **a** of the crystal. The wave vector for excitons is always parallel to the basis vector. Therefore, band $\mathscr{E}_1(\mathbf{k})$ can be called the transverse-exciton band and band $\mathscr{E}_2(\mathbf{k})$ can be called the longitudinal-exciton band. The effective mass of transverse excitons is negative and the effective mass of longitudinal excitons is positive. Light propagated along a crystal can excite only transverse excitons.

If the one-dimensional crystal contains one molecule in each unit cell, then each intramolecular electronic excitation will be associated with one quasi-continuous band of excited states. The energies of the sublevels in this band are determined by

$$E(\mathbf{k}) = \Delta\varepsilon + D + L(\mathbf{k}), \tag{5.13}$$

where $\Delta\varepsilon$ is the excitation energy of a molecule; D the change in the interaction energy of one molecule with all the rest in its transition to an excited state; and

$$\left.\begin{aligned} L(\mathbf{k}) &= M_1 \sum_{n=1}^{\infty} \frac{\cos \mathbf{ka}\, n}{n^3} \approx M_1 \cos \mathbf{ka}, \\ M_1 &= \frac{e^2\hbar F}{m\omega a^3}(1 - 3\cos^2\delta). \end{aligned}\right\} \tag{5.13a}$$

At small k and $\delta < 54°44'$, the mass of an exciton is positive; at $\delta > 54°44'$, it is negative.

The first theoretical calculations of the energy levels of excitons in complicated three-dimensional anisotropic molecular crystals for one fixed sense of k were made, on the basis of the theory in [7, 8], in [12, 45-49]. In these papers, the sums of the resonance interactions between molecules were made only for k = 0. The calculations were made numerically by summation of the resonance interactions of a certain molecule of the crystal with all of the molecules lying in spheres of ever increasing radius.[2] Usually consideration was confined to nearest molecules [45] or to molecules within a sphere of 20-30 Å radius.

This approximation is valid only for crystal exciton states that are associated with dipole-forbidden intramolecular transitions. But with dipole-allowed intramolecular transitions, this approximation is unsatisfactory from two standpoints. In the first place, as the radius R of the sphere increases, the sums of the resonance dipole–dipole interactions converge very poorly, because the number of terms increases as R^2 while the absolute value of each term decreases as R^{-3}. At small, nonzero k, therefore, the contribution to the sums made by molecules situated at a considerable distance is very great. This contribution results (at $R = \infty$) in nonanalytic terms in $L_{\alpha\beta}(k)$ [32]. Secondly, at finite R and k = 0, the resonance (dipole–dipole) sums are very small (in simple cubic crystals they are equal to zero) and cannot account for the experimentally observed values of exciton splitting at $k = Q \neq 0$. Because of this, in [46-50] it was postulated that, along with dipole–dipole resonance interactions, a considerable role could be played by octupole–octupole interactions, the introduction of which was difficult to justify independently.[3]

The elements (5.6) of the resonance-interaction matrix were calculated in the dipole–dipole approximation for all k by Davydov

[2]Allowance for an infinite number of molecules in calculating dipole sums was made by Ewald's method [53] in a paper by Craig and Walsh [51] also for k = 0.

[3]Calculations by Rice et al. [52] using the method of molecular orbits have shown that, in anthracene crystals, together with dipole – dipole interactions, short-range interactions of the octupole type can also be of great significance. It should be noted, however, that calculations of the molecular wave functions by the method of molecular orbits are very crude, so such numerical estimates are not sufficiently convincing.

and Sheka [33] for the exciton states in an anthracene crystal that are associated with the first electronic excitation of a molecule (27,570 cm^{-1}). A unit cell of an anthracene crystal is formed [44] by three basis vectors **a**, **b**, **c**, whose lengths are a = 8.56 Å, b = 6.04 Å, and c = 11.16 Å. In this case, **ab** = **cb** = 0, and **ac** = ac cos 125α. The positions of the centers of gravity of the molecules $\mathbf{n}\alpha$ are determined by radius vectors (4.15). The molecules that occupy lattice positions corresponding to α = 1 are oriented such that their median (M) and long (L) axes form with the basis vectors **a**, **b**, and **c'** = [**a** × **b**] the following angles:

	M	*L*
a	71.3°	119.7°
b	26.6°	97.0°
c'	71.8°	30.6°

The orientation of the molecules with α = 2 can be obtained by reflection in the plane **ac** of the molecules with α = 1.

Simple differentiation shows that (5.6) is transformed to

$$\widetilde{L}_{\alpha\beta}(\mathbf{k}) = -\{(d_\alpha \nabla)(d_\beta \nabla) S_{\alpha\beta}(\mathbf{k}, \mathbf{r})\}_{r=0}, \tag{5.14}$$

where

$$S_{\alpha\beta}(\mathbf{k}, \mathbf{r}) = \sum_{\mathbf{n}}{}' \frac{\exp(i\mathbf{k}\mathbf{r}_\mathbf{n})}{|\mathbf{r}_\mathbf{n} - \mathbf{r}|}. \tag{5.15}$$

Sum (5.15) can be calculated for infinite crystals by Ewald's method [53]. For this, using the integral representation

$$(|\mathbf{r}_\mathbf{n} - \mathbf{r}|)^{-1} = \frac{2}{\sqrt{\pi}} \int_0^\infty \exp(-\gamma^2 [\mathbf{r}_\mathbf{n} - \mathbf{r}]^2)\, d\gamma$$

and dividing the integration range into two intervals, we transform (5.15) to

$$S_{\alpha\beta}(\mathbf{k}, \mathbf{r}) = \gamma_0 \sum_{\mathbf{n}}{}' e^{i\mathbf{k}\mathbf{r}_\mathbf{n}} G(\xi) - \frac{2\delta_{\alpha\beta}}{\sqrt{\pi}} \int_0^{\gamma_0} e^{-\gamma^2 \mathbf{r}^2}\, d\gamma + \\ + \int_0^{\gamma_0} F_\gamma(\mathbf{r}) e^{i\mathbf{k}\mathbf{r}}\, d\gamma - \frac{2\delta_{\alpha\beta}}{\sqrt{\pi}} \int_0^{\gamma_0} e^{-\gamma^2 \mathbf{r}^2}\, d\gamma, \tag{5.16}$$

where

$$F_{\gamma}(\mathbf{r}) \equiv \frac{2}{\sqrt{\pi}} \sum_{\mathbf{n}} \exp\{-\gamma^2(\mathbf{r}-\mathbf{r_n})^2 - i\mathbf{k}(\mathbf{r}-\mathbf{r_n})\}$$

is a periodic function which has the lattice periods (constants), so it can be represented as a sum over the reciprocal-lattice vectors:

$$F_{\gamma}(\mathbf{r}) = \frac{2\pi}{v\gamma^3} \sum_{\mathbf{g}} \exp\left\{i\mathbf{g}\mathbf{r} - \left(\frac{\mathbf{g}+\mathbf{k}}{2\gamma}\right)^2\right\}, \tag{5.17}$$

where $v = \mathbf{a}[\mathbf{b} \times \mathbf{c}]$ is the volume of a unit cell;

$$\mathbf{g} = 2\pi \sum_{i=1}^{3} \nu_i \mathbf{b}_i, \qquad \nu_i = 0, \pm 1, \ldots; \tag{5.18}$$

and $\mathbf{b}_i$ are the basis vectors of the reciprocal lattice, such that

$$\mathbf{b}_1 = \frac{1}{v}[\mathbf{b} \times \mathbf{c}], \;\; \mathbf{b}_2 = \frac{1}{v}[\mathbf{c} \times \mathbf{a}], \;\; \mathbf{b}_3 = \frac{1}{v}[\mathbf{a} \times \mathbf{b}].$$

Summation in (5.17) is over all possible values of the vectors $\mathbf{g}$ in accordance with (5.18).

If we substitute (5.17) into (5.16), integrate, and introduce the function

$$G(\xi) \equiv \xi^{-1}[1 - \Phi(\xi)], \tag{5.19}$$

where

$$\xi = \gamma_0 |\mathbf{r} - \mathbf{r_n}|;$$

$$\Phi(\xi) \equiv \frac{2}{\sqrt{\pi}} \int_0^{\xi} e^{-z^2} dz \tag{5.20}$$

is the error function, we transform (5.16) to

$$S_{\alpha\beta}(\mathbf{k}, \mathbf{r}) = \frac{2}{\sqrt{\pi}} \sum_{\mathbf{n}}{}' e^{i\mathbf{k}\mathbf{r_n}} \int_{\gamma_0}^{\infty} e^{-\gamma^2(\mathbf{r_n}-\mathbf{r})^2} d\gamma +$$

$$+ \frac{4\pi}{v} \sum_{\mathbf{g}} (\mathbf{g}+\mathbf{k})^2 \exp\left\{-\left(\frac{\mathbf{g}+\mathbf{k}}{2\gamma_0}\right)^2 + i(\mathbf{g}+\mathbf{k})\mathbf{r}\right\}. \tag{5.21}$$

The proper selection of γ_0 ensures the rapid convergence of both series in (5.21). After substituting (5.21) into (5.14) and differentiating with allowance for (5.20), we obtain a final expression for the elements of the resonance-interaction matrix for an infinite crystal:

$$\widetilde{L}_{\alpha\beta}(\mathbf{k}) = \frac{4\pi d^2}{v} \sum_{\mathbf{g}} \frac{(\mathbf{s}_\alpha [\mathbf{g}+\mathbf{k}])(\mathbf{s}_\beta [\mathbf{g}+\mathbf{k}])}{(\mathbf{g}+\mathbf{k})^2} \exp\left\{-\left(\frac{\mathbf{g}+\mathbf{k}}{2\gamma_0}\right)^2\right\} - \\ - \gamma_0^3 d^2 \left\{ \sum_{\mathbf{n}}' \left[(s_\alpha s_\beta) \frac{1}{\xi_{\mathbf{n}}} \frac{\partial G(\xi_{\mathbf{n}})}{\partial \xi_{\mathbf{n}}} + \right.\right. \\ \left.\left. + \frac{(\mathbf{s}_\alpha \mathbf{r}_{\mathbf{n}})(\mathbf{s}_\beta \mathbf{r}_{\mathbf{n}})}{\mathbf{r}_{\mathbf{n}}^2} \left(\frac{\partial^2 G(\xi_{\mathbf{n}})}{\partial \xi_{\mathbf{n}}^2} - \frac{1}{\xi_{\mathbf{n}}} \frac{\partial G(\xi_{\mathbf{n}})}{\partial \xi_{\mathbf{n}}} \right) \right] e^{i\mathbf{k}\mathbf{r}_{\mathbf{n}}} - \frac{4\delta_{\alpha\beta}}{3\sqrt{\pi}} \right\}, \quad (5.22)$$

where

$$\mathbf{s}_\alpha = \frac{\mathbf{d}_\alpha}{|\mathbf{d}_\alpha|}, \quad \mathbf{s}_\beta = \frac{\mathbf{d}_\beta}{|\mathbf{d}_\beta|}, \quad \xi_{\mathbf{n}} = \gamma_0 |\mathbf{r}_{\mathbf{n}}|. \quad (5.23)$$

From the first sum of (5.22) we can extract the term corresponding to $\mathbf{g} = 0$, which has the form

$$\frac{4\pi}{v\mathbf{k}^2} (\mathbf{d}_\alpha \mathbf{k})(\mathbf{d}_\beta \mathbf{k}) e^{-k^2/4\gamma_0^2}. \quad (5.24)$$

This expression is not an analytic function of k at small k, since its limiting value (when $\mathbf{k} \to 0$) is a function of the sense of k. The remaining part of the matrix element $\widetilde{L}_{\alpha\beta}(\mathbf{k})$ is an analytic function of k. In particular, the diagonal elements can be written as

$$\widetilde{L}_{\alpha\alpha}(\mathbf{k}) = \frac{4\pi}{v} \frac{(\mathbf{d}_\alpha \mathbf{k})^2}{\mathbf{k}^2} e^{-k^2/4\gamma_0^2} + \Phi_{\alpha\alpha}(\mathbf{k}), \quad (5.25)$$

where the function $\Phi_{\alpha\alpha}(\mathbf{k})$ is not a function of α. Expression (5.25) is also valid for other crystals with a center of symmetry that contain σ identical molecules in a unit cell. The first term in (5.25) is a function of the orientation of the vector k relative to the sense $\mathbf{d}_\alpha$ of the dipole moment of a quantum transition in molecule α. Generally speaking, the molecules that make up one unit cell have different orientations and different senses of $\mathbf{d}_\alpha$. For an arbitrary sense of k, therefore, the scalar product $\mathbf{kd}_\alpha$ is not equal to the scalar product $\mathbf{kd}_\beta$. For all senses of k that are perpendicular to or parallel to the plane of symmetry of the crystal, however,

$$(\mathbf{kd}_\alpha) = (\mathbf{kd}_\beta) \quad \text{for all} \quad \alpha, \beta = 1, 2, \ldots, \sigma.$$

In this case,

$$\widetilde{L}_{\alpha\alpha}(\mathbf{k}) = \widetilde{L}_{\beta\beta}(\mathbf{k}) \quad \text{for } \alpha, \beta = 1, 2, \ldots, \sigma. \quad (5.26)$$

In all of the preceding formulas, only one excited state of a molecule was taken into account. But other excited states affect the exciton energy. As Davydov and Myasnikov have shown [34], this influence is effectively allowed for by decreasing the operator of intermolecular dipole–dipole interaction by the value ε_0. Here, ε_0 is the dielectric constant of the crystal in the frequency range of the f-th intramolecular transition (which is taken into account in (5.22)), which is determined by all of the electronic states of the molecules with the exception of the f-th state. A formally rough allowance for this effect can be made in formulas (5.22), (5.24), and (5.25) by substituting for the dipole moment of the transition $\mathbf{d}_\alpha$ the effective value

$$\mathbf{d}_{\text{eff}} \approx [\varepsilon_0]^{-1/2}\,\mathbf{d}_\alpha. \tag{5.27}$$

For anisotropic crystals, the tensor nature of ε_0 and, therefore, the possible noncoincidence of the senses of $\mathbf{d}_{\text{eff}}$ and $\mathbf{d}$ should be taken into account.

Davydov and Sheka [33] have made numerical calculations of matrix elements (5.22) without allowance for the effect of ε_0. Matrix elements (5.22) were calculated for the following values of the wave vectors, which satisfy conditions (5.26): a) the vectors $\mathbf{k}_c$, which are perpendicular to the plane **ab**, i.e., are directed along the vector $[\mathbf{a} \times \mathbf{b}]$ and take all values in the range $-\pi \leq \mathbf{ck}_c \leq \pi$; the vectors $\mathbf{k}_a$, which are directed along the vector $[\mathbf{b} \times \mathbf{c}]$ and take all values in the range $-\pi \leq \mathbf{ak}_a \leq \pi$; and c) the vectors $\mathbf{k}_b$, which are directed along the vector $[\mathbf{c} \times \mathbf{a}]$ and take all values in the range $-\pi \leq \mathbf{bk}_b \leq \pi$. In calculating matrix elements (5.22), let the parameter $\gamma_0 = 2.65 \times 10^7$, which ensured sufficiently fast convergence of the sums over the lattice and reciprocal-lattice vectors. The calculated values of $\widetilde{L}_{\alpha\beta}(\mathbf{k})$ for these wave-vector senses were used to determine the law of dispersion in the two exciton bands $E_1(\mathbf{k})$ and $E_2(\mathbf{k})$, which are given by the expressions

$$E_\mu(\mathbf{k}) - \Delta\varepsilon - D = \widetilde{L}_{11}(\mathbf{k}) - (-1)^\mu \widetilde{L}_{12}(\mathbf{k}), \qquad \mu = 1,\ 2.$$

The obtained values, with the correction for ε_0, which we let equal 2.5, are shown in Fig. 4. We let the oscillator strength of the molecular transition in anthracene equal 0.23. The curves for the range $-\pi$, 0 are obtained from those in Fig. 4 by rotation by 180° about the vertical axis, which passes through the point $\mathbf{k} = 0$. The directions of the dipole moments of the quantum transitions to the corresponding exciton states are shown for $\mathbf{k} \approx 0$. In particular,

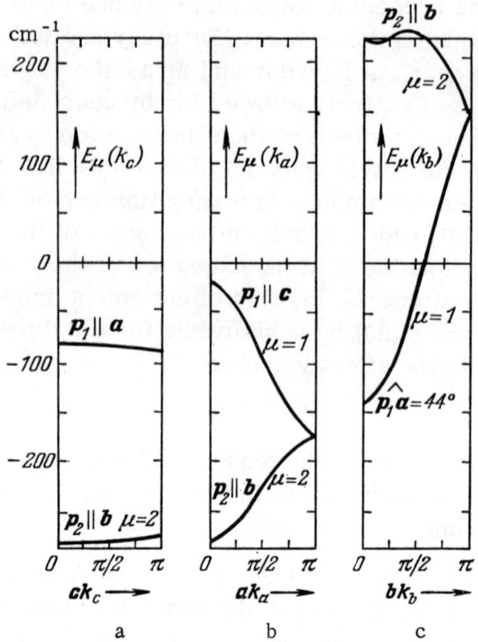

Fig. 4. Energies of two exciton bands (displaced by $\Delta\varepsilon + D$) in anthracene crystal for three wave-vector senses: $\mathbf{k}_c$ perpendicular to plane **ab**; $\mathbf{k}_a$ perpendicular to plane **bc**; and $\mathbf{k}_b$ perpendicular to plane **ac**. The vector $\mathbf{p}_\mu$ determines the direction of the dipole moment of the transition to the corresponding exciton state.

the curves with $\mu = 2$ pertain to one exciton band and correspond to the dipole moment that is directed along the **b** axis of the crystal. The exciton energy in this band can be approximated by the formula

$$E_2(\mathbf{k}) = E_0 + A\cos^2\varphi + M_a\cos(\mathbf{ka}) + M_b\cos(\mathbf{kb}) + M_c\cos(\mathbf{kc}),$$

where φ is the angle between vectors **k** and **b**; $E_0 = 25{,}240\ \text{cm}^{-1}$; $A = 220\ \text{cm}^{-1}$; $M_a = -23.2\ \text{cm}^{-1}$; $M_c = -1\ \text{cm}^{-1}$; and $M_b = 15\ \text{cm}^{-1}$. The transverse ($\mathbf{k} \perp \mathbf{b}$) and longitudinal ($\mathbf{k} \| \mathbf{b}$) excitons in this band have positive and negative effective masses, respectively.

In crystals composed of molecules whose excited states are triply degenerate, light excites only transverse excitons ($\mathbf{kd} = 0$) and the nonanalytic terms (5.24) of matrix elements (5.22) play a small role. But in anisotropic molecular crystals, as a rule, $\mathbf{kd} \neq 0$, and nonanalytic terms (5.24) take great values, causing considerable gaps in the exciton bands at the point $\mathbf{k} = 0$ (see Fig. 4).

The presence of nonanalytic terms in (5.22) is due to mathematical idealization of the problem, which comes down to extension of sums (5.6) to an infinite number of distant molecules. Because of the poor convergence of sums (5.6), the contributions from a large number of distant molecules are very considerable. Each term in (5.6) is an analytic function. Real crystals contain a very large but finite number of molecules. The sums of a finite number of analytic functions reduce to analytic functions.

The first calculations of (5.6) (at $\alpha = \beta = 1$) for a finite but rather large number of molecules in simple cubic lattices containing one molecule in each unit cell were made by Cohen and Keffer [54]. They showed that when allowance is made for the interaction of a molecule with all molecules located in a sphere of radius R of a simple cubic crystal with lattice constant a at $kR \ll 1$, sums such as (5.6) are determined by the expression

$$L^{(R)}(\mathbf{k}) = \frac{4\pi \mathbf{d}^2}{a^3}\left[\frac{(\mathbf{sk})^2}{\mathbf{k}^2} - \frac{1}{3}\right]\left(1 - \frac{3j_1(kR)}{kR}\right), \tag{5.28}$$

where $\mathbf{s} = \mathbf{d}/|\mathbf{d}|$, and $j_1(x)$ is a spherical Bessel function. Considering that when $kR \leq 1$,

$$\left(1 - \frac{3j_1(kR)}{kR}\right) \approx 0{,}1\,(kR)^2,$$

we can see that at finite R and when $\mathbf{k} \to 0$

$$\lim L^{(R)}(\mathbf{k}) - 0.$$

On the other hand, when $x \geq 1$, $j_1(x) = -\cos x/x$; therefore, at small but nonzero k and when $R \to \infty$

$$L^{\infty}(\mathbf{k}) = \frac{4\pi \mathbf{d}^2}{a^3}\left[\frac{(\mathbf{sk})^2}{\mathbf{k}^2} - \frac{1}{3}\right]. \tag{5.29}$$

In optically isotropic cubic crystals, when $ka \ll 1$, for any sense of k two types of excitons are possible: *transverse excitons*, $(\mathbf{dk}) = 0$, and *longitudinal excitons*, $\mathbf{d} \parallel \mathbf{k}$. From relations (5.13) and (5.29) it follows that when $R \to \infty$ and $ka \ll 1$, the energies of transverse ($E^{\perp}$) and longitudinal ($E^{\parallel}$) excitons will be

$$E_{\infty}^{\perp} = \Delta\varepsilon + D - \frac{4\pi d^2}{3a^2}, \quad E_{\infty}^{\parallel} = \Delta\varepsilon + D + \frac{8\pi d^2}{3a^2}. \qquad (5.30)$$

At finite R at the point **k** = 0 the energies of both types of excitons will be $\Delta\varepsilon + D$. As **k** increases, the energies of transverse and longitudinal excitons approach (the more rapidly the greater R) the corresponding values (5.30). Thus, the dimensions of the region of interaction between molecules have a substantial effect on $L^{(R)}$(**k**) at small **k**.

Calculations of (5.6) for monoclinic anthracene crystals when interaction is cut off at distance R_0 have been made by Davydov and Sheka [33]. The results of calculations of $E_{\mu}^{(R)}$(**k**) with allowance for the resonance interaction of each molecule of the crystal with the molecules surrounding it in a sphere of radius R_0 are shown in Fig. 5 for the wave vector $\mathbf{k} = \mathbf{k}_c$. The exciton energy (with allowance for the effect of ε_0) at small $\mathbf{k}_c$ is shown for three radii R_0 of the effective region of interaction: curves a_1 and a_2 for $R_0 = 1\ \mu$; curves b_1 and b_2 for $R_0 = 0.1\ \mu$; and curves c_1 and c_2 for $R_0 = 0.05\ \mu$. The maxima of the absorption curves must correspond to their intersections with the vertical arrows cQ_a and cQ_b, which correspond to the $c\mathbf{Q}_c$ values for the wave vectors of the polarization photons *a* and *b*. If we compare the energy differences

$$\Delta = E(Q_a) - E(Q_b)$$

with the experimental [55] value of 220 cm^{-1}, we see that the theoretical calculations are in approximate agreement, if we assume that the radius of the region of effective intermolecular interaction is greater than or equal to 0.1 μ. It was concluded in [33], in which the role of ε_0 was not taken into account, that the radius of the region of effective interaction was on the order of 0.05 μ.

6. Using Group Theory for Qualitative Interpretation of the Properties of Exciton States

In the preceding sections, we considered the exciton states that are associated with the nondegenerate excited states of a free molecule. In these cases, calculation of the energy and the wave

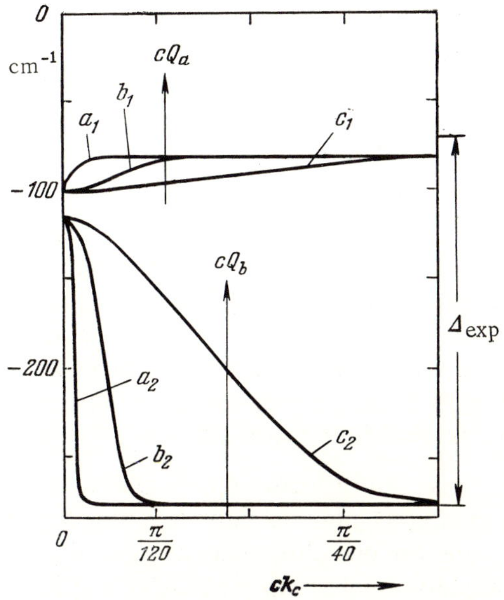

Fig. 5. Exciton energies for a small range of values of the wave vector $\mathbf{k}_c$ with allowance for the finite radius R_0 of the effective region of interaction between molecules.

functions of excitons comes down to solving equation system (3.14). If the molecule has degenerate excited states and the crystal has complicated symmetry, such calculations become cumbersome even for selected senses of the wave vector. A number of qualitative properties of exciton states in crystals can, however, be established without resorting to explicit calculations of the wave functions of the crystal, but taking into account the symmetry properties of the crystal and the molecule with group-theory methods. Group theory also considerably facilitates direct calculations. Group theory has been used by Davydov [8], Winston [37], Lubchenko and Rashba [56], and a number of others for qualitative interpretation of the properties of exciton states in molecular crystals.

The wave functions $\varphi_{n\alpha}$ of the steady states of one molecule apply to various irreducible representations of the symmetry group of a molecule in a crystal. In general, this symmetry group can differ from the symmetry group of a free molecule. Small distortions in the form of the molecules in a crystal are difficult to calculate. We shall assume that the functions $\varphi_{n\alpha}$ already take into account all distortions and study the collective effects determined by the proper arrangement of the molecules in the crystal and their resonance interaction. Winston [37] proposed that the symmetry group of a molecule in a crystal be called a local group, i.e., the symmetry group of the place in the crystal occupied by the molecule.

The use of group theory [57-61] to study exciton states in crystals involves taking into account the symmetry of the crystals. The principal symmetry elements of crystals of infinite dimensions include: parallel transitions in certain directions for specific distances, rotations about certain directions by specific angles, mirror reflections, screw axes, and slip planes.[4] The set of symmetry elements of a crystal form a space group.

In all, there are 230 different space groups [57] for three-dimensional crystals, 17 plane groups for two-dimensional crystals, and two line groups for one-dimensional crystals. Each space group has a translation subgroup that is formed by all of the lattice vectors $\mathbf{n} = \sum_{i=1}^{3} n_i \mathbf{a}_i$. If we take an arbitrary point in a crystal and make at it the translation defined by all of the vectors $\mathbf{n}$, we obtain a set of points that is called a simple Bravais lattice, i.e., the set of all transformations (inversion, rotations, and reflections[5]) that convert any vector $\mathbf{n}$ of a subgroup into the vector $\mathbf{n}'$ of the same subgroup forms a point group.[6] There are seven such point groups (they all contain the operation of inversion). Accordingly, the subgroup of translations is subdivided by directional-symmetry properties into seven systems: triclinic (S_2), monoclinic (C_{2h}), rhombic or orthorhombic (D_{2h}), rhombo-

[4]The elements of crystal symmetry that represent rotations or reflections that are associated with translations by a certain part of the lattice constant are called screw axes and slip planes. In the case of screw transformation, the displacement is parallel to the rotation axis, and in the case of the slip plane, it is parallel to the reflection plane. For example, the twofold screw axis represents rotation by 180° and displacement along the rotation axis by one-half of the lattice constant. Screw axes and slip planes occur, for example, in the space lattice of diamond, in hexagonal lattices, and in a number of molecular crystals.

[5]Only rotations by angles that are multiples of 60° and 90° about various axes and combinations of these rotations with inversion are possible.

[6]A point group contains transformations that leave at least one point fixed.

hedral or trigonal (D_{3d}), tetragonal (D_{4h}), hexagonal (D_{6h}), and cubic (O_h). Along with the crystals with which simple Bravais lattices are associated, there are crystals with more complicated Bravais lattices, which, according to the symmetry properties of the translation group, are classed with the same seven systems. The triclinic, rhombohedral, and hexagonal groups have only one (simple) Bravais lattice each. The other systems have more complicated Bravais lattices: one for monoclinic, three for rhombohedral, one for tetragonal, and two for cubic. Thus, there are only 14 Bravais lattices in all.

Besides translational symmetry, crystals possess a directional symmetry (rotations by angles that are multiples of 60° and 90°, inversion, and combinations of both). The set of symmetry elements that convert each direction in the crystal into its equivalent forms a *point group*. The point group of a Bravais lattice (point group of translation subgroup) is, as a rule, above the point group of the crystal. For example, the point group of a Bravais lattice always has a center of symmetry (inversion), but the point group of a crystal cannot have one.

The point group of a crystal contains all of the rotational parts of the symmetry elements of the complete space group. But the symmetry elements of the point group of a crystal do not necessarily coincide with the symmetry elements of the space group of the crystal. For example, a 2-fold screw axis in a space group corresponds to 180° rotation and obligatory subsequent translation by one-half of the lattice constant. The corresponding element in the point group of a crystal contains only 180° rotation. The symmetry elements of the point group of a crystal convert each direction in the crystal into its equivalent. On the other hand, the symmetry elements of the space group of a crystal must convert all of the points of the crystal into their equivalents. There exist 32 point groups, in accordance with which crystals are subdivided into 32 *crystal classes* [57].

All of the symmetry elements of the space group of a crystal can be represented as the product of the translations and the symmetry elements of a certain group, which is called the *factor group* of the space group with respect to the translation subgroup. It is very important that the factor group of the space group of a crystal be, with respect to the translation subgroup, isomorphic[7] with the point group of the crystal, which contains the rotational part of all of the operators of the complete space group [59, 60]. Because of isomorphism, any representation of the point group of a crystal is automatically a representation of the factor group. When investigating irreducible representations, therefore, instead of the factor group we can examine the point group of the crystal. The space groups of crystals that do not contain screw axes or slip planes are called simple space groups. In all, there are 73 simple space groups. Simple space groups are noteworthy in that their point groups are subgroups of the complete space group [60].

According to the general principles of quantum mechanics [43, 59, 61], the wave functions of the steady states of a quantum system are the bases for the irreducible representations of the group of symmetry operations that leave the Hamiltonian operator of the system unchanged. The Hamiltonian operator of a crystal is invariant under all of the symmetry operations of the space group of the crystal. It follows from this that the wave functions of the steady states—in particular, wave functions (3.11) of the exciton states—form the basis for various irreducible representations of the space group of the crystal. We can, therefore, denote and describe the states of a crystal that have a specific energy by indicating the irreducible representations of the space group of the crystal that are associated with them.

Irreducible representations of space groups are examined in Koster's book [60]. If we know the irreducible representations of the space groups of a crystal, we can establish the symmetry properties of the wave functions of the exciton states, determine the selection rules, and simplify the calculations. Since the space group contains the translation subgroup, the steady states can be classified according to the irreducible representation of the translation subgroup. Owing to the mutual commutation of the symmetry elements of the translation subgroup, it is an Abelian group and all of its irreducible representations are one-dimensional. Each irreducible representation of the translation subgroup is characterized by the wave vector **k**.

In a group-theory study of the properties of the excited states

[7]Two groups are isomorphic if there is a one-to-one correspondence between their elements.

of a crystal, it is convenient to use the first Brillouin zone (see Section 1 of Chapter II) to determine the reduced wave vectors that characterize the irreducible representations of the translation subgroup.

The wave function of the ground state of a crystal has k = 0 and belongs to the totally symmetric representation of the space group of the crystal. In Section 3, the functions of the ground (3.3) and excited (3.4) states of the crystal were expressed in terms of the wave functions of the molecules. The wave functions of the molecules are classified according to the irreducible representations of the symmetry group of the molecules, or, more precisely, according to the irreducible representations of the "local group."

The local symmetry group of a crystal characterizes the symmetry properties of the place in which a molecule is found in the crystal. The local group is a subgroup of the factor group of the crystal and of the point group of the crystal isomorphic with it. Therefore, the local group of a crystal contains all of the symmetry elements of the molecule that coincide with the symmetry elements of the point group. The local symmetry group of the molecules takes into account the Bethe splitting of molecular energy levels by the mean stationary field of the other molecules. In some cases, the disruption of the symmetry of a molecule in a crystal is small, so that the symmetry of a free molecule is preserved for all practical purposes and its excited states are characterized by the symmetry of the molecule.

As an example, let us consider the local symmetry of the molecules in an anthracene crystal. The point symmetry of an anthracene crystal is C_{2h} (see Table 2) and the symmetry group of a free molecule of anthracene is D_{2h} (Table 3). The identity element E and inversion are coincident symmetry elements in both groups. The local symmetry group in an anthracene crystal is therefore C_i. The group singlet state of a molecule corresponds to the function $\varphi^0_{n\alpha}$, which is classified with the totally symmetric representation of this group. If the f-th excited state $\varphi^f_{n\alpha}$ of a molecule belongs to the one-dimensional irreducible representation Γ_f of the symmetry group of the molecule, then the molecular energy level will not be degenerate, and, according to (3.11), the exciton states with the wave vector k are represented by the functions

$$\Phi^f_\mu(\mathbf{k}) = \sum_\alpha \psi^f_\alpha(\mathbf{k})\, u^*_{\alpha\mu}(\mathbf{k}), \tag{6.1}$$

where

$$\psi_\alpha^f(\mathbf{k}) = \frac{1}{\sqrt{N}} \sum_{\mathbf{n}} \psi_{\mathbf{n}\alpha}^f e^{i\mathbf{kn}}. \tag{6.2}$$

It follows directly from the form (3.4) of the function $\psi_{\mathbf{n}\alpha}^f$ that it also belongs to the irreducible representation of the symmetry group of the molecule. Function (6.2) belongs to the **k** representation of the translation subgroup and characterizes the excitations that are concentrated in the molecules α in each unit cell.

If an excited state of a molecule relates to the irreducible representation Γ_f, which has the dimension **s**, then the molecular energy level f corresponds to s wave functions $\psi_{\mathbf{n}\alpha}^{f\gamma}$, which we shall distinguish by the superscript γ. In this case, the exciton states in a crystal are represented by the functions

$$\Phi_\mu^f(\mathbf{k}) = \sum_{\alpha,\gamma} \psi_\alpha^{f\gamma}(\mathbf{k})\, u_{\alpha\mu}^{*\gamma}(\mathbf{k}), \tag{6.3}$$

where

$$\psi_\alpha^{f\gamma}(\mathbf{k}) = \frac{1}{\sqrt{N}} \sum_{\mathbf{n}} \psi_{\mathbf{n}\alpha}^{f\gamma} e^{i\mathbf{kn}}, \tag{6.4}$$

$$\psi_{\mathbf{n}\alpha}^{f\gamma} = \varphi_{\mathbf{n}\alpha}^{f\gamma} \prod_{\mathbf{m},\beta} \varphi_{\mathbf{m}\beta}^0, \qquad \mathbf{m}\beta \neq \mathbf{n}\alpha. \tag{6.4a}$$

Exciton states (6.1) and (6.3) are defined for each value of the wave vector **k** of the first Brillouin zone, which characterizes the corresponding irreducible representation of the translation subgroup.

Let us see how wave functions (6.1) and (6.3) are transformed under the symmetry operations of the space group. The translations brought about by the operator $T_\mathbf{n}$ do not have the wave vector k. Therefore, the only change in the wave functions with translation is the appearance of a phase factor,

$$T_\mathbf{n} \Phi_\mu^f(\mathbf{k}) = e^{i\mathbf{kn}} \Phi_\mu^f(\mathbf{k}).$$

Under the other symmetry operations of the space group that do not amount to simple translation, the function $\Phi_\mu^f(\mathbf{k})$ is transformed to the new function $\Phi_\mu^f(\mathbf{k}')$, which has another wave vector k' and which is obtained from the vector k by applying to it the symmetry element that corresponds to the symmetry element of the factor group.

Any element of the point group transforms each wave vector in the first Brillouin zone into a certain number of wave vectors, which, together with the initial vector, form the star of the k representation [57, 62]. Thus, each vector of the k-representation star is transformed by any symmetry element of the point group into a vector that also belongs to the star. If the vector **k** does not lie at singular points of the Brillouin zone (the axes of symmetry, symmetry planes, and boundary of the zone), then the number of vectors of the star is equal to the number of symmetry elements of the point group of the crystal. Such a star is called nondegenerate. If the end of the vector falls into a singular point of the zone, then two or more vectors of the star will be identical. In this case, the star is called degenerate. In particular, the star consisting of the one point $\mathbf{k} = 0$ corresponds to the center of the Brillouin zone.

To illustrate this, let us consider a monoclinic crystal with the point group C_{2h}. This group contains, in addition to the identity element, three symmetry elements: inversion, 180° rotation about the monoclinic axis, and reflection in a plane perpendicular to the monoclinic axis. In general, therefore, the star of the **k** representation for such a crystal consists of four vectors: $\mathbf{k}_1$, $\mathbf{k}_2$, $\mathbf{k}_3 = -\mathbf{k}_1$, and $\mathbf{k}_4 = -\mathbf{k}_2$, where $\mathbf{k}_2$ is obtained from $\mathbf{k}_1$ by 180° rotation about the monoclinic axis. With reflection in the plane perpendicular to the monoclinic axis, the vector $\mathbf{k}_2$ is transformed to $-\mathbf{k}_1$. The degenerate star of the **k** representation associated with the wave vector **k**, which lies along the monoclinic axis or in a plane perpendicular to it, consists of only two wave vectors $\mathbf{k}_1$ and $\mathbf{k}_2 = -\mathbf{k}_1$.

Let us assume that the star of the **k** representation contains l vectors $\mathbf{k}_1, \mathbf{k}_2, \ldots, \mathbf{k}_l$. Then, owing to the equivalence (with respect to the symmetry properties of the potential field of the crystal) of the wave vectors that belong to one star, the energies of the excited states (at fixed μ and f) for all vectors of a given star will be the same,

$$E^f_\mu(\mathbf{k}_1) = E^f_\mu(\mathbf{k}_2) = \ldots = E^f_\mu(\mathbf{k}_l). \tag{6.5}$$

In particular, in crystals with a center of symmetry wave vectors **k** and $-\mathbf{k}$ belong to a single star of the **k** representation, and so

$$E^f_\mu(\mathbf{k}) = E^f_\mu(-\mathbf{k}).$$

The wave functions $\Phi^f_\mu(\mathbf{k}_i)$ $(i = 1, 2, \ldots, l)$, which pertain to the different wave vectors of the star, are obtained from one another under the symmetry operations of the point group. According to (6.5), they all correspond to one energy (if there is no accidental degeneracy) and form the basis of the irreducible representation of the point group of the crystal. In other words, the wave functions $\Phi^f_\mu(\mathbf{k}_1), \ldots, \Phi^f_\mu(\mathbf{k}_l)$ form an irreducible vector space that is invariant under the symmetry operations of the point group. The symmetry operations of the point group transform these functions one into the other.

For the senses of $\mathbf{k}_1$ that correspond to degenerate stars, the irreducible representation that pertains to the level $\mathbf{E}^\mu_f(\mathbf{k}_1)$ can be one-dimensional. Then, all of the functions $\Phi^f_\mu(\mathbf{k}_i)$ for the vectors $\mathbf{k}_i$ that belong to the same degenerate star can differ from one another only by a phase factor, i.e.,

$$|\Phi^f_\mu(\mathbf{k}_1)| = |\Phi^f_\mu(\mathbf{k}_i)|.$$

In particular, in crystals with a center of symmetry, if the vector $\mathbf{k}_1$ belongs to a degenerate star, then the vector $-\mathbf{k}_1$ will also belong to it. Therefore, the functions that belong to one-dimensional irreducible representations satisfy the equation

$$|\Phi^f_\mu(\mathbf{k}_1)| = |\Phi^f_\mu(-\mathbf{k}_1)|.$$

In Section 1 of Chapter II it was pointed out that the wave vectors $\mathbf{k}$ and $\mathbf{k} + \mathbf{g}$ were equivalent if the vector $\mathbf{g}$ was determined by the basis vectors $\mathbf{b}_l$ of the reciprocal lattice with the aid of relations (1.4). A consequence of this equivalence is the equation

$$E^f_\mu(\mathbf{k}) = E^f_\mu(\mathbf{k} + \mathbf{g}). \tag{6.5a}$$

Using (6.5a), we can extend the function $\mathbf{E}^\mu_f(\mathbf{k})$ over the entire $\mathbf{k}$ space. The thus-formed function $\mathbf{E}^\mu_f(\mathbf{k})$ will be a periodic function with periods $2\pi\mathbf{b}_l$ $(l = 1, 2, 3)$.

Let us assume that the wave vectors $\mathbf{k}_1$ and $\mathbf{k}_2$ are parts of one $\mathbf{k}$-representation star. Then, according to (6.5), we have

$$E^f_\mu(\mathbf{k}_1) = E^f_\mu(\mathbf{k}_2). \tag{6.5b}$$

On the other hand, according to (6.5a), the relations

$$E^f_\mu(\mathbf{k}_1) = E^f_\mu(\mathbf{k}_1 \pm 2\pi\mathbf{b}_l). \tag{6.5c}$$

must be satisfied. From the structure of the first Brillouin zone (see Section 1 of Chapter II) it follows that when the end of one of the vectors of the star approaches the boundary (surface) of the Brillouin zone, the ends of all of the other vectors $\mathbf{k}_i$ of the same star will also approach the boundary of the Brillouin zone. If the vector $\mathbf{k}_1$ lies within the Brillouin zone, the vectors $\mathbf{k}_1 \pm 2\pi\mathbf{b}_l$ will lie outside it. When the end of the vector $\mathbf{k}_1$ approaches the boundary of the Brillouin zone from within, the end of one of the vectors $\mathbf{k}_1 \pm 2\pi\mathbf{b}_l$ will approach the vector $\mathbf{k}_2$ from without. Therefore, Eqs. (6.5b) and (6.5c) lead us to the conclusion that the function $\mathbf{E}_\mu^f(\mathbf{k})$ has the same values on both sides of the boundary of the Brillouin zone. In other words, the function $\mathbf{E}_\mu^f(\mathbf{k})$ passes through a maximum or a minimum when it intersects the boundary of the zone.

Explicit calculation of the coefficients $u_{\alpha\beta}(\mathbf{k})$ of functions (6.1) requires the solution of a system of σ (σ is the number of molecules in a unit cell) linear equations (3.10), which contain complex matrix elements $\mathscr{L}_{\alpha\beta}(\mathbf{k})$, whose explicit form is usually unknown. If the degeneracy multiplicity of the level f is equal to δ, then to determine the coefficients $u_{\alpha\mu}^{f\gamma}(\mathbf{k})$ it is necessary to solve a system of $\sigma\delta$ linear equations.

The problem of determining the coefficients $u_{\alpha\mu}^{f\gamma}(\mathbf{k})$ and the energies $\mathbf{E}_\mu^f(\mathbf{k})$ is considerably simplified if we allow for the fact that only exciton states with small wave vectors are of vital importance in the processes of absorption and emission of longwave (as compared with the lattice constant) light. The most important part of the Brillouin zone, therefore, is its central region near $\mathbf{k} = 0$.

From the point of view of group theory, calculation of the coefficients in functions (6.3) for values of the wave vectors $\mathbf{k}_i$ that pertain to a particular star comes down to finding the combinations of the functions (6.4) that correspond to the irreducible representation Γ_f of the local group of a molecule that would form irreducible representations of the space group of the crystal. If the wave vectors $\mathbf{k}_i$ are part of a nondegenerate star, all of the irreducible representations that pertain to this star will have the same dimension. The use of symmetry, therefore, does not divide the system of equations for the coefficients $u_{\alpha\mu}^{f\gamma}$ into simpler, independent subsystems. The irreducible representations connected with a degenerate star have a smaller dimension. Therefore, the construction of combinations of the functions $\psi_\alpha^{f\gamma}(\mathbf{k}_i)$, that have been transformed through irreducible representations that pertain to a degenerate star divides the system of equations into simpler subsystems and makes it easier to find the energies of the excited states.

A star has maximum degeneracy at k = 0. The value k = 0 corresponds to the totally symmetric representation of the translation subgroup. At k = 0, therefore, the irreducible representations of the space group of a crystal coincide with the irreducible representations of the factor group.

It should be borne in mind that at the point k = 0, the matrix elements $\mathscr{L}_{\alpha\beta}$ (k) in Eqs. (3.10) are not analytic functions of the wave vector (see Section 5). Their limiting values when $k \to 0$ are functions of the sense of k. The results obtained for k = 0, therefore, cannot, generally speaking, be transferred even to the case of very small but nonzero wave vectors. Evidently, this cannot be done mainly for senses of the wave vectors that belong to irreducible stars. Direct calculations show that in the case of monoclinic crystals, classification of the states according to the irreducible representations of the factor group gives correct results for small wave vectors parallel or perpendicular to the monoclinic axis, i.e., for wave vectors belonging to degenerate stars consisting of pairs of the wave vectors k and −k. In this case, the coefficients $u^{f\gamma}_{\alpha\mu}$ are not functions of the wave vectors and are uniquely determined from the symmetry conditions.

Now let us construct the wave functions that correspond to the irreducible representations of the factor group or of the point group isomorphic with it. If the exciton states can be classified according to the irreducible representations of the factor group, it is easy to determine [37] how many energy levels are produced in the crystal from one molecular level and what the symmetry properties of their wave functions are. To solve this problem, we must find the irreducible representations of the factor group that result from the given irreducible representations of the local group, which characterize the excited states of the molecules.

Let the f-th excited state of a molecule be characterized by the wave function $\varphi^{f\gamma}_{\alpha\mu}$ which is associated with the irreducible representation Γ_f of the local symmetry group of the molecule. The superscript γ of the function indicates the molecular functions that form the basis of the irreducible representation Γ_f of the local group. The number of different γ values (degeneracy multiplicity) equals the dimension of the representation Γ_f. Assume that we also know the characters $x^f(R_m)$, i.e., the sums of the diagonal elements of the matrices corresponding to the symmetry elements R_m in the irreducible representation Γ_f of the local group (see, for example, [43, 56]). If the local group is Abelian, i.e., its symmetry elements commute, then all of the representations of the local group are one-dimensional, coincide with the characters, and for the various symmetry elements have values of either 1 or −1. The functions of the ground state of molecules $\varphi^0_{n\alpha}$ pertain to the totally symmetric representation of the local group. According to definition (6.4a), therefore, the function pertains to the same irreducible representation Γ_f of the local group as does the function $\varphi^{f\gamma}_{n\alpha}$. The local symmetry group of a molecule is a subgroup of the factor group. Here, some of the symmetry elements of the local group coincide with the symmetry elements of the factor group. If a certain symmetry element R* of the factor group does not coincide with any symmetry element of the local group R_m, then R* will contain a translation by a part of the lattice constant. Under operation R*, in this case, the function is replaced by the function $\psi^{f\gamma}_{\beta}$, where $\beta \neq \alpha$. Consequently, the matrix of element R* of the corresponding representation of the factor group does not have nonzero diagonal elements. Hence, the characters of the factor-group symmetry elements that do not coincide with the symmetry elements of the local group are zero in all of the representations of the factor group. If the symmetry operation R* of the factor group coincides with the symmetry operation R_m of the local group, then it also coincides with the corresponding symmetry operation R of the point group of the crystal. In this case ($R = R_m$), there is a simple relationship between the characters of the corresponding representations in both groups: $\chi^f(R) = \sigma x^f(R_m)$, where σ is the number of molecules in a unit cell. Thus, the character $\chi^f(R)$ of any symmetry element R of the point group, which corresponds to the molecular energy level of the irreducible representation Γ_f of the local group, is calculated by the simple formula

$$\chi^f(R) = \sigma \sum_{R_m} x^f(R_m)\,\delta_{RR_m}, \tag{6.6}$$

where δ_{ij} is the Kronecker symbol; summation is over all symmetry elements of the local group; and σ is the number of molecules in a unit cell.

In general, the calculated characters $\chi^f(R)$ pertain to reducible representations of the point group. If we use the orthogonality relations of the characters of the irreducible group representations, we can decompose the character of the irreducible representation $\chi^f(R)$ in the characters $\chi_l(R)$ of the irreducible representations of the point group:

$$\chi^f(R) = \sum_l n_l^f \chi_l(R), \tag{6.7}$$

where summation is over all irreducible representations of the point group. The coefficients of the expansion n_l^f are determined by the formula

$$n_l^f = \frac{1}{h} \sum_R \chi_l^*(R)\chi^f(R), \tag{6.8}$$

where h is the number of symmetry operations in the group, and summation is over all elements of the point group.

As an example of how the obtained relations are used, let us consider an anthracene crystal. As has already been noted, this crystal belongs to the monoclinic system with the basis vectors **a**, **b**, **c**, with **a**⊥**b**. A unit cell contains two molecules. The basis vector **b** determines the direction of the monoclinic axis. Light that is normally incident on the plane **ab** of the crystal is used in experiments. Thus, we shall be concerned with small wave vectors that are perpendicular to the monoclinic axis, which belong, therefore, to a degenerate star of the k representation. The factor group of the space group is isomorphic with the point group C_{2h}. The characters of the irreducible representations of this group are given in Table 2.

The symbols for the irreducible representations are given in the first column, and the last column indicates the x, y, z projec-

TABLE 2. Characters of Irreducible Representations of Group C_{2h}

Irreducible representations	E	C_2^b	i	σ^b	**r**
A_g	1	1	1	1	
A_u	1	1	−1	−1	z
B_g	1	−1	1	−1	
B_u	1	−1	−1	1	x, y

TABLE 3. Characters of Irreducible Representations of Group D_{2h}

	E	C_2^ξ	C_2^η	C_2^ζ	i	σ^ξ	σ^η	σ^ζ	r
A_{1g}	1	1	1	1	1	1	1	1	
B_{1g}	1	—1	—1	1	1	—1	—1	1	
A_{2g}	1	1	—1	—1	1	1	—1	—1	
B_{2g}	1	—1	1	—1	1	—1	1	—1	
A_{1u}	1	1	1	1	—1	—1	—1	—1	
B_{1u}	1	—1	—1	1	—1	1	1	—1	ζ
A_{2u}	1	1	—1	—1	—1	—1	1	1	ξ
B_{2u}	1	—1	1	—1	—1	1	—1	1	η

tions of an arbitrary vector **r** onto the axes of a Cartesian system in which the z axis is directed along the monoclinic axis of the crystal.

An anthracene molecule has the symmetry corresponding to point group D_{2h}. The characters of the irreducible representations of this group are shown in Table 3. The last column gives the projections of the vector **r** onto the ξ, η, ζ axes of the coordinate system of a molecule. In rough approximation, an anthracene molecule can be represented as a general ellipsoid. We shall assume that the coordinate axis ζ is directed along the long axis and that the ξ axis is directed along the shortest axis of a molecule.

Let us examine the energy level that in a free molecule belongs to the irreducible representation B_{2u}. The electric dipole moment directed along the median axis is associated with the excitation B_{2u}. The local symmetry group of a molecule in an anthracene crystal is Abelian. It has only the two symmetry elements E and i and two irreducible representations: A_g with the characters $x_{A_g}(E) = x_{A_g}(i) = 1$ and A_u with the characters $x_{A_u}(E) = -x_{A_u}(i) = 1$. Using formula (6.6) and the characters of the irreducible representation A_u of the local group, which corresponds to the irreducible representation B_{2u} of a free molecule, we find the characters

$$\chi(E) = 2\chi_{B_{2u}}(E) = 2; \qquad \chi(C_2^b) = 0;$$

$$\chi(i) = 2\chi_{B_{2u}}(i) = -2; \qquad \chi(\sigma^b) = 0.$$

Then, using (6.8), we obtain expansion (6.7) as

$$\chi = \chi_{A_u} + \chi_{B_u}. \tag{6.9}$$

It follows from (6.9) that the molecular energy level of the irreducible representation B_{2u} in a crystal is associated with two types of excitons, which pertain to the irreducible representations A_u and B_u.

Since the ground state of a crystal belongs to the totally symmetric representation A_g, the electric dipole moment of the transitions to exciton states is transformed just as the exciton wave functions are. The irreducible representations corresponding to exciton wave functions, therefore, characterize the directions of the dipole moments of transitions to exciton states. In the case of anthracene, the exciton state that belongs to irreducible representation A_u is related to a dipole moment directed along the monoclinic axis, while the B_u exciton state is characterized by a moment lying in the plane xy, i.e., in a plane perpendicular to the monoclinic axis.

7. Experimental Confirmations of the Presence of Exciton States in Crystals

Splitting of nondegenerate molecular energy levels in the absorption spectra of polarized light by crystals has been observed by a number of researchers [10-13, 63]. Some results of experimental studies are reviewed in [9, 14, 15, 41, 64]. In this section, to illustrate the theory we shall consider only the basic characteristics of the absorption spectra of molecular crystals of anthracene, naphthacene, pentacene, naphthalene, and benzene.

Molecules of aromatic hydrocarbons usually consist of six-membered rings of carbon atoms with hydrogen atoms situated on the periphery. Anthracene ($C_{14}H_{10}$), naphthacene ($C_{18}H_{12}$), and pentacene ($C_{22}H_{14}$) molecules are planar and have symmetry D_{2h}. The carbon atoms form, respectively, three, four, and five rings:

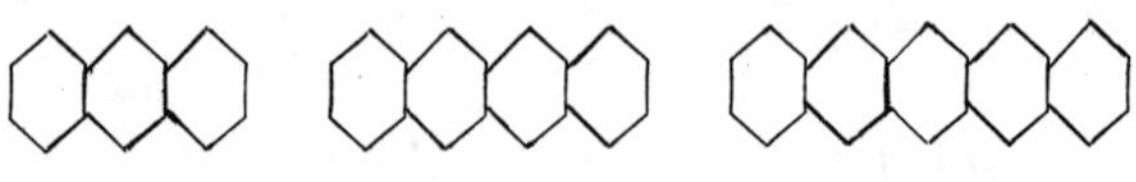

The parameters of unit cells of crystals formed from these molecules are shown in Table 4, in which **a**, **b**, and **c** are the basis vec-

TABLE 4. Parameters of Unit Cells of Anthracene, Napthacene, and Pentacene Crystals

	Anthracene	Napthacene	Pentacene
a	8.56	7.98	7.93
b	6.04	6.14	6.14
c	11.16	13.57	16.03
α	90°	101.3°	101.9°
β	124.7°	113.2°	112.6°
γ	90°	87.5°	85.8°

tors of a unit cell; α is the angle between **a** and **b**; β is the angle between **a** and **c**; and γ is the angle between **b** and **c**. Each unit cell contains two molecules. Anthracene crystals belong to the monoclinic system, and naphthacene and pentacene crystals belong to the triclinic system. To be sure, the deviations from the monoclinic system are very slight.

Anthracene Crystals. The first electron transition in an anthracene molecule (27,570 cm^{-1}) is associated with an intense dipole transition. Therefore, resonance interactions play a great role in the absorption spectrum of an anthracene crystal.

The spectrum of light absorption by anthracene single crystals has been studied by many researchers [13, 55, 63, 65-68]. The structure of the absorption spectrum for polarized light that is normally incident on the plane **ab** has been studied. The spectral re-

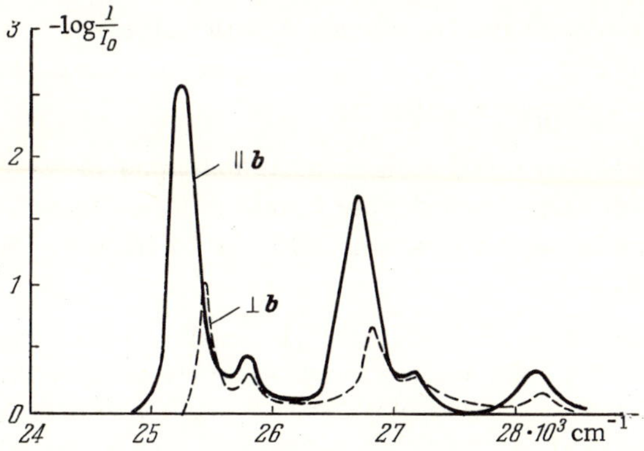

Fig. 6. Absorption spectra of polarized light in anthracene single crystal 0.1 μ thick at 20°K.

TABLE 5. Principal Bands of Anthracene Absorption Spectrum

Transition	Molecule excitation		Crystal excitation, 20°K			
	transition type	freq., cm^{-1}	b comp., cm^{-1}	*a* comp., cm^{-1}	ΔE	$P\left(\frac{b}{a}\right)$
1	00	27 570	25 213	25 432	220	5
2	00+1430	29 000	26 680	26 830	150	3
3	00+2·1430	30 430	28 200	28 280	80	3

gion beginning with frequencies near 25,000 cm^{-1} has been comparatively well investigated. The most precise data were obtained by Brodin and Marisova [55], who eliminated some procedural errors made in earlier papers. Figure 6 shows curves that they obtained for light absorption by an anthracene crystal 0.1 μ thick at 20°K. The solid curve corresponds to polarization in which the electric-field strength of the wave was parallel to the **b** axis (b component). The broken curve corresponds to the *a* component. The three principal maxima of one curve are shifted relative to the corresponding maxima of the other, forming the components of the resonance-splitting doublet. Table 5, which was compiled on the basis of the data in [55], gives the molecular-excitation frequencies and their corresponding frequencies in the crystal, the resonance splitting ΔE, and the polarization ratio P(b/*a*), i.e., the ratio of the integrated intensities of the two components of the doublet.[8]

It should be noted that the resonance splittings in Table 5 are the difference between the frequency maxima of both absorption bands. These bands are comparatively wide, and it is still not known exactly where the frequencies of the electron transitions without phonons are located in them. Therefore ΔE reflects only relatively the resonance splitting.

The second and third molecular transitions in the table are associated with single and double combinations of the totally symmetric intramolecular vibration 1430 cm^{-1} with the electron tran-

[8]Anthracene has very strong absorption, so single crystals 0.07-0.3 μ thick must be used. As Brodin has shown [55, 69], in such thin crystals the absorption of radiation of certain wavelengths violates the linear dependence of $\log(I/I_0)$ upon the thickness of the crystal. The data in Table 5 are for a thickness of 0.1 μ. They are somewhat different for other thicknesses.

sition. All three molecular transitions have a dipole moment lying in the "plane" of the molecule along the short axis (irreducible representation B_{2u}, Table 3). The presence of resonance splitting in all of the principal absorption bands in the frequency range $(25-30) \times 10^3$ cm^{-1} indicates that this absorption is due to the exciton states of the crystal, i.e., excitations distributed over a comparatively large region of the crystal and, therefore, reflecting its structure. Excitations localized at individual molecules would not be split and not be polarized in this way (see Fig. 1).

It should be noted that in many earlier experimental studies (e.g., [13, 63]), the resonance splitting of an electron transition was assumed to be 20 cm^{-1}, and not 220 cm^{-1} as in Table 5. As Brodin has shown [55], the incorrect splitting value was obtained because insufficiently strictly polarized radiation was used in measuring the b component. A slight bit of radiation with polarization along the *a* component causes considerable deformation of the spectrum of the b component. This is illustrated qualitatively in Fig. 7, in which curves 1 and 3 are the absorption curves of the b and *a* components. Line 2 shows the profile of the absorption band of the b component when a slight amount of the *a* component is present in the incident light. We see that the *a*-polarization light

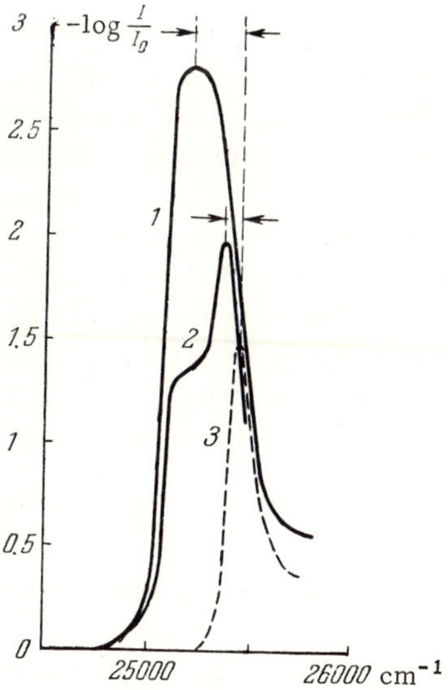

Fig. 7. Diagram of deformation of absorption curve when light that is not strictly polarized is used.

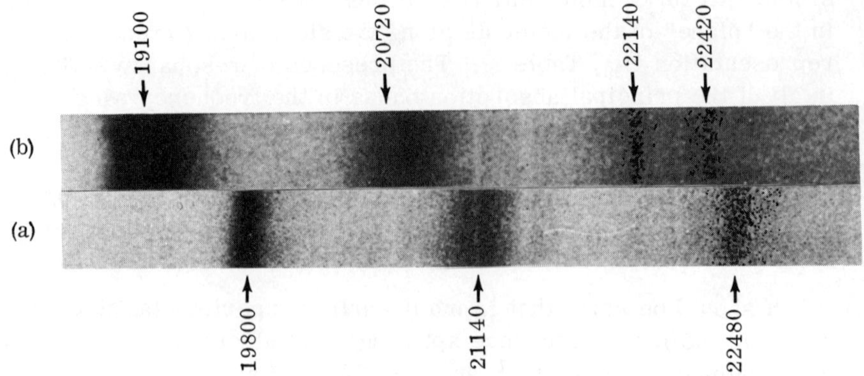

Fig. 8. Region of absorption spectrum of polarized light for naphthacene single crystal 0.27 μ thick at 20°K.

that has "broken through" on the left of the b-absorption band, being not absorbed for all practical purposes in the crystal, considerably deforms the absorption band and considerably reduces the distance between the maxima of the observed curves. Studies [70] made after Brodin's work with special precautions about the polarization confirmed the higher value of resonance splitting of an electron transition in anthracene.

Naphthacene and Pentacene Crystals. The first electron transition in a naphthacene molecule (~22,200 cm^{-1}) also pertains to the irreducible representation B_{2u} of group D_{2h} (Table 3). The comparatively high oscillator strength of the transition (~0.26) and the favorable arrangement of the molecules in the crystal lead to strong resonance interactions; this manifests itself as high doublet splittings, which are observed even at 77°K. Table 6 shows frequency maxima of absorption curves for single crystals at 77°K found by Eichis [71] and frequencies of the excited states of molecules obtained in a study of naphthacene vapors (in a layer 15 cm thick at 20°C).

As Prikhot'ko and Skorobogat'ko have shown [72], a temperature reduction results in narrowing of the absorption band and a clearer determination of the splitting (about 700 cm^{-1}) in the first electron transition. Figure 8 shows a region of the absorption spectrum of polarized light for a naphthacene single crystal 0.27 μ

TABLE 6. Principal Bands of Napthacene Absorption Spectrum

Transition	Molecule excitation		Crystal excitation, 77°K		
	transition type	freq., cm^{-1}	a comp., cm^{-1}	b comp., cm^{-1}	ΔE, cm^{-1}
1	0—0	22 200	19 810	19 235	575
2	00+1425	23 625	21 145	20 715	430
3	00+2800	25 000	22 455	22 320	225

thick at 20°K obtained in [72]. The corresponding absorption curves are shown in Fig. 9. Prikhot'ko and Skorobogat'ko have also studied [73] the spectrum of polarized-light absorption by pentacene single crystals. A spectral region for a pentacene single crystal 1 μ thick at 77°K obtained by them is shown in Fig. 10, and the corresponding absorption curves are given in Fig. 11. According to these data, the splitting of the first electronic excitation in a pentacene crystal is about 1000 cm^{-1}.

The sharp polarization of the polarized-light absorption bands for single crystals of anthracene, naphthacene, and pentacene undoubtedly indicates that the corresponding crystal excitations are collective excitations, i.e., exciton states.

The results of numerical calculations [33] of the structure of the exciton bands in anthracene crystals, which were discussed in Section 5, allow us to make a qualitative comparison between theory and experiment. If we ignore the effect of the interaction

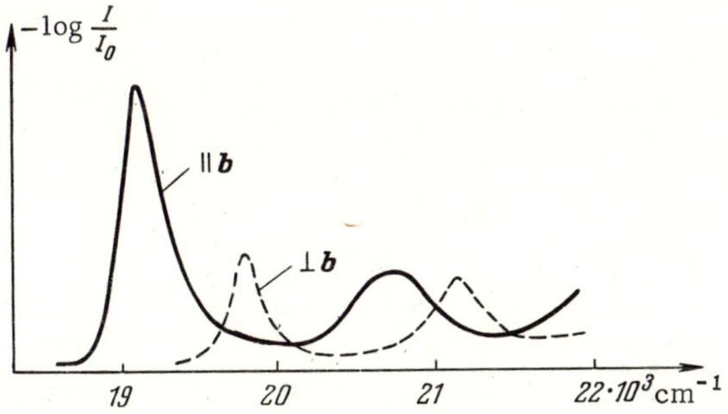

Fig. 9. Absorption curves of polarized light for naphthacene single crystal 0.27 μ thick at 20°K.

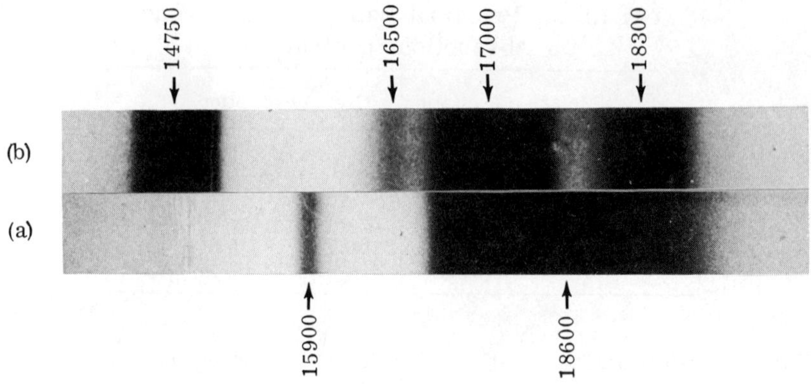

Fig. 10. Region of absorption spectrum of polarized light for pentacene single crystal 1 μ thick at 77°K.

of excitons with phonons, which will be examined below, then the production of excitons by light of polarization $\mathbf{E}_0$ and frequency ω with the wave vector $\mathbf{Q}$ will occur when the polarization condition $\mathbf{E}_0 \| \mathbf{p}_\mu$ and the laws of conservation of energy and quasi-momentum are satisfied:

$$\hbar\omega = E_\mu(\mathbf{k}), \qquad \mathbf{k} = \mathbf{Q}, \tag{7.1}$$

where $|\mathbf{Q}| = \omega n/c$, and n is the refractive index for light of frequency ω and polarization $\mathbf{E}_0$. Since the refractive index itself is a function of frequency, relations (7.1) can be satisfied in a certain range of ω and $\mathbf{k}$. In view of this, bands may be observed in the absorption spectrum. Allowance for exciton–phonon interaction results in their further broadening.

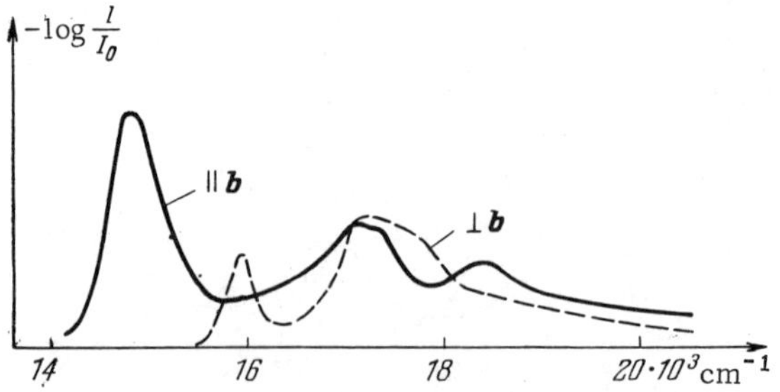

Fig. 11. Absorption curves of polarized light for pentacene single single crystal 1 μ thick at 77°K.

Thus, it may be expected that the light absorption band corresponds to the effective part of the exciton states that are associated with an interval of k in the range $|\mathbf{k}| = \omega n/c = Q$, where n is the refractive index corresponding to the frequency of maximum absorption. The two exciton bands have different refractive indices. Therefore, the energy difference between the maxima of the absorption bands (resonance splitting) is

$$\Delta E = \left| E_1^{(R_0)}\left(\frac{\omega_1}{c} n_1\right) - E_2^{(R_0)}\left(\frac{\omega_2}{c} n_2\right) \right|. \tag{7.2}$$

In the experiments of Brodin and Marisova [55], doublet bands were observed when light struck an anthracene single crystal perpendicular to the plane **ab**. In the first doublet, one of the components is a wide band with a maximum of about 25,213 cm^{-1} for the case when the electric vector of the light wave is parallel to the **b** axis in the crystal (b component). The second component of the doublet corresponds to the *a* component of the spectrum (the electric vector of the light wave is parallel to the **a** axis) and is a band with a maximum of about 25,432 cm^{-1}. The refractive indices at the maxima of the b-component and *a*-component bands are 3 and 2, respectively.

The case of light perpendicular to the plane **ab** is shown in Figs. 4a and 5. According to (4.16), in this case, the polarization vectors $\mathbf{p}_1$ and $\mathbf{p}_2$ of the two components of the doublet splitting are strictly perpendicular. Figure 4a shows that the longwave component of the absorption spectrum must have polarization parallel to the vector **b** (b component). This conclusion is in good agreement with experimental observations. The splitting can be calculated by formula (7.2) for each value of the assumed region of interaction (radius R_0) if the experimental values (3 and 2) of the refractive indices are used. Figure 5 shows with arrows the positions of $cQ_b = 3cQ_0$ and $cQ_a = 2cQ_0$, where $\mathbf{Q}_0$ is the wave vector of the light wave in a vacuum for a photon of 25,300 cm^{-1}. According to Fig. 5, we obtain the following splitting values as functions of the radius R_0 of the effective region of interaction between molecules:

$$\Delta E \approx \begin{cases} 210\ \text{cm}^{-1} & R_0 = 1\,\mu; \\ 200\ \text{cm}^{-1} & R_0 = 0.1\,\mu; \\ 100\ \text{cm}^{-1} & R_0 = 0.05\,\mu; \end{cases}$$

If we compare these with the experimental values (~220 cm^{-1}), we see that the region of effective molecular interaction in an anthracene crystal is on the order of 0.1 μ. Figure 4, b and c, shows the absorption characteristics for light perpendicular to the planes **bc** and **ac**. In the first case (Fig. 4, b), two absorption bands must be observed. The band with the lower energy is polarized along the **b** axis, and its energy is close to the energy of the absorption band of the b component of the spectrum that is observed when the light is perpendicular to the plane **ab**. The polarization of the absorption band with the higher energy is along the **c** axis of the crystal. If the wave vector of the light is perpendicular to the plane **ac**, i.e., along the monoclinic axis, one band must be observed in the absorption spectrum (transition to the second band is forbidden) (see Fig. 4c). The intensity maxima of this absorption band must be observed when the electric vector of the light wave makes an angle of ~44° with the **a** axis.

Because of its great absorption, all of the experimental data on the absorption spectra of light for anthracene are obtained with thin (about 0.07-0.3 μ) single crystals whose developed planes coincide with the plane **ab**. It has not been possible to prepare thin single crystals with developed planes **bc** or **ac**. Marisova recently studied [74] the spectra of reflected polarized light from faces **bc** and **ac** of anthracene single crystals. The obtained data confirm the theoretical conclusions about the number and polarization of the absorption bands.

When comparing experimental data with the above theory of the excited states of crystals, it should be borne in mind that the theory was developed under the assumption that the molecules were rigidly fixed at the points of an ideal crystal lattice. As we shall see below, allowance for the interactions of excitons with phonons, i.e., with molecular vibrations in the crystal, can result not only in quantitative but also in qualitative changes in the energy states of the crystal. The theory of exciton–phonon interaction is still in the development stage. Some aspects of this theory will be examined in Chapter IV. But here we shall discuss the qualitative considerations that are evidently confirmed by experimental data. Familiarity with these experimental data will be useful in the further development of the theory as well.

Owing to the interaction of excitons with phonons in a crystal, together with the transitions in which phonons do not participate (direct transitions), excited states will appear that are associated with the simultaneous formation of an exciton and a phonon (or of several phonons). Transition from the ground state to such excited states is possible when: a) the law of conservation of energy

$$E_{\mu}(\mathbf{k}) + \hbar\Omega_{\nu}(\mathbf{q}) = \hbar\omega, \tag{7.3}$$

where $\hbar\Omega_{\nu}(\mathbf{q})$ is the energy of the phonon of branch ν and $\mathbf{q}$ is its wave vector; b) the law of conservation of quasi-momentum

$$\mathbf{k} + \mathbf{q} = \mathbf{Q} \tag{7.4}$$

and c) specific polarization relations are satisfied simultaneously.

According to (7.3) and (7.4), transitions in which phonons participate may go not to one but to a number of sublevels of the exciton bands. This results in broadening of the absorption bands. The intensity distribution in such bands is a function of the exciton-phonon coupling with the various phonon-vibration branches. The theory of this phenomenon was developed on the basis of models in papers by Davydov, Rashba, and Lubchenko [75-77] and Toyozawa [78] for semiconductors and dielectrics with high dielectric constants. In Chapter IV we shall discuss the results of the further development of the theory of exciton–phonon interaction in molecular crystals.

Besides broadening of the absorption bands, the possibility of displacement of molecules in some cases has a considerable effect on the nature of the excited states of a crystal. In fact, with transition of a molecule to an excited state, the energy of its van der Waals interaction with the surrounding molecules changes. As a result of the interaction change, there is a tendency toward lattice deformation in the area of the excited molecule. On the other hand, resonance interaction between identical molecules leads to "blurring" of the molecular excitation over a certain region of the crystal. If the resonance interaction is more considerable (small D_f and heavy molecules), an exciton state is formed in the crystal. In this case, roughly speaking, the excitation moves from one molecule to another so rapidly that local lattice deformation does not have time to occur. Conversely, with weak resonance interactions, high D_f, and an easily deformable lattice, local deformation occurs along with excitation. This excited state is conveniently called a *local* excitation, since its movement through the crystal is very slow.

Local excitations are similar in their properties to excitations of impurity molecules in a crystal when resonance interaction of the impurity molecule with the molecules of the solvent is absent (a low impurity concentration and the impurity-excitation energy considerably removed from the fundamental-absorption energy).

Unfortunately, numerical estimates of the values determining the possibilities of realization of one or another excited state in real crystals are very complicated. The resonance interaction is characterized by the matrix L(**k**) with matrix elements (3.7). This matrix determines the width of the exciton bands. In general, it is proportional to the oscillator strength of the corresponding transition in a molecule. The value D_f is calculated by formula (3.6a). The lattice deformability can be estimated from the phonon-excitation frequencies.

Thus, various types of excited states are possible in real crystals, depending upon the relative values of L(**k**) and D_f and the deformability of the lattice. In the two limiting cases, these states are either the exciton states examined in Sections 2-6 or the localized excitations that will be considered in Sections 3 and 4 of Chapter V. Intermediate cases, too, can be realized for some excited states of molecules. The two above-mentioned limiting types of excited states have considerably different properties, so as we shall see below, they are easily identified in experiments. In fact, the absorption caused by exciton production reflects the collective properties of the excitations, which are primarily manifested as resonance (Davydov) splitting of the molecular absorption bands and their particular polarization. The absorption associated with the formation of localized excitations differs little from the polarization described by the oriented-gas model. The corresponding absorption bands are called *molecular-absorption bands*.

Experiments show that exciton states in molecular crystals are usually associated with intramolecular electronic excitations and their combinations with certain totally symmetric intramolecular vibrations. As a rule, the molecular energy levels that consist of combinations of an electronic excitation with not totally symmetric vibrations of the atoms in a molecule form only localized excitations in the crystal. The light absorption spectra for naphthalene and benzene single crystals at low temperatures can serve to illustrate these principles.

<u>Naphthalene Crystals.</u> A naphthalene free molecule ($C_{10}H_8$) has symmetry D_{2h}. A crystal belongs to the monoclinic system, with the point group C_{2h}^5. A crystal has two molecules in each unit cell, which is formed by the three basis vectors **a**, **b**, **c**. The vector **b** coincides with the monoclinic axis. Naphthalene is crystallized in the form of thin wafers with the developed plane **ab**. X-ray structural analysis [79] has established that the molecules in a crystal have their long axes almost parallel to the **c** axis of the crystal, and their median axes are at an angle of 29.5° to the monoclinic axis. In a crystal, the molecules are slightly deformed (the length of some C–C bonds is changed by ~0.006 Å). Their local symmetry group becomes C_i.

Unlike anthracene and naphthacene molecules, the first electron transition in a free molecule of naphthalene (frequency, 32,020 cm^{-1}) is evidently oriented along the long axis of the molecule [80] and is very weak (oscillator strength, $\sim 10^{-3}$). The perturbations caused by not totally symmetric vibrations also change the electronic state of the molecule. For example, the excitation corresponding to the combination of an electron transition with a not totally symmetric vibration of 438 cm^{-1} relates to the irreducible representation B_{2u} and has a comparatively large dipole moment oriented along the median axis of the molecule. Roughly speaking, a not totally symmetric molecular vibration results in a new molecular state whose wave function contains a small part of the wave function of the second electron transition. Deformation of a naphthalene molecule in a crystal also strengthens and changes the orientation of the electric dipole moment of the first electron transition.

The spectrum of naphthalene crystals was studied in [12, 81-83]. The characteristics of the principal transitions in naphthalene molecules and crystals at 20°K are shown in Table 7. A region of the absorption spectrum of polarized light for a naphthalene single crystal is shown in Fig. 12.

It follows from Table 7 that molecular transitions 1 and 3 in a crystal correspond to exciton states, while transitions 2 and 4 correspond to localized excitations. The total intensity of the second-transition band in a crystal is almost equal to the total intensity of the bands corresponding to purely electronic transition 1. For a free molecule, the intensity of the second transition exceeds by a factor of almost 10 the intensity of the first transition.

TABLE 7. Principal Extended States in Naphthalene

Transition	Molecule excitation: transition type	Molecule excitation: freq., cm^{-1}	Crystal excitation, 20°K: a comp., cm^{-1}	Crystal excitation, 20°K: b comp., cm^{-1}	Crystal excitation, 20°K: ΔE, cm^{-1}	$P\left(\frac{b}{a}\right)$
1	00	32 020	31 476	31 623	147	160
2	00+433	32 458	31 960	31 960	0	6,6
3	00+702	32 722	32 231	32 261	30	15
4	00+911	32 931	32 413	32 413	0	1

This intensity distribution is evidently due to molecule deformation in the crystal.

The high polarization ratio of the two components of the doublet associated with the first electron transition is of particular interest. As Prikhot'ko and Soskin have shown [83], the 31,623 cm^{-1} band in the b component is more intense by a factor of 160 than the 31,476 cm^{-1} band in the a component. This great difference in intensity engendered doubt about whether they belong to a single electronic excitation of a free molecule. This doubt was completely eliminated by Sheka [84], who gave a clear picture of the formation of an exciton doublet in a crystal.

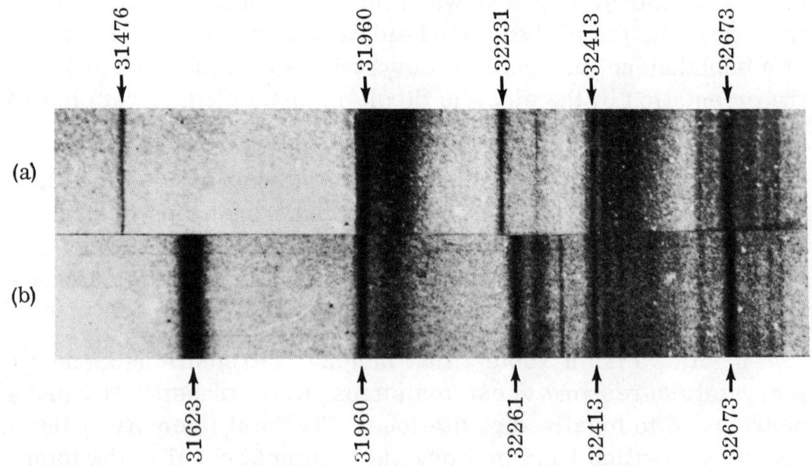

Fig. 12. Region of polarized-light absorption spectrum for naphthalene single crystal at 20°K. a) Polarization along vector **a**; b) polarization along vector **b**. Frequencies given in cm^{-1}.

Sheka studied the absorption spectrum of solid solutions of naphthalene in deuteronaphthalene ($C_{10}D_8$) in a wide range of concentrations. The spectrum of the excited states of the deuteronaphthalene molecules is shifted 115 cm^{-1} in the shortwave direction relative to the spectrum of ordinary naphthalene. At low naphthalene concentrations, there is no resonance interaction between its molecules and one absorption band with frequency 31,541 cm^{-1} is associated with an intramolecular electron transition in a crystal. When the naphthalene concentration is increased, resonance interactions begin to appear and this band is divided into two bands, the distance between which increases as the concentration of naphthalene molecules increases. When the crystal is pure naphthalene (100% concentration), these bands form the exciton doublet shown in Fig. 7.

Benzene Crystals. A benzene molecule (C_6H_6) has symmetry D_{6h}. Light absorption in the near ultraviolet region is associated with transition of a free molecule from the ground state to the electronic excited state of the irreducible representation B_{2u} of the group D_{6h}.[9] A purely electronic transition of this type is forbidden in the dipole approximation. It corresponds to a frequency of 38,089 cm^{-1}. Transition to the state formed by the combination of an electronic excitation with not totally symmetric vibration E_{2g} (frequency, 520 cm^{-1}) is allowed.

The absorption spectrum of polarized light in benzene single crystals has been studied by Broude, Medvedev, and Prikhot'ko [11, 86, 87]. According to x-ray studies [88], a benzene crystal belongs to the rhombic system, with point group V_h^{15}, with four molecules in a unit cell, which is formed by the three mutually perpendicular vectors **a**, **b**, **c**, whose absolute values are 7.28, 9.45, and 6.73 Å, respectively. In a crystal, a benzene molecule is slightly distorted (the change in some C–C bonds is on the order of 0.005 Å) and the first electron transition (38,089 cm^{-1}) becomes allowed.

Table 8 gives some results of studies [11, 86, 87] at 20°K of the absorption spectra of variously oriented benzene single crys-

[9]We use the symbols of irreducible representations and the table of characteristics of group D_{6h} from [85]. The lines of intersection with the benzene ring of the nodal planes of the wave function of the irreducible representation B_{2u} pass through the carbon atoms [14].

TABLE 8. Principal Extended States in Benzene

Transition	Molecule excitation		Crystal excitation, cm^{-1}			
	transition type	freq., cm^{-1}	*a* component	c component	b component	ΔE
1	00	38,089	37,803	37,843*	–	40
2	00+523	38,612	from 38,351 to 38,485	from 38,351 to 38,485	from 38,351 to 38,585	0
3	00+925	39,014	38,724	38,768	–	44
4	00+523+925	39,537	from 39,275 to 39,372	from 39,275 to 39,372	from 39,275 to 39,372	0
5	00+2× 925	39,939	39,667*	39,690*	–	23
6	00+3×925	40,864	40,600	40,615	–	15

tals. The states marked with asterisks are doublet states (with a distance of 1-7 cm^{-1}). The second and fourth molecular levels in a crystal correspond to localized excitations with a complicated structure. The structure of the second and fourth bands remains the same in all three components of the spectrum. A difference is observed only in their relative intensity. The crystal absorption bands corresponding to intramolecular transitions 1, 3, 5, and 6 are associated with exciton excitations, so they appear in different components of the spectrum with different frequencies.

In a study of the spectra of solid solutions of benzene in deuterobenzene, Broude and Onoprienko [89] detected a gradual development of resonance splitting as the concentration of molecules of one type was increased. At very low benzene concentrations, the absorption band associated with the intramolecular electron transition of 38,089 cm^{-1} is alone. But at a concentration as low as 10-15%, marked splitting of this band into two components is observed: the splitting is ~3-4 cm^{-1}. At a 50% concentration, the resonance splitting reaches 20 cm^{-1}. As the concentration increases, the splitting increases almost linearly, reaching 40 cm^{-1} at 100% concentration.

Thus, the example of the absorption spectra of anthracene, naphthacene, pentacene, naphthalene, and benzene crystals shows us that the polarized-light absorption bands in crystals at low temperatures can be roughly divided into two types: a) the bands associated with exciton production, which appear in the different spectrum components with different intensities and different frequencies; and b) the bands associated with localized excitations. In the near ultraviolet region, only exciton production is observed in the spectra of anthracene and naphthacene. Both excitons and localized excitations occur in the spectra of naphthalene and benzene. Localized excitations are associated with the intramolecular energy levels formed by a combination of an electronic excitation with not totally symmetric vibrations of the atoms in a molecule.

The form and nature of the exciton absorption bands in the spectra of molecular crystals are quite varied. In the spectra of anthracene and naphthacene, they are wide, intense bands, which is in accord with the high oscillator strengths of the molecular transitions. In the benzene spectrum, the exciton absorption bands are comparatively narrow. In the naphthalene spectrum, the bands of the first exciton doublet differ considerably in intensity. But in other cases, the intensities of the doublet components are of the same order of magnitude. There is still no theory that explains all of these diverse cases.

8. Exciton Luminescence

The excitons and localized excitations produced by light in a crystal are nonequilibrium excitations. The energy of such excitations is completely or partially converted to radiant heat energy (luminescence) or causes a photoelectric effect, photochemical reactions, etc. The most characteristic property of exciton states is the fact that they embrace large (as compared with the lattice constant) regions of the crystal and move, generally speaking, from one region to another. This migration of excitation energy is very important. It is the basis for a number of phenomena such as, for example, sensitized luminescence or quenching, when the excitation energy of the base material of the crystal is imparted to its impurities, which emit it or convert it to heat, chemical, or other forms of energy.

The nature of these processes is determined by the structure of the exciton energy bands and their interaction with lattice vibrations, crystal defects, and impurities. The kinetics of these processes is greatly dependent upon temperature. In this section, we shall discuss the qualitative description of the phenomena connected with luminescence.

Luminescence is one of the important competing processes of conversion of a part or all of the energy of the electronic-vibrational excitation of the molecules of the crystal into other forms of energy. Of particular importance among these competing processes are those of nonradiative conversion of all or a part of the excitation energy into thermal lattice-vibration energy, i.e., into phonon energy. A quantitative theory of nonradiative conversion of excitation energy into phonon energy has not yet been developed. Mainly experimental data must be used for a qualitative description of these processes.

Nonradiative processes in solids are responsible for the establishment of thermal equilibrium, i.e., of a statistical energy distribution with respect to all degrees of freedom of the crystal. The probability of nonradiative transitions and, therefore, the rate of establishment of thermal equilibrium are functions of the relationship between the corresponding excitations and the lattice-vibration phonons. Evidently, the relationship between intramolecular vibrations and lattice vibrations is considerably greater than the relationship between electronic excitations of molecules and lattice vibrations. Because of this, at low temperatures the electronic-vibrational excitations lose comparatively rapidly ($\sim 10^{-12}$–10^{-13} sec) the excitation energy associated with intramolecular vibrations. Therefore, luminescence proceeds from the lowest state of electronic excitation to the ground or vibrational excitations in the electronic ground state. With light absorption, the transitions are from the ground state to states of electronic or electronic-vibrational excitation.

Such transitions in impurity molecules of solid solutions are illustrated in Fig. 13 for low temperatures. The absorption spectrum begins with a band of purely electronic transition (frequency ω_0') and continues into the shortwave region (frequencies ω_1', ω_2',...). The frequency differences are equal to the frequencies of intramolecular vibrations in an excited state of a molecule. The luminescence spectrum begins with the electron-transition frequency ω_0 and continues into the longwave region (frequencies ω_1, ω_2,...). The frequency differences $\omega_i - \omega_0 = \Omega_i$ correspond to the intramolecular vibrations in the ground state of a molecule. If the coupling of an electronic excitation with the lattice phonons is small, then $\omega_0 \approx \omega_0'$ and the absorption and luminescence spectra will be joined at ω_0. In the ground as well as in the electronic excited

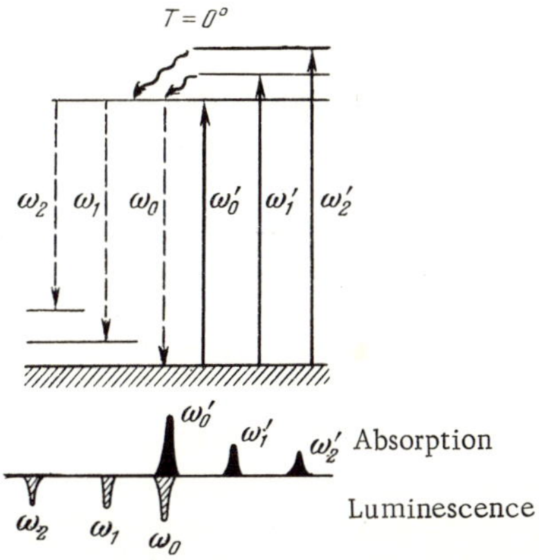

Fig. 13. Absorption and luminescence spectra of impurity molecules at low temperature.

states, with an increase in temperature the intramolecular vibrational degrees of freedom will be "populated" in proportion to the Boltzmann factor

$$\exp(-\hbar\Omega_i/kT),$$

therefore, the absorption and luminescence spectra will overlap.

It has been found that, in addition to the establishment of an equilibrium energy distribution between the lattice phonons and the intramolecular vibrational degrees of freedom, a quasi-equilibrium distribution of electronic excitations of various kinds is established in a crystal during the lifetime of electronic excitations (10^{-8}-10^{-9} sec) if their energy difference does not very greatly exceed the mean phonon energy. For example, in the presence of two types of excited states with energies E_1 and E_2, the ratio of the numbers of excited states

$$\frac{N(2)}{N(1)} = \frac{p_2}{p_1} \exp\left(-\frac{E_2 - E_1}{kT}\right), \tag{8.1}$$

where p_i is the multiplicity of the corresponding state, and T is temperature. After an equilibrium distribution has been established, when $E_2 - E_1 \gg kT$, there are only excitations of type E_1 in the crystal. The sublevels of any exciton band of the crystal obey rule (8.1).

A light wave with frequency ω and wave vector **Q** excites in the crystal a state with energy $E(\mathbf{k}) = \hbar\omega$ at $\mathbf{k} = \mathbf{Q}$. As a result of the establishment of a quasi-equilibrium distribution, at $T \approx 0$ the exciton excitation "rolls down" to a sublevel with a k value at which $E(\mathbf{k})$ has the minimum value E_0. At $T \neq 0$, the sublevels of an exciton band are "populated" in accordance with the Boltzmann factor

$$\exp\left\{\frac{E(\mathbf{k}) - E_0}{kT}\right\}. \tag{8.2}$$

Owing to the quasi-equilibrium distribution during the lifetime of an electronic excitation, luminescence begins with the band that is associated with the electronic excitation of least energy.[10]

Now let us examine the characteristics of the luminescence of molecular crystals in which exciton states play a great role.

Luminescence of Pure Crystals. First let us consider the luminescence of ideal crystals, which do not contain impurities or defects and which have fairly large dimensions. Take an anthracene crystal in which light excites only exciton states. As we shall see below, the nature of the luminescence of such crystals is greatly dependent upon the temperature and structure of the exciton band.

Assume that light produces an exciton only from the lower band $E(\mathbf{k})$. The range of absolute values of the wave vector k is roughly defined by the interval $0, \pi/a$, where a is the mean lattice constant. Owing to the law of conservation of quasi-momentum, only the sublevel of the exciton band with $\mathbf{k} = \mathbf{Q}$, where $|\mathbf{Q}| = 2\pi/\lambda$, is excited. Since $\lambda \gg a$, the sublevel of the exciton band corresponding to the center ($\mathbf{k} \approx 0$) of the Brillouin zone is excited.

1) The value of $E(\mathbf{Q})$ is found near the bottom of the exciton band with energy E_0 (the case of a positive exciton effective mass),

[10]Here we shall consider only singlet excited states. Aromatic molecules also have triplet states of lower energy. Under ordinary conditions, in molecular crystals the transition of singlet to triplet excitations is strongly forbidden and, as a rule, requires more time than the lifetime of a singlet excited state. It has also been indicated [90] that in a solid solution of azulene in naphthalene, luminescence proceeds not from the first but from the second excited state of an azulene molecule.

so that[11]

$$E(\mathbf{Q}) - E_0 \ll kT. \tag{8.3}$$

2) The value of $E(\mathbf{Q})$ is found at a distance from the bottom of the band E_0 such that

$$E(\mathbf{Q}) - E_0 \gg kT. \tag{8.4}$$

Consider the case when inequality (8.3) is satisfied. Here, a band with frequency

$$\omega_1 = \hbar^{-1} E(\mathbf{Q})$$

and quite definite polarization is observed in the absorption spectrum. After light absorption, a quasi-equilibrium distribution is established and excitons fill a region of the exciton band near $\mathbf{k} \approx \mathbf{Q}$; the higher the temperature, the wider is this region. In view of the law of conservation of quasi-momentum, however, luminescence with direct transition of the crystal to the ground state is possible only from the sublevel $\mathbf{k} = \mathbf{Q}$. Therefore, a polarized band that is resonance-coincident with the absorption band must be observed in the luminescence spectrum.

Luminescence from exciton-band states with wave vectors $\mathbf{k} \neq \mathbf{Q}$ in an ideal lattice is possible also with the participation of phonons with momentum $\mathbf{q}$ that satisfies the equation $\mathbf{k} + \mathbf{q} = \mathbf{Q}$. Transitions in which phonons participate are unlikely. They cause a weak background about the fundamental frequency ω_1. As the temperature increases, the intensity of the background grows (the

---

[11]If the effective mass is positive, i.e., the exciton energy increases with an increase in quasi-momentum, then in an isotropic crystal for transverse excitons

$$E(\mathbf{Q}) - E_0 = \frac{\hbar^2 Q^2}{2m^*}.$$

Then, if we substitute the value of (2.20) for the effective mass, we find

$$E(\mathbf{Q}) - E_0 = {}^1/_4 (Qa)^2 \Delta L, \tag{8.3a}$$

where ΔL is the width of the exciton band, and a is the lattice constant. For a quantum transition with energy $\sim 3 \cdot 10^4$ cm^{-1} and $a \sim 10^{-7}$ cm, we have

$$E(\mathbf{Q}) - E_0 \approx 10^{-4} \Delta L.$$

If the exciton effective mass is negative, then $E(\mathbf{Q}) - E_0 \approx \Delta L$.

probability of the transitions increases) and it expands to the short-wave side, since the region of filling of states with $\mathbf{k} \neq \mathbf{Q}$ increases. Luminescence from states with $\mathbf{k} \neq \mathbf{Q}$ is also possible at places where the translational symmetry of the crystal is violated (lattice defects, the crystal surface). Transitions of this type are also unlikely. They create a background about the ω_1 band whose intensity is proportional to the defect concentration and is independent of temperature when the lattice defects are not caused by thermal motion. The background produced by a violation of translational symmetry cannot have strict polarization.

Also possible are transitions from exciton states not directly to the ground state but to vibration sublevels of the ground state, which correspond to "vibration bands," i.e., the bands produced by intramolecular vibrations when molecules are joined in a crystal. Each intramolecular vibration in a crystal that contains σ molecules in a unit cell is associated with σ vibration bands with energies $\hbar\Omega_i(\mathbf{q})$, $i = 1, 2, \ldots, \sigma$, where $\mathbf{q}$ is the wave vector. The allowed electron transitions combine with the vibration-band states corresponding to totally symmetric molecular vibrations. The widths of these vibration bands and the distances between them are practically equal to zero, i.e., $\Omega_i(\mathbf{q}) \approx \Omega$, since resonance interactions are absent with these totally symmetric intramolecular vibrations.[12]

Quantum transitions from exciton states to vibration-band states with photon emission $(\omega, \mathbf{Q})$ are possible when the selection rules

$$\hbar\omega = E_\mu(\mathbf{k}) - \hbar\Omega(\mathbf{q}), \quad \mathbf{k} + \mathbf{q} = \mathbf{Q}. \tag{8.5}$$

are satisfied. These selection rules can be satisfied for any $\mathbf{k}$. Therefore, transitions from all occupied sublevels of the exciton band are probable.

A second characteristic of transitions to vibration-band states (in crystals containing not less than two molecules in a unit cell) is that they result in the emission of unpolarized radiation. We

[12]The vibration bands associated with intramolecular vibrations that appear in the infrared spectrum have a finite width, due to resonance interaction between the dipole moments of the transitions in the molecules. Excitations of this type have the properties of exciton states for electronic excitations. In particular, resonance splitting of the infrared-absorption bands can be observed in crystals with several molecules in a unit cell.

can show this by the example of crystals containing two molecules in a unit cell with symmetry corresponding to the point group C_{2h} (see Table 2). The two vibration bands of such a crystal, which are associated with a single totally symmetric vibration of the irreducible representation A_{1g} of the symmetry group D_{2h} (see Table 3), are characterized by the irreducible representations A_g and B_g. This can be seen with the aid of formulas (6.6) and (6.7). The dipole moment of the transition from the exciton state A_u to the vibration band A_g (just as the transition to the ground state of the crystal) is associated with the irreducible representation $A_u \times A_g = B_u$, i.e., is directed along the monoclinic axis. But the dipole moment of the transition from this same state to the vibration band B_g pertains to the irreducible representation $A_u \times B_g = B_u$, i.e., is perpendicular to the monoclinic axis.

Thus, in crystal luminescence, along with a polarized band of frequency ω_1 for transition to the ground state, we can have unpolarized bands in the frequency range $\omega_i = \omega_1 - \Omega_i$, where Ω_i are the frequencies of the intramolecular vibrations in the electronic ground state of a molecule. With an increase in temperature, the electron vibration luminescence bands are broadened (a greater number of sublevels of the exciton band are occupied) and their intensity increases.

Figure 14 shows the energy levels of the crystal when inequality (8.3) is satisfied. Only one intramolecular vibration is

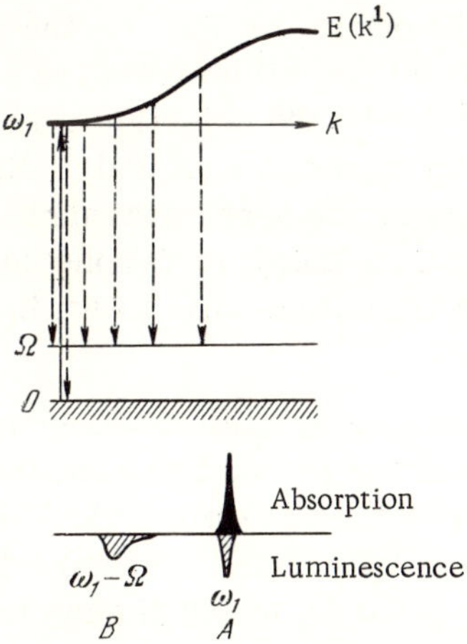

Fig. 14. Absorption and luminescence spectra of pure crystal produced by excitation band with positive exciton effective mass.

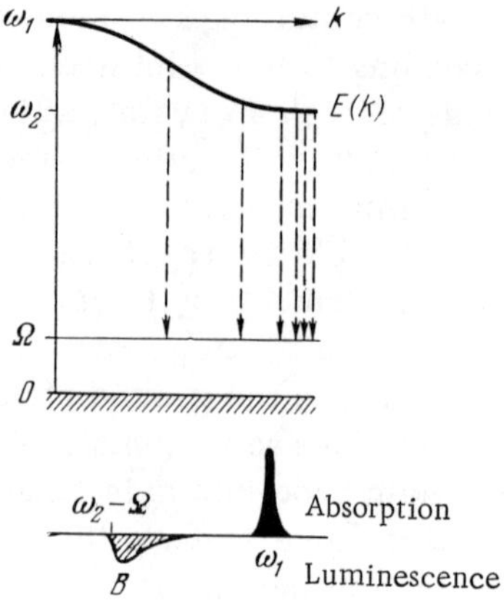

Fig. 15. Absorption and luminescence spectra of pure crystal produced by excitation band with negative exciton effective mass.

taken into account. The solid arrow indicates the transition for absorption. The broken arrows indicate the transitions for luminescence. Band A pertains to the transition to the ground state. It is in resonance-coincidence with the absorption band. Band B characterizes the transitions to the vibrational sublevels of the ground state.

Now let us consider the case when inequality (8.4) is satisfied. Figure 15 shows the energy band in which E(Q) is close to the upper edge of the band (negative effective mass). When light is absorbed, a band is observed in the frequency range $\omega_1 = \hbar^{-1}E(\mathbf{Q})$. After light absorption and a quasi-equilibrium distribution is established in the crystal, excitons with k values corresponding to an energy close to the bottom of the exciton band will occur. In Fig. 15, these values correspond to $|\mathbf{k}| = \pi/a$. A direct transition from the states to the crystal ground state is forbidden by the selection rule $\mathbf{k} = \mathbf{Q}$, and transitions in which phonons participate are unlikely. Nonradiative transitions of excitation energy to heat and transitions to vibration bands associated with intramolecular vibrations compete with these transitions. The former transitions are shown in Fig. 15 by broken arrows.

Thus, when inequality (8.4) is satisfied, luminescence from exciton states is difficult, since the more probable luminescence band that is resonance-coincident with the absorption band is either absent or very weak (due to low occupancy of the corresponding sublevel of the exciton band). Only B bands, which pertain to transitions with excitation of intramolecular vibrations, are present in the luminescence spectrum. These bands are situated on the longwave side of the absorption band at distances

$$\Omega_i + \frac{E(Q) - E(\pi/a)}{\hbar}.$$

When temperature increases, the widths of the bands and their intensities increase according to the most complete occupation of the sublevels of the exciton band, reaching ΔL when $\Delta L \approx kT$. At sufficiently high temperatures, therefore, the width of the electron-vibration luminescence bands roughly characterizes the width of the exciton band when the latter does not overlap other bands.

If the crystal has several exciton bands and local excitations, then, according to the selection rules, several absorption bands of specific polarization must be observed in the absorption spectrum. At low temperatures, after a quasi-equilibrium distribution has been established, only excitons with energies close to the energy E_0 of the bottom of the first exciton band will remain. In accordance with the above, the nature of the luminescence in this case will depend upon the ratio of kT and $E(\mathbf{Q}) - E_0$. At sufficiently high temperatures, providing there is a quasi-equilibrium distribution, excited states will also be present in the second exciton band. If $E_2(\mathbf{Q}) - E_0 \ll kT$, in this case, then two polarized bands that are resonance-coincident with the absorption bands must appear in the luminescence spectrum. If $E_2(\mathbf{Q}) - E_0 \gg kT$ and $\min(E_2(\mathbf{k}) - E_0) \ll kT$, luminescence from the states of the second band is possible only with transitions to the vibrational sublevels of the ground state.

<u>Luminescence of Impure Crystals.</u> As a rule, even well-purified crystals contain impurities. Sometimes impurities are introduced artificially. Various kinds of lattice defects and thermal density fluctuations also represent kinds of impurities. Henceforth, local inclusions of any kind will be considered impurity molecules.

If the energy of the excited states of the impurity molecules is somewhat lower than the energy of the first exciton band, when a quasi-equilibrium distribution is established the excitation energy of the crystal at low temperatures will become the excitation energy of the impurity molecules. In this case, if the probability of nonradiative transitions of the excitation energy into heat is less than the probability of radiation, the impurity molecules will be sources of luminescence. Although under these conditions the impurity molecules are centers of luminescence, the role of excitons in luminescence is great. They impart to the impurity molecules the energy that is absorbed by the molecules of the base material.

The exciton energy moves through the crystal and excites the impurity molecules when it reaches them. As they move through the crystal, the excitons are scattered by phonons. If the lifetime of the excitons considerably exceeds the mean time between two collisions of them with phonons, the exciton distribution in the crystal can be described with the aid of a diffusion equation.[13]

The notion of the diffusional nature of exciton motion has been employed by Agranovich and Faidysh [93] and a number of others [94-100].

The migration of excitation energy in a crystal has been proven by many experimental studies. In 10^{-8} sec, an exciton has time to cover a distance of tens and hundreds of thousands of lattice constants. The migration and transfer of excitation energy from the base material to trace impurities, which become centers of luminescence, explain the fact that at low temperatures the major portion of the luminescence comes from the levels of the impurity molecules, even when the impurity concentration is not over a hundredth or thousandth of weight percent.

Lipsett and Dekker [101], for example, have shown that at a 10^{-6} mole concentration of naphthacene in crystalline anthracene,

---

[13]A reabsorption mechanism in which the base material emits radiation that is absorbed by the base material and the impurities is possible along with the exciton mechanism of energy migration. Wolf has shown [91] that the latter mechanism is less effective by several orders of magnitude than the former mechanism in naphthalene crystals with an anthracene impurity. In some cases, the effect of reabsorption is very great. The relative roles of both energy-migration mechanisms has been studied theoretically by Agranovich [92]. The theory of the diffusion motion of excitons considers an exciton as a point formation. Considering the considerable spatial extent of an exciton ($\sim 10^3$ lattice constants), the movement of an exciton excitation to impurity molecules does not necessarily involve its migration — it is sufficient that the impurity be in the region of the crystal embraced by the excitation. This effect becomes even more important when it is considered that an impurity molecule has a "binding" influence on an exciton.

the intensity of the naphthacene luminescence is comparable with the intensity of luminescence of the anthracene molecules. In this case, the naphthacene molecules directly receive only 10^{-6} of the energy absorbed by the crystal. For the same concentrations of naphthacene molecules in a liquid or solid (vitreous) solution, naphthacene luminescence is absent. Borisov and Vishnevskii have shown [102] that appreciable anthracene luminescence appears in a naphthalene crystal even when the distance between the anthracene molecules is $\sim 0.1\,\mu$.

Very small amounts of impurities are practically impossible to detect by chemical methods of analysis. For this reason, impurity luminescence has sometimes been mistaken for the luminescence of a pure crystal. For example, it was assumed for a long time that the main luminescence spectrum of a naphthalene crystal, which begins with a band with a frequency of 31,062 cm^{-1}, was due to naphthalene molecules. Not until 1958 did Prikhot'ko and Shpak [103] prove conclusively that this luminescence was due to the presence in naphthalene of trace amounts of β-methylnaphthalene, α-naphthol, and other impurities.

Shpak and Sheka have made the most complete study of the impurity and natural luminescence of naphthalene [104-106]. They showed that in naphthalene crystals containing a small β-methylnaphthalene impurity, the absorbed luminous energy (at 20.4°K) is transferred by excitons migrating through the crystal to the impurity molecules, which become centers of luminescence. As the crystal is purified, there begins to appear in the a component a luminescence band with a frequency of 31,480 cm^{-1}, which is very close to the band with a 31,476 cm^{-1} maximum in the absorption spectrum (see Table 7). The polarization of the luminescence and the resonance coincidence with the band of the absorption spectrum indicate that this luminescence band is associated with direct transitions from exciton states to the ground state. As was shown above, this luminescence is possible if the sublevel $E_a(\mathbf{Q})$ is close to the bottom of the band. Thus, the nature of the luminescence indicates that the excitons in the a band of a naphthalene crystal in the area of $\mathbf{k} \approx 0$ have a positive effective mass.

At 20°K, the value of kT ($\sim$14 cm^{-1}) is smaller by a factor of about 10 than the energy difference between the bottom E_0 of the first band in the a component and $E_b(\mathbf{Q})$ (31,623 cm^{-1}) in the corresponding band of the b component. Therefore, the sublevels

$E_b(Q)$ in a state of quasi-equilibrium are not occupied and do not participate in the luminescence. When the temperature of the crystal is increased to 77°K, however, another band, with a frequency of 31,623 cm^{-1}, which is polarized in the b direction, appears in the luminescence spectrum of naphthalene. The position of this band coincides with the band observed in the b component of the absorption spectrum. This change in the luminescence spectrum is easily understood when one considers that at 77°K, the value of kT (54 cm^{-1}) is smaller than $E_b(Q) - E_0$ only by a factor of 2.7, and the probability of transitions to the ground state from the level $E_b(Q)$ is greater by a factor of 160 than the probability of transition from the level $E_a(Q)$.

Figure 16 shows microphotograms of luminescence spectra obtained by Shpak and Sheka [105]. The luminescence of a naphthalene crystal containing a small β-methylnaphthalene impurity at 20.4°K is shown in Fig. 16a. A luminescence band with a 31,060 cm^{-1} maximum is clearly visible; it corresponds to the luminescence of the impurity. Microphotograms of the luminescence of a very pure naphthalene crystal at 20.4 and 77°K are given in Fig. 16b. The measurements were made without allowance for polarization. At 20.4°K there is an exciton luminescence band of 31,480 cm^{-1}, and at 77°K there are two exciton luminescence bands, which correspond to the resonance doublet in the absorption spectrum. The exciton nature of these luminescence bands is confirmed by a study of their polarization (Fig. 16c). A broad unpolarized luminescence band with a maximum of about 30,970 cm^{-1} can also

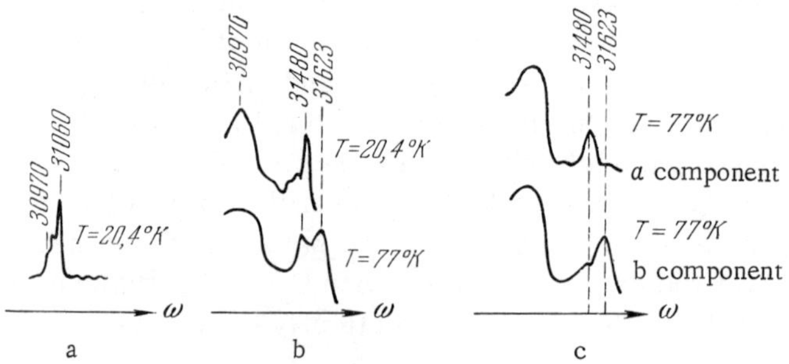

Fig. 16. Microphotograms of luminescence spectra of naphthalene single crystal: a) crystal containing small β-methylnaphthalene impurity; b) luminescence of pure crystal without allowance for its polarization; c) polarized luminescence of pure crystal.

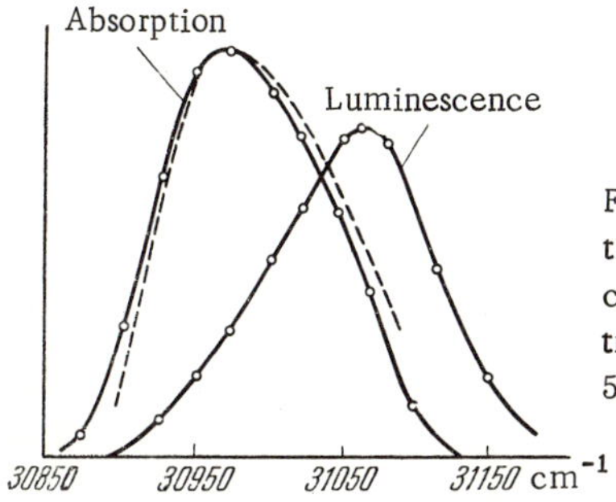

Fig. 17. Absorption and luminescence spectra of naphthalene single crystal at 100°K, corresponding to transitions between electronic excited state and vibration level of 512 cm^{-1} in the molecule ground state.

be seen in Fig. 16, b and c. This band corresponds to transition from the sublevels of the exciton band to the vibration sublevels of the ground state, which are associated with the totally symmetric vibration of 512 cm^{-1}, well known from Raman scattering. As might be expected, this electron-vibration band is unpolarized. Its width is about 200 cm^{-1}. Figure 17, which was borrowed from a paper by Broude, Shpak, and Sheka [107], shows (solid curves) the profiles of such an electron-vibration luminescence band and of the absorption band corresponding to inverse transitions from the vibration sublevels of the ground state to an electron band.[14]

The link between electron transitions and phonons in a naphthalene crystal is weak. In the first approximation, therefore, the broadening of the bands due to interaction with phonons may be ignored. In this approximation, the form of the absorption band must be determined by the curve

$$F_{\text{abs}}(E) = \rho(E - E_0)\, w(E, \Omega) \exp\left(-\frac{\hbar\Omega}{kT}\right), \tag{8.6}$$

where $\rho(E - E_0)$ is the probability density function of the sublevels with energy E in the exciton band; and $w(E, \Omega)$ is the probability of a dipole transition between the sublevels of the exciton band and the vibrational state Ω. If it is assumed that the probability of nonradiative transitions of excitation energy to heat from all sublevels of the exciton band is the same, then the shape of the lu-

[14]In a naphthalene crystal, the exciton bands of both components of the resonance splitting overlap partially [107]. In reality, therefore, the transitions occur between the sublevels of this compound band and the vibration sublevels of the ground state.

minescence band must be expressed by the curve

$$F_{\text{lum}}(E) = AF_{\text{abs}}(E)\exp\left(-\frac{E-E_0}{kT}\right). \quad (8.7)$$

In Fig. 17, the broken curve represents the function $F_{\text{lum}}(E)$, which was obtained using (8.7) from the absorption curve $F_{\text{abs}}(E)$.

Thus, by studying the polarization and temperature variation of the intensity of the luminescence bands of sufficiently pure single crystals and comparing them with the absorption spectrum, in a number of cases we can uniquely determine their nature. Sharply polarized luminescence bands whose polarization and position coincide with the polarization and position of the components of the resonance doublets in the absorption spectrum pertain to transitions from exciton bands directly to the ground state. At low temperatures, only the longwave component of the doublet is observed in the luminescence spectrum. Comparatively wide, blurred, and unpolarized bands in the luminescence spectrum are associated with transitions from exciton bands to the vibration sublevels of the ground state. Their width increases greatly with temperature. They are easily distinguished from the comparatively narrow luminescence bands of impurities and local excitations. The widths of such bands vary little with temperature.

The above-mentioned characteristics of the various bands of the luminescence spectrum are well confirmed by studies of the luminescence spectrum of an anthracene crystal. The luminescence of anthracene that is usually observed at low temperatures is due, as Prikhot'ko and Fugol' have shown [108], to impurities. The luminescence of very pure anthracene at 4.2°K has been studied by Alexander et al. [109]. The most complete studies of the polarized luminescence of anthracene at temperatures of 20.4, 77, and 290°K were made by Shpak and Sheremet [110]. The main results of these studies are as follows. It was shown that at 20.4°K the luminescence spectrum of very pure anthracene crystals in the b component begins with a broad intense band (~60 cm^{-1} wide) with a maximum in the area of 25,055 cm^{-1}. The polarization and the position of the shortwave wing of this band coincide with the polarization and position of the band in the b component of the absorption spectrum. On the longwave side of the first luminescence band are broad luminescence bands, which are evidently associated with transitions from the exciton band to the vibration sublevels of the ground state, which correspond to totally symmetric intramolecular vibrations of 394, 1167, 1262, 1402, 1558, and 1644 cm^{-1} in the Raman spectrum. These same bands are observed in the *a* component of the luminescence spectrum.

At 77°K, besides the above-mentioned bands, a luminescence band in the area of 25,200 cm^{-1} appears in the *a* component of the spectrum. The shortwave edge of this band reaches a frequency of 25,300 cm^{-1} and partially overlaps the absorption band corresponding to the component of the resonance doublet (with a maximum of 25,432 cm^{-1}). This luminescence band is absent in the b component of the spectrum.

The width of the luminescence bands associated with transitions from the sublevels of the exciton bands to the ground state is determined by the interaction of electron transitions with lattice vibrations. As a result of this interaction, transitions from the various sublevels of the exciton band become allowed, providing $k + q = Q$. In anthracene the exciton bands are wider than in naphthalene or benzene. This is due to the considerable interaction of an electronic excitation (high oscillator strength) with lattice vibrations. The comparatively strong interaction of an electronic excitation with lattice vibrations is also indicated by the considerable broadening of the luminescence bands as temperature is increased (to 500 cm^{-1} at 290°K) and the only partial overlapping of the exciton luminescence and absorption bands.

Chapter III

Theory of Exciton States in the Second-Quantization Representation (Fixed Molecules)

1. The Energy Operator of a Crystal with Fixed Molecules

In Chapter II, we considered the elementary theory of the exciton states of a crystal on the basis of perturbation theory in a form proposed by Heitler and London. We shall call this the Heitler–London approximation. As is well known, systems consisting of a large number of identical interacting subsystems (atoms, molecules, particles, etc.) are more conveniently studied in a second-quantization representation. In this chapter, we shall study exciton states by second quantization, which will enable us to obtain the results discussed in Chapter II more simply and to study higher approximations of the theory. The second-quantization method has been used in the theory of exciton states of molecular crystals by Agranovich [1] and others [2-4].

In the coordinate representation, the energy operator of a crystal with fixed molecules can be written as

$$H_{\text{III}} = \sum_{n} H_n + \frac{1}{2} \sum_{n,\ m}' V_{n,m}, \tag{1.1}$$

where n = $\{\mathbf{n}, \alpha\}$; m = $\{\mathbf{m}, \beta\}$; **n** and **m** are lattice vectors; and the prime on the summation indicates that there are no terms with $\mathbf{n}\alpha = \mathbf{m}\beta$ in the sum. Operator (1.1) is defined in the function space $\psi(\ldots\xi_n\ldots)$ of the internal variables ξ_n of all of the molecules of the crystal.

We proceed to the second-quantization representation by selecting a complete set of orthonormal functions that characterize the states of an individual subsystem (molecule). In our case, we can use the eigenfunctions $\varphi_{nf}(\xi_n)$ of the operator H_n, which correspond to the eigenvalues ε_f. As eigenfunctions of the operator H_n, the functions φ_{nf} satisfy strict orthogonality conditions

$$\int \varphi_{nf}^* \varphi_{nf'} \, d\xi = \delta_{ff'}.$$

We shall be concerned with singlet excited states of low (as compared with the ionization energy) energy. In molecular crystals made up of aromatic molecules, the overlap integrals of the functions that pertain to the various molecules are also very small:

$$\int \varphi_{nf}^* \varphi_{mf} \, d\xi < 10^{-2} - 10^{-3}, \quad \text{if } n \neq m.$$

When the concentration of Wannier excitons in semiconductors and dielectrics is not very great, their wave functions do not overlap for all practical purposes. As a simplification, therefore, it may be assumed that the general orthonormality conditions are satisfied:

$$\int \varphi_{nf}^* \varphi_{mf'} \, d\xi = \delta_{ff'} \delta_{nm}.$$

This simplification is not fundamental, and it need not be used.

In representation of the occupation numbers corresponding to the eigenfunctions φ_{nf} of the operators H_n, the state of the crystal is characterized by functions of the occupation numbers N_{nf}, where the first subscript n indicates a molecule of the crystal and the second subscript f indicates its state. The value $f = 0$ denotes the ground state. The occupation numbers N_{nf} are equal to either zero or unity. If $N_{nf} = 1$, for example, the molecule n is in the state f; if $N_{nf} = 0$, the molecule n is not in the state f. Since each molecule can be in only one state, the occupation numbers N_{nf}, which characterize the steady state of a molecule, must satisfy the condition

$$\sum_{f} N_{nf} = 1, \tag{1.2}$$

where summation is over all possible states of the molecule n. Considering that the total number of molecules in a crystal containing N unit cells with σ molecules in each unit cell is equal to σN, the occupation numbers must also satisfy the condition

$$\sum_{n,\ f} N_{nf} = \sigma N. \tag{1.3}$$

The wave function, which represents the various states of the crystal, is, in the occupation-number representation, a function of all of the numbers N_{nf} (their number is infinite), of which σN numbers are 1 and the remaining are 0. We shall denote these functions by the symbol

$$\psi(\ldots N_{nf} \ldots).$$

All of the operators in the second-quantization representation are defined in the function space of the occupation numbers. The simplest operator is the number-of-states operator $\hat{N}_{nf}$. In the occupation-number representation, this operator is diagonal and its eigenvalues are the numbers N_{nf}, which equal 0 or 1, i.e.,

$$\hat{N}_{nf}\, \psi(\ldots N_{nf} \ldots) = N_{nf}\, \psi(\ldots N_{nf} \ldots). \tag{1.4}$$

The operator $\hat{N}_{nf}$ is Hermitian. It can be expressed in terms of two non-Hermitian operators b^{+}_{nf} and b_{nf} by the relation

$$\hat{N}_{nf} = b^{+}_{nf} b_{nf}. \tag{1.5}$$

In order that operator (1.5) satisfy Eq. (1.4) and have eigenvalues of 0 or 1, it is sufficient that the effect of the new operators on the occupation-number functions be determined by the equations

$$\left.\begin{aligned} b^{+}_{nf}\psi(\ldots N_{nf} \ldots) &= (1 - N_{nf})\,\psi(\ldots, N_{nf}+1, \ldots), \\ b_{nf}\psi(\ldots N_{nf} \ldots) &= N_{nf}\psi(\ldots, N_{nf}-1, \ldots). \end{aligned}\right\} \tag{1.6}$$

Equations (1.6) allow us to call b^{+}_{nf} the creation operator of the state nf, i.e., the transition operator of the molecule n to the state f. If $N_{nf} = 0$, the operator b^{+}_{nf} transforms the crystal state to a state in which $N_{nf} = 1$. But the operator b^{+}_{nf} cannot act on a state in which $N_{nf} = 1$. The operator b_{nf} annihilates the state nf.

It follows from Eqs. (1.6) that operators b_{nf}^{+} and b_{nf} that have the same subscripts nf must satisfy the commutation relations

$$b_{nf}b_{nf}^{+}+b_{nf}^{+}b_{nf}=1;\quad b_{nf}b_{nf}=0. \tag{1.7}$$

Operators b_{nf}^{+} and b_{nf} that pertain to different n or f values act on the various variables N_{nf} of the wave functions $\psi(\ldots N_{nf}\ldots)$. Therefore, they must commute, i.e.,

$$\left.\begin{array}{c} b_{nf}b_{n'f'}^{+}=b_{n'f'}^{+}b_{nf},\ b_{n'f'}b_{nf}=b_{nf}b_{n'f'}, \\ \text{if } n\neq n' \text{ or } f\neq f'. \end{array}\right\} \tag{1.8}$$

Thus, creation b_{nf}^{+} and annihilation b_{nf} operators that are associated with a certain state f of the molecule n satisfy the Fermi commutation rules. But if the operators refer to different molecules or to different excited states of the same molecule, they satisfy the Bose commutation relations.

Let us introduce the operators

$$\left.\begin{array}{l} \hat{\Psi}(\xi)=\sum\limits_{n,\,f} b_{nf}\,\varphi_{nf}(\xi), \\ \hat{\Psi}^{+}(\xi)=\sum\limits_{n,\,f} b_{nf}^{+}\,\varphi_{nf}^{*}(\xi), \end{array}\right\} \tag{1.9}$$

where $\varphi_{nf}(\xi)$ are the eigenfunctions of the operators $H_n(\xi)$. Then, transformation from the operators $\hat{F}(\ldots\xi_n\ldots)$ of the coordinate representation, which are defined in the function space $\psi(\ldots\xi_n\ldots)$ of the internal variables of all of the molecules of the crystal, to the operators of the occupation-number representation $\hat{F}(\ldots b_{nf}, b_{nf}^{+}\ldots)$, which are defined in the function space of the occupation numbers N_{nf}, is accomplished simply:

1) If $\hat{F}(\ldots\xi_n\ldots)=\sum\limits_{n} V_n(\xi_n)$, then

$$\hat{F}(\ldots b_{nf}\ldots)=\int \hat{\Psi}^{+}(\xi)\,V(\xi)\,\hat{\Psi}(\xi)\,d\xi=\sum_{n,\,f,\,g} b_{nf}^{+}\langle\varphi_{nf}|V|\varphi_{ng}\rangle b_{ng}, \tag{1.10a}$$

where

$$\langle\varphi_{nf}|V|\varphi_{ng}\rangle\equiv\int \varphi_{nf}^{*}(\xi)\,V(\xi)\,\varphi_{ng}(\xi)\,d\xi.$$

2) If $\hat{F}(\ldots\xi_n\ldots)=\sum\limits_{n,\,m}{}' W_{n,m}(\xi_n,\xi_m)$, then

$$\hat{F}(\ldots b_{nf}\ldots) = \int \hat{\Psi}^{+}(\xi)\,\hat{\Psi}^{+}(\eta)\,W(\xi,\eta)\,\hat{\Psi}(\eta)\,\Psi(\xi)\,d\xi\,d\eta = \\ = \sum_{n,f,g,h,i} b_{nf}^{+}\,b_{ng}^{+}\,\langle\varphi_{nf}\varphi_{ng}\,|\,W\,|\,\varphi_{nh}\varphi_{ni}\rangle\,b_{nh}b_{ni}. \qquad (1.10b)$$

3) If $\hat{F}(\ldots\xi_n\ldots) = \sum'_{n,m,p} W(\xi_n,\xi_m,\xi_p)$, then

$$\hat{F}(\ldots b_{nf}\ldots) = \int \hat{\Psi}^{+}(\xi)\,\Psi^{+}(\eta)\,\Psi^{+}(\zeta)\,W(\xi,\eta,\zeta)\,\Psi(\zeta)\,\Psi(\eta)\,\times \\ \times\,\Psi(\xi)\,d\xi\,d\eta\,d\zeta, \qquad (1.10c)$$

etc.

Let us consider the simplest examples of transformations (1.10).

a) The energy operator of the noninteracting molecules in the coordinate representation

$$H_0 = \sum_n H_n(\xi_n),$$

$$[H_n(\xi_n) - \varepsilon_f]\,\varphi_{nf}(\xi_n) = 0,$$

is, according to (1.10), transformed in the occupation-number representation to

$$H_0 = \int \hat{\Psi}^{+} H_n \Psi\,d\xi = \sum_{n,f} b_{nf}^{+} b_{nf}\varepsilon_f. \qquad (1.11)$$

Energy operator (1.11) commutes with the occupation-number operator $\hat{N}_{nf} = b_{nf}^{+}b_{nf}$. The quantum numbers $N_{nf} = 0, 1$, therefore, are motion integrals and the functions $\psi(\ldots N_{nf}\ldots)$ are eigenfunctions of operator (1.11), i.e., represent the steady states that have a definite energy. Assume that in a crystal containing σN molecules the molecule n' is in the f-th excited state while all of the other molecules are in the ground state, i.e., $N_{n'f} = 1$, $N_{n0} = 1$, $n \neq n'$; all of the other quantum numbers $N_{nf} = 0$. In this state, the crystal energy

$$E = (\sigma N - 1)\,\varepsilon_0 + \varepsilon_f.$$

b) Any operator V, which is given in the coordinate representation by the sum of the operators $\sum_n V_n(\xi_n)$, each of which acts only on the functions of the internal variables of one molecule, has, according to (1.10a), in the occupation-number representation the form

$$V = \sum_{n, f, g} b_{ng}^{+} b_{nf} \langle g | V_n(\xi_n) | f \rangle, \tag{1.12}$$

where

$$\langle g | V_n(\xi_n) | f \rangle = \int \varphi_{ng}^{*} V_n(\xi_n) \varphi_{nf} \, d\xi_n. \tag{1.13}$$

c) Any operator W, which in the coordinate representation is given by the operator sum

$$W = \sum_{n, m} V_{n,m}(\xi_n, \xi_m),$$

each operator of which acts only on the functions of the internal variables of two molecules, has, according to (1.10), in the occupation-number representation the form

$$W = \sum_{n, m} \sum_{f, f'} \sum_{g, g'} b_{nf'}^{+} b_{mg'}^{+} b_{mg} b_{nf} \langle f'g' | W_{nm} | gf \rangle. \tag{1.14}$$

If, using rules (1.10), we transform the energy operator (1.1) of a crystal with fixed molecules, we obtain this operator in the occupation-number representation

$$H = \sum \varepsilon_f b_{nf}^{+} b_{nf} + \frac{1}{2} \sum{}' b_{nf'}^{+} b_{mg'}^{+} b_{mg} b_{nf} \langle f'g' | V_{nm} | gf \rangle, \tag{1.15}$$

where f, f', g, g' are the quantum numbers that characterize all of the steady states of the molecules. In the first sum of (1.15), summation is over all n and f. The second summation is over all n, m, f, f', g, and g' at $n \neq m$;

$$\langle f'g' | V_{nm} | gf \rangle = \int \varphi_{nf'}^{*} \varphi_{mg'}^{*} V_{nm} \varphi_{mg} \varphi_{nf} \, d\xi \, d\eta. \tag{1.16}$$

The operators V_{nm} are symmetric in n and m. Therefore, matrix elements (1.16) have the symmetry conditions

$$\langle f'g' | V_{nm} | gf \rangle = \langle g'f' | V_{nm} | fg \rangle. \tag{1.17}$$

Energy operator (1.15) does not commute with the number-of-states operator $\hat{N}_{nf}$. Therefore, the quantum numbers N_{nf} are not motion integrals. In other words, in the presence of interactions between molecules, which are characterized by the second sum in Hamiltonian (1.15), the localization of an excitation at one or more isolated molecules does not correspond to steady states. Calculation of the steady states of the quantum system described by energy operator (1.15) comes down to finding a canonical transformation that diagonalizes operator (1.15). Particular cases of this diagonalization will be considered in the following sections.

Although none of the occupation numbers N_{nf} is a motion integral of the system with Hamiltonian (1.15), their sum is, in accordance with (1.3), a motion integral equal to the number of molecules in the crystal. The operator for the total number of states in the crystal has, in the occupation-number representation, the form

$$\hat{N} = \int \Psi^{+}(\ldots \xi_n \ldots)\, \Psi(\ldots \xi_n \ldots)\, d\xi = \sum_{nf} b_{nf}^{+} b_{nf}.$$

Since the number of molecules in the crystal is fixed and equals σN, the operators b_{nf} are defined in the function space for which are valid the operator equations

$$\sigma N = \sum_{n,\, f} b_{nf}^{+} b_{nf} \equiv \sum_{nf} \hat{N}_{nf}, \tag{1.18}$$

$$\sum_{f} \hat{N}_{nf} = 1. \tag{1.19}$$

2. The Heitler – London Approximation in the Theory of Excitons

In molecular crystals, the first singlet intramolecular electronic excitations often do not have similar energies. In this case, when calculating the energy of singlet excitons in a crystal, i.e., in diagonalization of operator (1.15), it is sufficient to take into account only the ground state of a molecule and one of its excited states. In this approximation, in the sums over g and f of energy operator (1.15) we should retain only terms for g and f equal to 0 and f, i.e., consider the matrix elements

$$\langle f'g' \,|\, V_{nm} \,|\, gf \rangle = \int \varphi_{nf'}^{*}\, \varphi_{mg'}^{*} V_{nm} \varphi_{mg} \varphi_{nf}\, d\xi\, d\eta, \tag{2.1}$$

for f', g', g, and f equal to 0 or f. In the nondegenerate singlet states, the functions φ_{n0} and φ_{nf} are real; hence, the corresponding matrix elements (2.1) are real.

When the concentration of excited molecules is small, in operator (1.15) we may ignore the interactions between molecules that are characterized by the matrix elements

$$\langle ff \mid V_{nm} \mid ff \rangle = \int \varphi_{nf}^2 V_{nm} \varphi_{mf}^2 \, d\xi \, d\eta,$$

and take into account only the following:

1) the interactions between molecules in the ground state. Such interactions are characterized by the matrix elements

$$\langle 00 \mid V_{nm} \mid 00 \rangle = \int \varphi_{n0}^2 V_{nm} \varphi_{m0}^2 \, d\xi \, d\eta; \tag{2.2}$$

2) the interactions of an excited molecule with unexcited molecules. These interactions are characterized by the matrix elements

$$\langle 0f \mid V_{nm} \mid f0 \rangle = \int \varphi_{nf}^2 V_{nm} \varphi_{m0}^2 \, d\xi \, d\eta; \tag{2.3}$$

3) the interactions associated with intramolecular-excitation transfer. Such interactions are characterized by the matrix elements

$$\langle 0f \mid V_{nm} \mid 0f \rangle = \int \varphi_{n0} \varphi_{mf} V_{nm} \varphi_{m0} \varphi_{nf} \, d\xi \, d\eta; \tag{2.4}$$

4) the interactions that are characterized by the matrix elements

$$\langle ff \mid V_{nm} \mid 00 \rangle = \langle 00 \mid V_{nm} \mid ff \rangle = \int \varphi_{n0} \varphi_{m0} V_{nm} \varphi_{mf} \varphi_{nf} \, d\xi d\eta. \tag{2.5}$$

Matrix elements (2.4) and (2.5) are equal to one another in our approximation (we ignore electron transfer between molecules). Further, we shall use the abbreviation

$$\langle 0f \mid V_{nm} \mid 0f \rangle = \langle ff \mid V_{nm} \mid 00 \rangle = M_{nm}^f. \tag{2.6}$$

If in (1.15) we retain only the above-mentioned matrix elements and take into account operator equation (1.18) and Eq. (1.19), which in our case reduces to

$$\hat{N}_{n0} = 1 - \hat{N}_{nf}, \tag{2.7}$$

then the energy operator takes the form

$$H = \mathscr{E}_0 + H_1 + H_2 + H_3, \tag{2.8}$$

where

$$\mathscr{E}_0 = N\sigma\varepsilon_0 + \frac{1}{2}\sum_{n,m}{}' \langle 00 | V_{nm} | 00 \rangle$$

is a constant term. Further,

$$H_1 = \sum_n \Big(\Delta\varepsilon_f + \sum_{m(\neq n)} D^f_{nm}\Big) \hat{N}_{nf}, \tag{2.9}$$

in which

$$\Delta\varepsilon_f = \varepsilon_f - \varepsilon_0$$

is the excitation energy of an isolated molecule, and

$$D^f_{nm} = \langle 0f | V_{nm} | f0 \rangle - \langle 00 | V_{nm} | 00 \rangle. \tag{2.10}$$

Therefore, $D_f = \sum_{m(\neq n)} D^f_{nm}$ determines the change in the interaction energy of all of the molecules of the crystal with one molecule in its transition to an excited state. The last two operators in (2.8) are defined by the expressions

$$H_2 = \sum_{n,m}{}' M^f_{nm} b^+_{n0} b^+_{mf} b_{m0} b_{nf}, \tag{2.11}$$

$$H_3 = \frac{1}{2}\sum_{n,m}{}' M^f_{nm} \{ b^+_{n0} b^+_{m0} b_{mf} b_{nf} + b^+_{nf} b^+_{mf} b_{m0} b_{n0} \}. \tag{2.12}$$

Energy operator (2.8) is further simplified by introducing the new operators

$$B_{nf} = b^+_{n0} b_{nf}, \qquad B^+_{nf} = b^+_{nf} b_{n0}. \tag{2.13}$$

The new operators have a simple physical meaning. The operator B_{nf} corresponds to transition of the molecule n from the f-th excited state to the ground state. The operator B^{+}_{nf} corresponds to the reverse process – transition of the molecule n from the ground state to an excited state.

The eigenvalues N_{nf} equal zero or unity; therefore, $\hat{N}^2_{nf} = \hat{N}_{nf}$. Considering (1.5) and (2.7), we have

$$\left.\begin{aligned} B^{+}_{nf}B_{nf} &= \hat{N}_{nf}(1-\hat{N}_{n0}) = \hat{N}^2_{nf} = \hat{N}_{nf},\\ B_{nf}B^{+}_{nf} &= \hat{N}_{n0} = 1-\hat{N}_{nf}. \end{aligned}\right\} \tag{2.14}$$

It also follows from (1.8) and (2.13) that $B_{nf}B^{+}_{n'f'} = B^{+}_{n'f'}B_{nf}$, if $n \neq n'$ or $f \neq f'$. From the obtained equations we find the commutation relations for operators (2.13):

$$\left.\begin{aligned} &B_{nf}B^{+}_{nf} - B^{+}_{nf}B_{nf} = 1 - 2\hat{N}_{nf};\\ &B_{n'f'}B^{+}_{nf} - B^{+}_{nf}B_{n'f'} = 0, \text{ if } n \neq n' \text{ or } f' \neq f. \end{aligned}\right\} \tag{2.15}$$

At ordinary densities of electromagnetic radiation, a small portion of the molecules in a crystal are excited, i.e., the mean value $\langle N_{nf}\rangle$ of the operator $\hat{N}_{nf}$ is considerably ($\sim 10^{-5}$–10^{-7}) smaller than unity. Under these conditions, commutation relations (2.14) reduce to the conventional commutation relations for bosons:

$$B_{n'f'}B^{+}_{nf} - B^{+}_{nf}B_{n'f'} = \delta_{nn'}\delta_{ff'}; \tag{2.16}$$

all other combinations of B_{nf} and $B^{+}_{n'f'}$ commute.

Using the first equation of (2.14), let us transform operator (2.9) to

$$H_1 = \sum_n (\Delta\varepsilon_f + D_f)\, B^{+}_{nf}B_{nf}. \tag{2.17}$$

Similarly, let us transform operators (2.11) and (2.12):

$$H_2 = \sum_{n,m}{}' M^f_{nm}B^{+}_{mf}B_{nf}, \tag{2.18}$$

$$H_3 = \frac{1}{2}\sum_{n,m} M^f_{nm}(B^{+}_{mf}B^{+}_{nf} + B_{mf}B_{nf}). \tag{2.19}$$

Now let us diagonalize operator (2.8), using various approximations. First, consider (2.8) without the operator H_3. Then, with allowance for (2.17) and (2.18), the excitation-energy operator takes the form

$$\Delta H = H - \mathscr{E}_0 = \sum_n (\Delta\varepsilon_f + D_f)\, B^+_{nf} B_{nf} + \sum_{n,\,m}{}' M^f_{nm} B^+_{mf} B_{nf}. \qquad (2.20)$$

This approximation corresponds to the Heitler–London approximation, which was considered in Chapter II. The conditions of applicability of the Heitler–London approximation will be examined below, in Section 3. Let us study the eigenvalues of ΔH separately for crystals with one molecule in a unit cell and in general.

Crystals with One Molecule in a Unit Cell. For crystals with one molecule in a unit cell, the subscripts n and m are replaced by the lattice vectors **n** and **m**. The excitation-energy operator

$$\Delta H = \sum_{\mathbf{n}} (\Delta\varepsilon_f + D_f)\, B^+_{\mathbf{n}f} B_{\mathbf{n}f} + \sum_{\mathbf{nm}}{}' M^f_{\mathbf{nm}} B^+_{\mathbf{m}f} B_{\mathbf{n}f} \qquad (2.21)$$

is diagonalized by converting from the operators B_{nf} to the new operators $B_f(\mathbf{k})$ by the unitary transformation

$$B_{\mathbf{n}f} = \frac{1}{\sqrt{N}} \sum_{\mathbf{k}} B_f(\mathbf{k}) \exp(i\mathbf{kn}), \qquad (2.22)$$

where **k** is the wave vector, which is defined by Eqs. (II.2.7). If we substitute (2.22) into (2.16), we see that the operators $B_f(\mathbf{k})$ satisfy the commutation relations

$$B_{f'}(\mathbf{k}')\, B^+_f(\mathbf{k}) - B^+_f(\mathbf{k})\, B_{f'}(\mathbf{k}') = \delta_{\mathbf{kk}'} \delta_{ff'}. \qquad (2.23)$$

Therefore, the operators $B^+_f(\mathbf{k})$ are the creation operators for states characterized by the quantum numbers f and the wave vectors **k**. The operators $B_f(\mathbf{k})$ annihilate these same states.

If we substitute (2.22) into operator (2.21), we have

$$\Delta H = \sum_{\mathbf{k},\,f} \{\Delta\varepsilon_f + D_f + L_f(\mathbf{k})\}\, B^+_f(\mathbf{k})\, B_f(\mathbf{k}), \qquad (2.24)$$

where

$$L_f(\mathbf{k}) \equiv \sum_{\mathbf{m}(\neq \mathbf{n})} M^f_{\mathbf{nm}} \exp\{i\mathbf{k}(\mathbf{n} - \mathbf{m})\}. \qquad (2.24a)$$

Operator (2.24) is diagonal in the operators $\hat{N}_f(\mathbf{k}) = B_f^+(\mathbf{k})B_f(\mathbf{k})$ of the occupation numbers of the exciton states $\mathbf{k}f$. Therefore, its eigenfunctions are

$$|\ldots N_f(\mathbf{k})\ldots\rangle,$$

where the quantum numbers $N_f(\mathbf{k}) = 0, 1, 2, \ldots$ indicate the number of exciton excitations of a particular type. The state of a crystal with one exciton excitation $\mathbf{k}f$ is determined by the set of quantum numbers $N_{f'}(\mathbf{k}') = \delta_{\mathbf{kk}'}\delta_{ff'}$. According to (2.24), this state is associated with the excitation energy

$$E_f(\mathbf{k}) = \Delta\varepsilon_f + D_f + L_f(\mathbf{k}). \tag{2.25}$$

Thus, the exciton states (which correspond to the intramolecular excitation f) in a crystal are characterized by the wave vector k and the energy $E_f(\mathbf{k})$. For crystals with infinite dimensions, the wave vector **k** takes continuous values in the **k** space, which is bounded by the first Brillouin zone. In crystals with finite dimensions, **k** takes discrete values in the first Brillouin zone. The number of these possible values is equal to the number of unit cells.

It follows directly from (2.24a) and (2.25) that

$$E_f(\mathbf{k}) = E_f(\mathbf{k} + \mathbf{g}), \tag{2.25a}$$

where

$$\mathbf{g} = 2\pi \sum_{i=1}^{3} m_i \mathbf{b}_i,$$

$\mathbf{b}_i$ are the reciprocal-lattice vectors, and $m_i = 0, \pm 1, \pm 2, \ldots$. Thus, the wave vectors **k** and **k** + **g** are equivalent. In particular, we shall assume

$$B_f(\mathbf{k}) = B_f(\mathbf{k} + \mathbf{g}). \tag{2.26}$$

Equations (2.25a) and (2.26) enable us to return to the first Brillouin zone even when, for example, as a result of interaction of an exciton with a phonon, its momentum falls outside the limits of the first Brillouin zone.

Crystals with Several Molecules in a Unit Cell. If a crystal contains σ molecules in a unit cell, then n and m in (2.20) are determined by the expressions

$$n \equiv (\mathbf{n}, \alpha), \quad m = (\mathbf{m}, \beta),$$

where **n** and **m** are vectors describing the position of the unit cell, and α and β are the numbers 1, 2, ..., σ, which determine the position and orientation of the molecules in the unit cell.

Energy operator (2.20) can be diagonalized in two steps. First, we make the canonical transformation

$$B_{\mathbf{n}f} = \frac{1}{\sqrt{N}} \sum_{\mathbf{k}} B_{\alpha f}(\mathbf{k}) \exp(i\mathbf{kn}), \tag{2.27}$$

where $B_{\alpha f}(\mathbf{k})$ are new operators that satisfy the commutation relations

$$B_{\alpha' f'}(\mathbf{k}') B^{+}_{\alpha f}(\mathbf{k}) - B^{+}_{\alpha f}(\mathbf{k}) B_{\alpha' f'}(\mathbf{k}') = \delta_{\mathbf{kk}'} \delta_{ff'} \delta_{\alpha\alpha'}. \tag{2.28}$$

Substituting (2.27) into (2.20), we find a new expression for the excitation-energy operator:

$$\Delta H = \sum_{\mathbf{k},\,\alpha} (\Delta\varepsilon_f + D_f)\, B^{+}_{\alpha f}(\mathbf{k}) B_{\alpha f}(\mathbf{k}) + \sum_{\mathbf{k},\,\alpha,\,\beta} L^{f}_{\alpha\beta}(\mathbf{k})\, B^{+}_{\beta f}(\mathbf{k}) B_{\alpha f}(\mathbf{k}), \tag{2.29}$$

where

$$L^{f}_{\alpha\beta}(\mathbf{k}) = \sum_{\mathbf{m}}{}' M^{f}_{\mathbf{n}\alpha,\,\mathbf{m}\beta} \exp\{i\mathbf{k}(\mathbf{n}-\mathbf{m})\}. \tag{2.30}$$

For simplicity, we introduce the symbols

$$\mathcal{L}^{f}_{\alpha\beta}(\mathbf{k}) \equiv (\Delta\varepsilon_f + D_f)\,\delta_{\alpha\beta} + L^{f}_{\alpha\beta}(\mathbf{k}). \tag{2.31}$$

Since we shall consider only one f-th excited state, the subscript f will be omitted below. In the new symbols, operator (2.29) has the form

$$\Delta H = \sum_{\mathbf{k},\,\alpha,\,\beta} \mathcal{L}_{\alpha\beta}(\mathbf{k})\, B^{+}_{\beta}(\mathbf{k}) B_{\alpha}(\mathbf{k}). \tag{2.32}$$

Operator (2.32) is diagonalized by means of the canonical transformation

$$B_{\alpha}(\mathbf{k}) = \sum_{\mu=1}^{\sigma} u^{*}_{\alpha\mu}(\mathbf{k})\, a_{\mu}(\mathbf{k}) \tag{2.33}$$

under the conditions

$$\sum_{\mu} u^{*}_{\alpha\mu} u_{\beta\mu} = \delta_{\alpha\beta}, \quad \sum_{\alpha} u^{*}_{\alpha\mu} u_{\alpha\nu} = \delta_{\mu\nu}. \tag{2.34}$$

If we substitute (2.33) into (2.27), we see that the new operators satisfy the commutation relations

$$a_\mu(\mathbf{k})\, a_\nu^+(\mathbf{k}') - a_\nu^+(\mathbf{k}')\, a_\mu(\mathbf{k}) = \delta_{\mathbf{k}\mathbf{k}'}\delta_{\mu\nu}. \qquad (2.35)$$

Since canonical transformation (2.33) diagonalizes the energy operator (2.32), then

$$\sum u^*_{\alpha\nu}(\mathbf{k})\, \mathcal{L}_{\alpha\beta}(\mathbf{k})\, u_{\beta\mu}(\mathbf{k})\, a_\mu^+(\mathbf{k})\, a_\nu(\mathbf{k}) = \sum E_\mu(\mathbf{k})\, a_\mu^+(\mathbf{k})\, a_\mu(\mathbf{k}). \qquad (2.36)$$

Equation (2.36) is satisfied if the matrix elements $u_{\alpha\mu}$ satisfy the system of equations

$$\sum_\beta \mathcal{L}_{\alpha\beta}(\mathbf{k})\, u_{\beta\mu}(\mathbf{k}) = E_\mu(\mathbf{k})\, u_{\alpha\mu}(\mathbf{k}). \qquad (2.37)$$

This system coincides with system (II.3.10). It follows from (2.37) that each fixed value of the wave vector k corresponds to σ exciton states that differ in μ. The wave function

$$|\ldots n_\mu(\mathbf{k}) \ldots\rangle,$$

where $n_\mu(\mathbf{k})$ is the number of excited states of type μ, **k** and is an eigenfunction of operator (2.36) and of the number-of-states operator

$$\hat{n}_\mu(\mathbf{k}) = a_\mu^+(\mathbf{k})\, a_\mu(\mathbf{k}).$$

The value

$$E_\mu(\mathbf{k}) = \sum_{\beta,\alpha} u^*_{\alpha\mu}(\mathbf{k})\, \mathcal{L}_{\alpha\beta}(\mathbf{k})\, u_{\beta\mu}(\mathbf{k})$$

determines the energy of each exciton $\mu\mathbf{k}$. In the second-quantization representation, the wave function of an excited state with one exciton $\mu\mathbf{k}$ is obtained from the ground-state function by the rule

$$|1_\mu(\mathbf{k})\rangle = a_\mu^+(\mathbf{k})\,|0\rangle. \qquad (2.38)$$

The wave function of the ground state $|0\rangle$ in the coordinate representation [see (II.3.3)] corresponds to the function

$$\psi^0 = \prod_{n\alpha} \varphi^0_{n\alpha}.$$

Let us transform function (2.38) to the coordinate representation. It follows from (2.33) and (2.27) that

$$a_\mu^+(\mathbf{k}) = \sum_\alpha u^*_{\alpha\mu}(\mathbf{k})\, B_\alpha^+(\mathbf{k}), \qquad B_\alpha^+(\mathbf{k}) = \frac{1}{\sqrt{N}} \sum_{\mathbf{n}} e^{i\mathbf{k}\mathbf{n}} B^+_{\mathbf{n}\alpha}.$$

Then, considering (2.13), we can write

$$a^{+}_{\mu}(\mathbf{k}) = \frac{1}{\sqrt{N}} \sum_{\mathbf{n},\,\alpha} u^{*}_{\alpha\mu}(\mathbf{k}) e^{i\mathbf{kn}} b^{+}_{n\alpha,f} b_{n\alpha,0}. \tag{2.39}$$

Therefore, the exciton function in the coordinate representation has the form

$$\Phi^{f}_{\mu}(\mathbf{k}) = \frac{1}{\sqrt{N}} \sum_{\mathbf{n},\,\alpha} u^{*}_{\alpha\mu} e^{i\mathbf{kn}} b^{+}_{n\alpha,f} b_{n\alpha,0}\, \psi^{0} = \frac{1}{\sqrt{N}} \sum_{\mathbf{n},\,\alpha} u^{*}_{\alpha\mu} e^{i\mathbf{kn}} \varphi^{f}_{n\alpha} \prod_{m\beta(\neq n\alpha)} \varphi^{0}_{m\beta}.$$

This result agrees with expression (II.3.11), which was found earlier.

3. The Theory of Excitons without the Heitler – London Approximation

In the preceding section, we found the excitation energy of a crystal in the Heitler–London approximation, i.e., without taking into account H_3 in energy operator (2.8). Now let us calculate the crystal energy without using this simplification (Agranovich [1]). Using (2.17)-(2.19), operator (2.8) can be written in the form (the fixed subscript f is omitted)

$$\Delta H = \sum_{n} (\Delta\varepsilon + D) B^{+}_{n} B_{n} + \frac{1}{2} \sum_{n,\,m}{}' M_{nm} (B^{+}_{m} B^{+}_{n} + 2B^{+}_{m} B_{n} + B_{m} B_{n}). \tag{3.1}$$

Operator (3.1) is diagonalized by the unitary transformation

$$B_{n} = \frac{1}{\sqrt{N}} \sum_{\mathbf{k},\,\mu} \{u_{\alpha\mu}(\mathbf{k}) e^{i\mathbf{kn}} a_{\mu}(\mathbf{k}) + v^{*}_{\alpha\mu}(\mathbf{k}) e^{-i\mathbf{kn}} a^{+}_{\mu}(\mathbf{k})\} \tag{3.2}$$

to the new Bose operator $a_{\mu}(\mathbf{k})$, which satisfies commutation relations (2.35). Transformation (3.2) is unitary when

$$\left.\begin{aligned} &\sum_{\mu} \{u^{*}_{\alpha\mu}(\mathbf{k}) u_{\beta\mu}(\mathbf{k}) - v^{*}_{\beta\mu}(\mathbf{k}) v_{\alpha\mu}(\mathbf{k})\} = \delta_{\alpha\beta}, \\ &\sum_{\mathbf{k}} \exp\{i\mathbf{k}(\mathbf{m}-\mathbf{n})\} = N\delta_{\mathbf{mn}}, \\ &\sum_{\alpha} u_{\alpha\mu}(\mathbf{k}) v_{\alpha\mu'}(\mathbf{k}) - u_{\alpha\mu'}(\mathbf{k}) v_{\alpha\mu}(\mathbf{k}) = 0. \end{aligned}\right\} \tag{3.3}$$

If we substitute (3.2) into (3.1), we see that this operator is transformed to the diagonal form

$$\Delta H = -\sum_{\mathbf{k},\,\mu,\,\alpha} E_{\mu}(\mathbf{k})\, v_{\alpha\mu}(\mathbf{k})\, v^{*}_{\alpha\mu}(\mathbf{k}) + \sum_{\mu,\,\mathbf{k}} E_{\mu}(\mathbf{k})\, a^{+}_{\mu}(\mathbf{k})\, a_{\mu}(\mathbf{k}),$$

if the coefficients $u_{\alpha\mu}$ and $v_{\alpha\mu}$ of unitary transformation (3.2) satisfy the system of equations

$$\left.\begin{aligned} &[\Delta\varepsilon + D - E_{\mu}(\mathbf{k})]\, u_{\alpha\mu}(\mathbf{k}) + \sum_{\beta} L_{\alpha\beta}(\mathbf{k})\,[u_{\beta\mu}(\mathbf{k}) + v_{\beta\mu}(\mathbf{k})] = 0,\\ &[\Delta\varepsilon + D + E_{\mu}(\mathbf{k})]\, v_{\alpha\mu}(\mathbf{k}) + \sum_{\beta} L_{\alpha\beta}(\mathbf{k})\,[u_{\beta\mu}(\mathbf{k}) + v_{\beta\mu}(\mathbf{k})] = 0,\end{aligned}\right\} \tag{3.4}$$

where $L_{\alpha\beta}(\mathbf{k})$ is determined by Eq. (2.30).

If the excitation energy of the crystal is read from the "vacuum" state $|\,0\rangle$, i.e., the state in which there are no excitons, then the excitation-energy operator takes the form

$$\Delta H = \sum_{\mu,\,\mathbf{k}} E_{\mu}(\mathbf{k})\, a^{+}_{\mu}(\mathbf{k})\, a_{\mu}(\mathbf{k}). \tag{3.5}$$

The state with one exciton of type μ, **k** has, therefore, the energy $E_{\mu}(\mathbf{k})$, which can be calculated by solving system (3.4) under conditions (3.3). Thus, the function $E_{\mu}(\mathbf{k})$ characterizes the exciton energy spectrum, i.e., the dependence of the exciton energy in each band μ upon the exciton quasi-momentum $\hbar\mathbf{k}$.

In the simple case when a unit cell contains one molecule, there is only one exciton band ($\alpha = \beta = \mu = 1$) and equation system (3.3) and (3.4) takes the form

$$\left.\begin{aligned} u^{2} - v^{2} &= 1,\\ [\Delta\varepsilon + D - E(\mathbf{k})]\, u + L(\mathbf{k})\,[u + v] &= 0,\\ [\Delta\varepsilon + D + E(\mathbf{k})]\, v + L(\mathbf{k})\,[u + v] &= 0.\end{aligned}\right\} \tag{3.6}$$

Solution of this system gives

$$\left.\begin{aligned} E(\mathbf{k}) &= \{[\Delta\varepsilon + D + L(\mathbf{k})]^{2} - L^{2}(\mathbf{k})\}^{1/2},\\ [\Delta\varepsilon + D + E(\mathbf{k})]\, v &= [\Delta\varepsilon + D - E(\mathbf{k})]\, u.\end{aligned}\right\} \tag{3.7}$$

If

$$L(\mathbf{k}) \ll \Delta\varepsilon + D, \tag{3.8}$$

then system (3.7) can be replaced by the approximate expressions

$$\left.\begin{aligned} E(\mathbf{k}) &\approx \Delta\varepsilon + D + L(\mathbf{k}) - \frac{L^2(\mathbf{k})}{2(\Delta\varepsilon + D)} \approx \Delta\varepsilon + D + L(\mathbf{k}), \\ v &\approx \frac{uL(\mathbf{k})}{2(\Delta\varepsilon + D) + L(\mathbf{k})} \approx 0, \quad u \approx 1. \end{aligned}\right\} \tag{3.9}$$

Solutions (3.9) agree with the values (2.25) obtained in the Heitler–London approximation. Thus, satisfaction of inequality (3.8) is a condition of applicability of the Heitler–London approximation. In molecular crystals, for states f corresponding to the first electronic excitation of free molecules, $|L^f(\mathbf{k})|$ rarely exceeds 10^3 cm^{-1}, and $\Delta\varepsilon + D_f$ is $\sim 3 \times 10^4$ cm^{-1}. Thus, inequality (3.8) is well satisfied, and calculations can be made in the Heitler–London approximation. In some molecular crystals, $|L^f(\mathbf{k})|$ values of $\sim 10^4$ cm^{-1} correspond to the second electronic excitation. In these cases, the corrections for the Heitler–London approximation can be considerable.

In order to obtain a general expression of the corrections for the Heitler–London approximation when the crystal has several molecules in a unit cell, we transform Eqs. (3.4) to the form

$$\left.\begin{aligned} v_{\alpha\mu}(\mathbf{k}) &= \frac{\Delta\varepsilon + D - E_\mu(\mathbf{k})}{\Delta\varepsilon + D + E_\mu(\mathbf{k})} u_{\alpha\mu}(\mathbf{k}), \\ [E_\mu(\mathbf{k}) - \Delta\varepsilon - D]\, u_{\alpha\mu} &= 2\left(1 + \frac{E_\mu(\mathbf{k})}{\Delta\varepsilon + D}\right)^{-1} \sum_\beta L_{\alpha\beta}(\mathbf{k})\, u_{\beta\mu}(\mathbf{k}). \end{aligned}\right\} \tag{3.10}$$

The Heitler–London approximation is obtained from (3.10) when $E^0_\mu(\mathbf{k}) = \Delta\varepsilon + D$ is substituted on the right of these equations. Then

$$v^{\text{H-L}}_{\alpha\mu}(\mathbf{k}) = 0, \tag{3.11}$$

$$\sum_\beta \{[E^{\text{H-L}}_\mu(\mathbf{k}) - \Delta\varepsilon - D]\,\delta_{\alpha\beta} - L_{\alpha\beta}(\mathbf{k})\}\, u^{\text{ГЛ}}_{\beta\mu}(\mathbf{k}) = 0. \tag{3.12}$$

If we substitute (3.11) into (3.3), we obtain in the Heitler–London approximation

$$\sum_\mu u^{*\text{H-L}}_{\alpha\mu}(\mathbf{k})\, u^{\text{H-L}}_{\beta\mu}(\mathbf{k}) = \delta_{\alpha\beta}. \tag{3.13}$$

As might be expected, Eqs. (3.12) and (3.13) agree, respectively, with Eqs. (II.3.10) and (II.3.9a), which determine the energy and the wave functions in the Heitler–London approximation.

To obtain the exciton energy in the next approximation, we substitute into the second equation of (3.10) the values $u_{\alpha\beta}^{H\text{-}L}(\mathbf{k})$ which satisfy condition (3.13):

$$\{E_\mu^2(\mathbf{k}) - (\Delta\varepsilon + D)^2\}\, u_{\alpha\mu}^{H\text{-}L}(\mathbf{k}) = 2\,[\Delta\varepsilon + D] \sum_\beta L_{\alpha\beta}(\mathbf{k})\, u_{\beta\mu}^{H\text{-}L}(\mathbf{k}).$$

According to (3.12), the sum on the right of the obtained equation can be expressed in terms of the energy $E_\mu^{H\text{-}L}(\mathbf{k})$. Thus, we obtain

$$\{E_\mu^2(\mathbf{k}) - (\Delta\varepsilon + D)\,[2E_\mu^{H\text{-}L}(\mathbf{k}) - (\Delta\varepsilon + D)]\}\, u_{\alpha\mu}^{H\text{-}L}(\mathbf{k}) = 0. \qquad (3.14)$$

It follows from Eq. (3.14) that

$$E_\mu(\mathbf{k}) = \{(\Delta\varepsilon + D)\,[2E_\mu^{H\text{-}L}(\mathbf{k}) - (\Delta\varepsilon + D)]\}^{1/2}. \qquad (3.15)$$

Equation (3.15) allows us to calculate the exciton energy more accurately if we know the solution $E_\mu^{H\text{-}L}(\mathbf{k})$ in the Heitler–London approximation.

We shall need a relationship between $u_{\alpha\mu}(\mathbf{k})$ and $u_{\alpha\mu}^{H\text{-}L}(\mathbf{k})$. For all α at fixed μ, these values differ from one another by the same factor, which can be found by comparing the normalization conditions for $u_{\alpha\mu}(\mathbf{k})$ and $u_{\alpha\mu}^{H\text{-}L}(\mathbf{k})$.

Using the first condition of (3.3) and the first relation of (3.10), we find that the normalization condition for $u_{\alpha\mu}(\mathbf{k})$ has the form

$$\sum_\mu \left[1 - \left(\frac{\Delta\varepsilon + D - E_\mu(\mathbf{k})}{\Delta\varepsilon + D + E_\mu(\mathbf{k})}\right)^2\right] u_{\alpha\mu}^*(\mathbf{k})\, u_{\beta\mu}(\mathbf{k}) = \delta_{\alpha\beta}. \qquad (3.16)$$

Now, comparing (3.16) and (3.13), we find that

$$u_{\alpha\mu}^{H\text{-}L}(\mathbf{k}) = 2\,\frac{\{E_\mu(\mathbf{k})\,[\Delta\varepsilon + D]\}^{1/2}}{\Delta\varepsilon + D + E_\mu(\mathbf{k})}\, u_{\alpha\mu}(\mathbf{k}). \qquad (3.17)$$

4. Exciton States with Allowance for Several Adjacent Molecular Levels

In the preceding sections, we investigated the exciton states of crystals under the assumption that only two steady states of a molecule (the ground 0 and the excited f) participated in their formation. This approximation is valid when the excitation level f is sufficiently removed from the other excitation levels of the molecule. If there is a group of adjacent levels, all such levels will have some part in the formation of an exciton. This case has been studied theoretically by Agranovich [5] in the second-quantization representation and by Craig [6, 7] in the Heitler–London approximation.[1] Below, we shall use the second-quantization representation.

If we retain in operator (1.15) only matrix elements of the type

$$\left.\begin{array}{c}\langle 00 | V_{nm} | 00\rangle, \quad \langle 0g | V_{nm} | f0\rangle, \\ M_{nm}^{gf} = \langle 0g | V_{nm} | 0f\rangle = \langle 0f | V_{nm} | 0g\rangle = \langle 00 | V_{nm} | gf\rangle,\end{array}\right\} \tag{4.1}$$

then, considering the normalization conditions

$$\sum_g N_{ng} + \sum_f N_{nf} + N_{n0} = 1,$$

where f and g number the excited states of the molecules, we obtain

$$H = \mathscr{E}_0 + H_1 + H_2 + H_3, \tag{4.2}$$

where

$$\mathscr{E}_0 = N\varepsilon_0 + \frac{1}{2} \sum_{n,m}' \langle 00 | V_{nm} | 00\rangle$$

is a constant term, and

$$H_1 = \sum_{n,f} (\Delta\varepsilon_f + D_f)\hat{N}_{nf};$$

$$H_2 = \frac{1}{2} \sum_{n,m,f,g}' M_{nm}^{fg} \{b_{n0}^+ b_{mg}^+ b_{m0} b_{nf} + b_{n0} b_{mf} b_{m0}^+ b_{ng}^+\},$$

$$H_3 = \frac{1}{2} \sum_{n,m,f,g}' M_{nm}^{fg} \{b_{n0}^+ b_{m0}^+ b_{mf} b_{mg} + b_{nf}^+ b_{mg}^+ b_{m0} b_{n0}\}.$$

[1] See also [1], where the problem is solved by the Green-function method.

Converting, by means of (2.13) and (2.14), to the operators B_{nf}, we transform energy operator (4.2) to

$$\Delta H = H - \mathscr{E}_0 = \sum_{n, f} (\Delta\varepsilon_f + D_f) B^+_{nf} B_{nf} +$$

$$+ \frac{1}{2} \sum_{n, m, g, f}' M^{fg}_{nm} (B^+_{mg} + B_{mg}) (B^+_{nf} + B_{nf}). \tag{4.3}$$

Energy operator (4.3) has been diagonalized by Agranovich [5]. Here, for simplicity, we shall confine ourselves to solving the problem in the Heitler–London approximation, i.e., we shall study only the operator

$$\Delta H = \sum_{n, f} (\Delta\varepsilon_f + D_f) B^+_{nf} B_{nf} + \sum_{n, m, g, f}' M^{fg}_{nm} B^+_{mg} B_{nf}. \tag{4.4}$$

In the summation, f and g take the values 1, 2,..., l, where l is the number of excited states taken into account, n = $(\mathbf{n}, \alpha)$, m = $(\mathbf{m}, \beta)$.

Operator (4.4) can be diagonalized in two steps. First, we make canonical transformation (2.22). Then we obtain

$$\Delta H = \sum_{\mathbf{k}, \alpha, \beta, g, f} \mathscr{L}^{fg}_{\alpha\beta} (\mathbf{k}) B^+_{\beta g} (\mathbf{k}) B_{\alpha f} (\mathbf{k}), \tag{4.5}$$

where

$$\mathscr{L}^{fg}_{\alpha\beta} (\mathbf{k}) \equiv (\Delta\varepsilon_f + D_f) \delta_{\alpha\beta} \delta_{fg} + L^{fg}_{\alpha\beta} (\mathbf{k}), \tag{4.6}$$

$$L^{fg}_{\alpha\beta} (\mathbf{k}) = \sum_{\mathbf{m}} M^{fg}_{\mathbf{n}\alpha, \mathbf{m}\beta} \exp \{i\mathbf{k} (\mathbf{n} - \mathbf{m})\}. \tag{4.7}$$

Then operator (4.5) is diagonalized by the canonical transformation

$$B_{\alpha f} (\mathbf{k}) = \sum_{\mu, g} u^*_{\alpha f, \mu g} (\mathbf{k}) a_{\mu g} (\mathbf{k}) \tag{4.8}$$

to the new operators $a_{\mu g}(\mathbf{k})$, which satisfy the commutation relations

$$a_{\mu f} (\mathbf{k}) a^+_{\nu g} (\mathbf{k}') - a^+_{\nu g} (\mathbf{k}') a_{\mu f} (\mathbf{k}) = \delta_{\mathbf{k}\mathbf{k}'} \delta_{gf} \delta_{\mu\nu}.$$

Here, the elements of the transformation matrix (u) must satisfy the system of equations

$$\sum_{\beta, g} \mathscr{L}^{fg}_{\alpha\beta}(\mathbf{k})\, u_{\beta g, \mu s}(\mathbf{k}) = E_{\mu s}(\mathbf{k})\, u_{\alpha g, \mu s}(\mathbf{k}), \tag{4.9}$$

where α, β, μ take the values 1, 2,..., σ, and g and s take the values 1, 2,..., l.

If we solve the system of σl equations (4.9) at fixed **k**, we determine σl roots $E_{\mu s}(\mathbf{k})$ and their corresponding coefficients $u_{\alpha f;\mu s}(\mathbf{k})$. It is clear that such solutions are very tedious when $l > 2$ and $\sigma > 2$. Therefore, we shall consider particular cases of system (4.9).

<u>One Molecule in a Unit Cell.</u> In this case, $\alpha = \beta = \mu = 1$ and system (4.9) takes the form

$$\sum_{g=1}^{l} \mathscr{L}^{fg}(\mathbf{k})\, u_{gs}(\mathbf{k}) = E_s(\mathbf{k})\, u_{fs}(\mathbf{k}), \tag{4.10}$$

where l is the number of levels taken into account; $f = 1, 2, ..., l$.

$$\mathscr{L}^{fg}(\mathbf{k}) = (\Delta\varepsilon_f + D_f)\,\delta_{fg} + L^{fg}(\mathbf{k}),$$

$$L^{fg}(\mathbf{k}) = \sum_{m\,(\neq n)} M^{fg}_{\mathbf{nm}} \exp\{i\mathbf{k}(\mathbf{n} - \mathbf{m})\},$$

$$M^{fg}_{\mathbf{nm}} = \int \varphi^0_{\mathbf{n}} \varphi^f_{\mathbf{m}} V_{\mathbf{nm}} \varphi^0_{\mathbf{m}} \varphi^g_{\mathbf{n}}\, d\tau.$$

According to (4.10), to find $E_s(\mathbf{k})$ and $u_{fs}(\mathbf{k})$ we must solve the system of homogeneous equations

$$\sum_{g=1}^{l} \mathscr{L}^{fg}(\mathbf{k})\, u_g(\mathbf{k}) = E(\mathbf{k})\, u_f(\mathbf{k}). \tag{4.11}$$

The condition of nontrivial solvability of system (4.11) comes down to an equation of degree l in $E(\mathbf{k})$,

$$\det\{\mathscr{L}^{fg}(\mathbf{k}) - E(\mathbf{k})\,\delta_{fg}\} = 0. \tag{4.12}$$

The l roots of this equation $E_s(\mathbf{k})$ also determine the l exciton bands, a portion of which may completely or partially overlap. If the l molecular levels in question correspond to an l-fold degenerate molecular level, it may be said that the degeneracy of this molecular level in the crystal is completely or partially removed (Bethe splitting) and each of the levels formed is converted to a quasi-continuous band.

When $l > 2$, the solution of (4.11) can be simplified by using group-theory methods based on the symmetry properties of the crystal.

In the particular case when only two excited states are significant, the roots of Eq. (4.12) can be written in explicit form:

$$E_s(\mathbf{k}) = \frac{1}{2}\{A_1 + A_2 - (-1)^s [(A_1 - A_2)^2 + 4 | L^{1,2}(\mathbf{k})|^2]^{1/2}\}, \quad (4.13)$$

where

$$A_f = \Delta\varepsilon_f + D_f + L^{f,f}(\mathbf{k}); \qquad f = 1, 2.$$

The coefficients u_{fs} here are determined with accuracy to normalization by the equations

$$u_{1s}(\mathbf{k}) = \frac{2 | L^{1,2}(\mathbf{k}) | u_{2s}(\mathbf{k})}{A_2 - A_1 - (-1)^s \sqrt{(A_1 - A_2)^2 + 4 | L^{1,2}(\mathbf{k}) |^2}}, \qquad s = 1, 2. \quad (4.14)$$

The wave functions of the excited states in the coordinate representation in this case have the form

$$\Phi^s(\mathbf{k}) = \frac{1}{\sqrt{N}} \sum_{nf} u^*_{fs}(\mathbf{k}) e^{i\mathbf{k}\mathbf{n}} \varphi^f_n \prod_m \varphi^0_m, \quad m \neq n.$$

Considering (4.11) and (4.12), we can see that the contribution of the second intramolecular excited state to the exciton excitation corresponding to the first intramolecular excitation (and vice versa) is significant only when

$$(A_2 - A_1)^2 < | L^{1,2}(\mathbf{k}) |^2.$$

If the reverse inequality is satisfied, the role of "mixing" of the various molecular states has little significance, and the approximation of one excitation level can be used.

Several Molecules in a Unit Cell. When there are σ molecules in a unit cell, solution of the σl equations (4.9) is very tedious. Therefore, it is convenient to diagonalize operator (4.4) in two steps. For this, we transform (4.4) to

$$\Delta H = \sum_f \Delta H_f + \sum_{n,m}' \sum_{f,g}' M_{nm}^{fg} B_{mg}^{+} B_{nf}, \tag{4.15}$$

where

$$\Delta H_f = \sum_n (\Delta\varepsilon_f + D_f) B_{nf}^{+} B_{nf} + \sum_{n,m}' M_{nm}^{ff} B_{mf}^{+} B_{nf} \tag{4.16}$$

is an energy operator that takes into account only one excitation level in the molecules of the crystal. Operator (4.16) coincides with operator (2.20). According to (2.27) and (2.33), therefore, this operator is diagonalized by the canonical transformation

$$B_{nf} = \frac{1}{\sqrt{N}} \sum_{\mathbf{k}} u_{\alpha\mu_f}^{*}(\mathbf{k})\, a_{\mu_f}(\mathbf{k}) \exp(i\mathbf{kn}) \tag{4.17}$$

to the form

$$\Delta H_f = \sum_{\mathbf{k},\mu_f} E_{\mu_f}(\mathbf{k})\, a_{\mu_f}^{+}(\mathbf{k})\, a_{\mu_f}(\mathbf{k}), \tag{4.18}$$

where $\mu_f = 1, 2, \ldots, \sigma$ are the exciton bands corresponding to the intramolecular excitation f; $a_{\mu f}(\mathbf{k})$ are new operators, which satisfy the commutation relations

$$a_{\mu_f}(\mathbf{k})\, a_{\nu_f}^{+}(\mathbf{k}') - a_{\nu_f}^{+}(\mathbf{k}')\, a_{\mu_f}(\mathbf{k}) = \delta_{\mathbf{kk}'}\delta_{\mu_f\nu_f}; \tag{4.19}$$

and the elements $u_{\alpha\mu_f}$ of the transformation matrix are determined by the system of σ equations

$$\left.\begin{aligned} &\sum_{\mu_f} u_{\alpha\mu_f}^{*} u_{\beta\mu_f} = \delta_{\alpha\beta}, \\ &\sum_{\beta} \mathcal{L}_{\alpha\beta}^{f}(\mathbf{k})\, u_{\beta\mu_f}(\mathbf{k}) = E_{\mu_f}(\mathbf{k})\, u_{\alpha\mu_f}(\mathbf{k}), \end{aligned}\right\} \tag{4.20}$$

where

$$\mathcal{L}_{\alpha\beta}^{f}(\mathbf{k}) = (\Delta\varepsilon_f + D_f)\,\delta_{\alpha\beta} + L_{\alpha\beta}^{ff}(\mathbf{k}), \tag{4.21}$$

$$L_{\alpha\beta}^{ff}(\mathbf{k}) = \sum_n M_{n\alpha,0\beta}^{ff} \exp(i\mathbf{kn}). \tag{4.22}$$

The solutions of system (4.20) were examined in Section 3 of Chapter II. If $E_{\mu_f}(\mathbf{k})$ and the coefficients $u_{\alpha\mu_f}(\mathbf{k})$ are known for each of the l excited states in question ($f = 1, 2, \ldots, l$), then total operator (4.15) can be transformed using (4.17).

Thus, we obtain

$$\Delta H = \sum_{f,\,\mathbf{k},\,\mu_f} E_{\mu_f}(\mathbf{k})\, a^{+}_{\mu_f}(\mathbf{k})\, a_{\mu_f}(\mathbf{k}) + \sum_{f,\,g}{}' \sum_{\mu_f,\,\mu_g} G^{fg}_{\mu_f\mu_g}(\mathbf{k})\, a^{+}_{\mu_f}(\mathbf{k})\, a_{\mu_g}(\mathbf{k}), \tag{4.23}$$

where

$$\left.\begin{aligned} G^{fg}_{\mu_f\mu_g}(\mathbf{k}) &= \sum_{\alpha,\,\beta} L^{fg}_{\alpha\beta}(\mathbf{k})\, u_{\beta\mu_f}(\mathbf{k})\, u^{*}_{\alpha\mu_g}(\mathbf{k}), \\ L^{fg}_{\alpha\beta}(\mathbf{k}) &= \sum_{\mathbf{n}} M^{fg}_{\mathbf{n}\alpha,\,0\beta} \exp(i\mathbf{k}\mathbf{n}). \end{aligned}\right\} \tag{4.24}$$

In operator (4.23), μ_f and μ_g take values of 1, 2,..., σ and number the exciton bands corresponding to the f-th and g-th excited states of a molecule. Each exciton band corresponds to a wave function that is the basis of the irreducible representation of the space group of the crystal. Matrix elements (4.24) differ from zero only when bands μ_f and μ_g pertain to a single irreducible representation of this group. Operator (4.23), therefore, is decomposed to a sum of independent operators that pertain to particular irreducible representations of the symmetry space group of the crystal. Each of these operators may be diagonalized independently.

5. Photoexcitons

In the preceding sections, when examining the spectrum of elementary excitations of a crystal, we took into account only Coulomb interaction between charges. According to quantum electrodynamics (see, for example, [8], Sections 1 and 6), the total interaction between charges is described by an electromagnetic field with the aid of the vector and scalar potentials. In the Coulomb calibration of the vector potential (div **A** = 0), along with the static Coulomb interaction, we must take into account the interaction of charges with the transverse electromagnetic field, whose quantum exchange between charged particles completely takes into account their delayed interaction. Thus, since the exciton states already allow for the Coulomb interaction between charges, in order to study the effect of the delayed interaction it is sufficient to consider the interaction of excitons with the electromagnetic field, which is defined by the vector potential in the Coulomb calibration.

In the Coulomb calibration, this electromagnetic field in a crystal without free charges is completely transverse, since div A = div D = 0.

In the infrared region, the role of the transverse electromagnetic field without allowance for absorption was studied long ago by Born and Ewald (see [9], for example) and by Tolpygo [10] according to the microtheory of crystal lattice dynamics and more recently [11-14] using the semiphenomenological approach.

Neamtan [15] and Fano [4] have attempted to construct a quantum theory of the delay effects for calculating the elementary excitations of a system of charges. They studied the longwave excitations in macroscopically homogeneous isotropic systems of atoms and molecules. The longwave elementary excitations of a coupled system of the atoms and oscillators of the field were calculated. In this case, according to Neamtan [15], the phenomenological dielectric constant of the medium $\varepsilon(\omega, \mathbf{k})$ was defined by the relation

$$\varepsilon(\omega, \mathbf{k}) = \frac{c^2\mathbf{k}^2}{\omega^2}, \tag{5.1}$$

where ω is the frequency and k is the real wave vector of the corresponding elementary excitation. Neamtan considered the case of weak coupling of the atoms with the electromagnetic field (the oscillator strengths of the quantum transitions were small) for which $\varepsilon(\omega, \mathbf{k}) \approx 1$.

Fano [4] did not assume the oscillator strengths to be small, but he took into account the Coulomb interaction between charges only partially by the "random phase" method. Fano studied the interaction of two external electric charges located in the coupled system of atoms and electromagnetic field and showed that the "direct" interaction between charges could be obtained by eliminating the normal longitudinal elementary excitations of this system, which interact with the "external" charges. This interaction is described by the Coulomb formula with the corresponding dielectric constant of the system.

Quantum-mechanical calculations of the delay effects in the theory of excitons using the Coulomb calibration for the vector potential have been made by Hopfield [16] (for cubic crystals) and by Agranovich [1] (for crystals with any symmetry). Absorption effects were also ignored in these papers.

Here, we shall examine the interaction of excitons with a transverse electromagnetic field on a model of a crystal with rigidly fixed atoms (molecules), i.e., also without allowance for absorption. We shall assume that the crystal is described by the three basis vectors $\mathbf{a}_1$, $\mathbf{a}_2$, $\mathbf{a}_3$ and occupies the volume $V = vN$ in the form of a parallelepiped with edges $\mathbf{L}_i = \mathbf{a}_i N_i$. Here, $i = 1, 2, 3$; $N = N_1 N_2 N_3$ is the number of unit cells; and v is the volume of a unit cell. Let each cell of the crystal contain one molecule, which can be found in excited states with energy ε_f, where f is the set of quantum numbers of the corresponding electronic states of the atoms (molecules). For simplicity, we shall assume that each of the excited states f is characterized by the electric dipole moment of the quantum transition $\mathbf{d}_f$ from the ground state. Then, each intramolecular excitation with the energy ε_f corresponds to an exciton band with the dispersion law $E_f(\mathbf{k})$, which is determined by expression (2.25). Here, $\mathbf{k}$ is the wave vector of an exciton, which, according to (II.2.7), takes N discrete values in the first cell of the $\mathbf{k}$ space. Below in this section, we shall use units in which $\hbar = 1$.

In the occupation-number representation, the operator that characterizes the exciton states of the crystal has the form [see (2.24) and (2.25)]

$$H_{\text{ex}} = \sum_{\mathbf{k},\, f} E_f(\mathbf{k})\, B^{+}_{\mathbf{k}f} B_{\mathbf{k}f}, \tag{5.2}$$

where $B^{+}_{\mathbf{k}f}$ and $B_{\mathbf{k}f}$ are the Bose creation and annihilation operators of excitons corresponding to the intramolecular electronic excitation E_f, the quasi-momentum k, and the electric dipole moment of the transition (from the ground state) d_f. Summation in (5.2) is over all k lying in the first cell of the k space, i.e., $-\pi < \mathbf{k}\mathbf{a}_i \leq \pi$.

Let us assume that the transverse electromagnetic field is enclosed in the volume of the crystal and satisfies the same cyclic boundary conditions as do the excitons.[2] Then, in the occupation-number representation for photons that have the momentum k and

[2]This assumption is very important. Only under this assumption is the interaction of excitons with the wave vector **k** accomplished by a photon with the same wave vector and the corresponding polarization. Therefore, new elementary excitations (photoexcitons) with the same wave vector are produced in a system of interacting excitons and photons. This case can be realized when the crystal has infinite dimensions. If the crystal occupies a finite space and the field, as usual, is in an infinite space, an exciton with the wave vector **k** will interact with an infinite number of degrees of freedom of transverse photons. The exciton energy, therefore, must be distributed over an infinite number of degrees of freedom of the field, which leads to radiation damping of the exciton states.

the polarization $\mathbf{u}_{\mathbf{k}\alpha}$, the energy operator of the electromagnetic field can be written as

$$H_\gamma = \sum_{\mathbf{k},\,\alpha}^{\infty} c\,|\mathbf{k}|\, a_{\mathbf{k}\alpha}^{+} a_{\mathbf{k}\alpha}; \qquad \alpha = 1,\, 2, \tag{5.3}$$

where

$$c\,|\mathbf{k}| = \omega(\mathbf{k}) \tag{5.3a}$$

is the photon energy (c is the velocity of light); and $a_{\mathbf{k}\alpha}^{+}$ and $a_{\mathbf{k}\alpha}$ are the Bose creation and annihilation operators for photons with energy $\omega(\mathbf{k})$, quasi-momentum $\mathbf{k}$, and polarization unit vectors $\mathbf{u}_{\mathbf{k}\alpha}$ such that $(\mathbf{u}_{\mathbf{k}_1}\mathbf{u}_{\mathbf{k}_2}) = 0$, $(\mathbf{k}\mathbf{u}_{\mathbf{k}\alpha}) = 0$ at $\alpha = 1, 2$. Summation in (5.3) is over the entire discrete k-space, which is defined by the equations (this is indicated by ∞ over the sum)

$$\mathbf{k}\mathbf{a}_i = \frac{2\pi}{N_i}\nu_i, \qquad \nu_i = 0, \pm 1, \pm 2, \ldots, \infty.$$

The vector-potential operator $\mathbf{A}(\mathbf{r})$ of the electromagnetic field (with Coulomb calibration, i.e., div $\mathbf{A} = 0$) is expressed in terms of the operators $a_{\mathbf{k}\alpha}$ by means of the formula (see, for example, [18], Section 134)

$$\mathbf{A}(\mathbf{r}) = \sum_{\mathbf{k},\,\alpha}^{\infty} \sqrt{\frac{2\pi c}{V\cdot|\mathbf{k}|}}\, \mathbf{u}_{\mathbf{k}\alpha}\gamma_{\mathbf{k}\alpha} \exp(i\mathbf{k}\mathbf{r}), \tag{5.4}$$

where

$$\gamma_{\mathbf{k}\alpha} = \gamma_{-\mathbf{k},\,\alpha}^{+} \equiv a_{\mathbf{k}\alpha} + a_{-\mathbf{k},\,\alpha}^{+}, \qquad \alpha = 1,\, 2. \tag{5.4a}$$

In the coordinate representation in the dipole approximation, the interaction operator of the electromagnetic field with the molecules of the crystal is given by the expression

$$H_{\text{ex-}\gamma} = -\frac{e}{mc}\sum_{\mathbf{n}} \mathbf{A}(\mathbf{n})\,\hat{\mathbf{p}}_{\mathbf{n}} + \frac{e^2 S}{2mc^2}\sum_{\mathbf{n}} \mathbf{A}^2(\mathbf{n}), \tag{5.5}$$

where $\hat{\mathbf{p}}_{\mathbf{n}}$ is the total momentum operator of all S optically active electrons of a neutral molecule situated at lattice point $\mathbf{n}$; and e and m are the electron charge and mass.

If $B_{\mathbf{n}f}^{+}$ is the transition operator (introduced in Section 2) of the $\mathbf{n}$-th molecule from the ground $|0\rangle$ to an excited $|f\rangle$ state, then in the second-quantization representation the operator $\hat{\mathbf{p}}_{\mathbf{n}}$ is transformed to

$$\hat{\mathbf{p}}_{\mathbf{n}} \Rightarrow \sum_{f(\neq 0)} \{\langle f | \hat{\mathbf{p}}_{\mathbf{n}} | 0 \rangle B_{\mathbf{n}f}^{+} + \langle 0 | \hat{\mathbf{p}}_{\mathbf{n}} | f \rangle B_{\mathbf{n}f}\}. \quad (5.6)$$

Let $H_{\mathbf{n}}$ be the Hamiltonian of a molecule in the crystal and let $\hat{\mathbf{d}}_{\mathbf{n}}$ be the operator of the total electric dipole moment of all S optically active electrons of the molecule. Then, the following commutation relations are valid (at $\hbar = 1$):

$$[\hat{\mathbf{d}}_{\mathbf{n}}, H_{\mathbf{n}}] = \frac{ie}{m} \hat{\mathbf{p}}_{\mathbf{n}};$$

$$[\mathbf{d}_{\mathbf{n}}^{x}, \mathbf{p}_{\mathbf{n}}^{y}] = ieS\delta_{xy}.$$

Using these commutation relations, we find

$$\langle f | \hat{\mathbf{p}}_{\mathbf{n}} | 0 \rangle = \frac{im\omega_f}{e} \mathbf{d}_f, \quad (5.7)$$

$$\sum_{f(\neq 0)} \omega_f d_f^x d_f^y = \frac{e^2}{2m} S\delta_{xy}, \quad (5.8)$$

where $\mathbf{d}_f$ is the electric dipole moment of the quantum transition of a molecule from the ground state to the f-th excited state. Equation (5.8) expresses the sum rule of the oscillators of the transition:

$$F_f^x = \frac{2m\,(d^x)^2}{e^2} \omega_f, \qquad \sum_f F_f^x = S.$$

If we substitute (5.4), (5.6), and (5.7) into operator (5.5), we find

$$H_{\text{ex-}\gamma} = -\frac{i}{c} \sum_{\mathbf{n},f} \omega_f (\mathbf{A}(\mathbf{n})\, \mathbf{d}_f)(B_{\mathbf{n}f}^{+} - B_{\mathbf{n}f}) + \frac{e^2 S}{2mc^2} \sum_{\mathbf{n}} \mathbf{A}^2(\mathbf{n}). \quad (5.9)$$

If now, using

$$B_{\mathbf{n}f} = \frac{1}{\sqrt{N}} \sum_{\mathbf{k}} B_f(\mathbf{k}) \exp(i\mathbf{k}\mathbf{n}),$$

we convert to the exciton creation and annihilation operators, we obtain a final expression for the exciton–photon interaction operator in the occupation-number representation for excitons and phonons:

$$H_{\text{ex-}\gamma}=\sum_{\mathbf{k},\,\alpha}\frac{\omega_0^2}{4c\,|\,\mathbf{k}\,|}\gamma_{\mathbf{k}\alpha}\gamma_{\mathbf{k}\alpha}^{+}+\sum_{\mathbf{k},\,\alpha,\,f}D(\mathbf{k},\,\alpha,\,f)\,\gamma_{\mathbf{k}\alpha}\,(B_{\mathbf{k}f}^{+}-B_{-\mathbf{k},\,f}),\qquad(5.10)$$

where

$$\left.\begin{aligned}D(\mathbf{k},\,\alpha,\,f)&=-i\omega_f\,(\mathbf{u}_{\mathbf{k}\alpha}\mathbf{d}_f)\sqrt{\frac{2\pi}{vc\,|\,\mathbf{k}\,|}},\\ \omega_0^2&=\frac{4\pi e^2S}{mv}=\frac{4\pi}{v}\sum_{f(\neq 0)}\omega_f\,(\mathbf{d}_f\mathbf{u}_{\mathbf{k}\alpha})^2.\end{aligned}\right\}\qquad(5.11)$$

In the derivation of expression (5.10) we ignored the interaction of excitons with x-ray and gamma-ray photons whose wave vectors lie outside the first cell of the **k** space. Summation over **k**, therefore, extends only to the wave vectors that lie within the first cell of the **k**-space. A more general study of the role of delayed interaction, with allowance for the shortwave components of the field that transmits the interaction between molecules, will be made in Section 7 of Chapter IV.

It follows from expression (5.10) that excitons ($\mathbf{k}$, $\mathbf{d}_f$) interact with photons ($\mathbf{Q}$, $\mathbf{u}_{Q\alpha}$) only when $\mathbf{Q} = \pm\mathbf{k}$ and $(\mathbf{d}_f\mathbf{u}_{\mathbf{k}\alpha}) \neq 0$. Longitudinal excitons, i.e., excitons for which the wave vector **k** is parallel to $\mathbf{d}_f$, do not interact with a transverse electric field, since for them $(\mathbf{d}_f\mathbf{u}_{\mathbf{k}\alpha}) = 0$.

Using operators (5.2) and (5.9) and the part of operator (5.3) with the vectors **k** satisfying the inequalities $-\pi < \mathbf{k}\mathbf{a}_i \leq \pi$, let us formulate the total Hamiltonian of the system of interacting excitons and photons. Providing $E_f(\mathbf{k}) = E_f(-\mathbf{k})$, we can write

$$H'=\frac{1}{2}\sum_{\mathbf{k}}H_{\mathbf{k}},\quad -\pi<\mathbf{k}\mathbf{a}_i\leqslant\pi,\qquad(5.12)$$

where

$$\begin{aligned}H_{\mathbf{k}}&=\sum_f E_f(\mathbf{k})\,[B_{\mathbf{k}f}^{+}B_{\mathbf{k}f}+B_{-\mathbf{k},\,f}^{+}B_{-\mathbf{k},\,f}]+\\&+\sum_{\alpha}c\,|\,\mathbf{k}\,|\left(1+\frac{\omega_0^2}{2c^2\mathbf{k}^2}\right)[a_{\mathbf{k}\alpha}^{+}a_{\mathbf{k}\alpha}+a_{-\mathbf{k},\,\alpha}^{+}a_{-\mathbf{k},\,\alpha}]+\frac{\omega_0^2}{2c\,|\,\mathbf{k}\,|}\sum_{\alpha}(a_{\mathbf{k}\alpha}a_{-\mathbf{k},\,\alpha}+\\&+a_{\mathbf{k}\alpha}^{+}a_{-\mathbf{k},\,\alpha}^{+})+\sum_{\alpha,f}D(\mathbf{k},\,\alpha,\,f)\{\gamma_{\mathbf{k}\alpha}\,(B_{\mathbf{k}f}^{+}-B_{-\mathbf{k},\,f})+\gamma_{-\mathbf{k},\,\alpha}\,(B_{-\mathbf{k},\,f}^{+}-B_{\mathbf{k}f})\}.\end{aligned}\qquad(5.13)$$

Operator (5.13) can be diagonalized with the Bogolyubov–Tyablikov canonical transformation (see [19], Chapter IV, Section 11). For this, from the four types of operators $a_{k\alpha}$, $a_{-k\alpha}$, $B_{k,f}$, $B_{-k,f}$, where k is fixed, $\alpha = 1, 2$, and f indicates ($f = 1, 2,\ldots$) all of the excited states of a molecule, let us move to the new Bose operators ξ_μ ($\mu = 1, 2,\ldots$) by means of the canonical transformations

$$\left.\begin{aligned} B_{\mathbf{k}f} &= \sum_\mu (\xi_\mu u_{\mathbf{k}f,\,\mu} + \xi_\mu^+ \overset{*}{v}_{\mathbf{k}f,\,\mu}), \\ a_{\mathbf{k}\alpha} &= \sum_\mu (\xi_\mu u_{\mathbf{k}\alpha,\,\mu} + \xi_\mu^+ \overset{*}{v}_{\mathbf{k}\alpha,\,\mu}). \end{aligned}\right\} \tag{5.14}$$

In order that transformations (5.14) be canonical, the functions $u_{l\mu}$ and $v_{l\mu}$ must satisfy the relations

$$\left.\begin{aligned} &\sum_l (u_{l\mu} \overset{*}{u}_{l\mu'} - v_{l\mu} \overset{*}{v}_{l\mu'}) = \delta_{\mu\mu'}, \\ &\sum_l (u_{l\mu} v_{l\mu'} - v_{l\mu'} u_{l\mu}) = 0, \\ &\sum_l (u_{l\mu} \overset{*}{v}_{l\mu'} - v_{l\mu'} u_{l\mu}) = 0. \end{aligned}\right\} \tag{5.15}$$

Summation in (5.15) is over all $l = \mathbf{k}f;\ -\mathbf{k},f;\ \mathbf{k}\alpha;\ -\mathbf{k},\alpha$ at fixed k.

Substituting (5.14) into operator (5.13), let us diagonalize it:

$$H_{\mathbf{k}} = H_{-\mathbf{k}} = \sum_\mu \omega_\mu(\mathbf{k}) \{\xi_\mu^+ \xi_\mu - \zeta_\mu(\mathbf{k})\}, \tag{5.16}$$

where

$$\zeta_\mu(\mathbf{k}) = \sum_l |v_{l\mu}|^2, \tag{5.16a}$$

if the functions of transformation (5.14) satisfy the system of equations

$$(\omega_\mu - c|\mathbf{k}|)\, u_{\mathbf{k}\alpha;\,\mu} = \\ = \sum_f D(\mathbf{k}, \alpha, f)\,[v_{-\mathbf{k},\,f;\,\mu} - u_{\mathbf{k}f;\,\mu}] + \frac{\omega_0^2}{2c|\mathbf{k}|}[u_{\mathbf{k}\alpha;\,\mu} + v_{-\mathbf{k},\,\alpha;\,\mu}], \tag{5.17a}$$

$$(\omega_\mu + c|\mathbf{k}|)\, v_{-\mathbf{k},\,\alpha;\,\mu} = \\ = \sum_f D(\mathbf{k}, \alpha, f)\,[u_{\mathbf{k}f;\,\mu} - v_{-\mathbf{k},\,f;\,\mu}] - \frac{\omega_0^2}{2c|\mathbf{k}|}[v_{\mathbf{k}\alpha;\,\mu} + u_{-\mathbf{k},\,\alpha;\,\mu}], \tag{5.17b}$$

$$(\omega_\mu - E_f(\mathbf{k}))\, u_{\mathbf{k}f;\,\mu} = \sum_\alpha D(\mathbf{k}, \alpha, f)\,[v_{-\mathbf{k},\,\alpha;\,\mu} + u_{\mathbf{k}\alpha;\,\mu}], \qquad (5.17c)$$

$$(\omega_\mu + E_f(\mathbf{k}))\, v_{-\mathbf{k},\,f;\,\mu} = \sum_\alpha D(\mathbf{k}, \alpha, f)\,[v_{-\mathbf{k},\,\alpha;\,\mu} + u_{\mathbf{k}\alpha;\,\mu}]. \qquad (5.17d)$$

If we compare (5.17a) with (5.17b) and (5.17c) with (5.17d), we find

$$\left.\begin{aligned} [\omega_\mu + c\,|\mathbf{k}|]\, v_{-\mathbf{k},\,\alpha;\,\mu} &= [c\,|\mathbf{k}| - \omega_\mu]\, u_{\mathbf{k}\alpha;\,\mu}, \\ [\omega_\mu + E_f(\mathbf{k})]\, v_{-\mathbf{k},\,f;\,\mu} &= [\omega_\mu - E_f(\mathbf{k})]\, u_{\mathbf{k}f;\,\mu}, \end{aligned}\right\} \qquad (5.18)$$

by means of which we can eliminate the functions $v_{l\mu}$ from Eqs. (5.17a) and (5.17c). Thus, we obtain the equations

$$\left.\begin{aligned} [\omega_\mu^2 - c^2\mathbf{k}^2 - \omega_0^2]\, u_{\mathbf{k}\alpha;\,\mu} &= \sum_f \frac{2E_f(\mathbf{k})\, D(\mathbf{k}, f, \alpha)\, E_f(\mathbf{k})}{\omega_\mu + E_f(\mathbf{k})}\, u_{\mathbf{k}f,\,\mu}, \\ [\omega_\mu - E_f(\mathbf{k})]\, u_{\mathbf{k}f;\,\mu} &= \sum_\alpha \frac{2D(\mathbf{k}, f, \alpha)\, c\,|\mathbf{k}|}{\omega_\mu + c\,|\mathbf{k}|}\, u_{\mathbf{k}\alpha;\,\mu}. \end{aligned}\right\} \qquad (5.18a)$$

Then, if we eliminate from this system the functions $u_{\mathbf{k}f;\mu}$, we find a system of equations that determine the energies of the new elementary excitations $\omega_\mu(\mathbf{k})$, which are sometimes called *polaritons* [20] or *photoexcitons*, since they correspond to a combination of excitons and photons:

$$[\omega_\mu^2 - c^2\mathbf{k}^2 - \omega_0^2]\, u_{\mathbf{k}\alpha;\,\mu} - \sum_\beta T_{\alpha\beta} u_{\mathbf{k}\beta;\,\mu} = 0, \qquad (5.19)$$

where

$$T_{\alpha\beta} = 8\pi \sum_f \frac{\omega_f^2 E_f(\mathbf{k})\, (\mathbf{u}_{\mathbf{k}\alpha}\mathbf{d}_f)\, (\mathbf{u}_{\mathbf{k}\beta}\mathbf{d}_f)}{v\,[\omega_\mu^2 - E_f^2(\mathbf{k})]} \qquad (5.20)$$

are matrix elements, which are functions of the unknown energy ω_μ. Using the approximate equation

$$\omega_f \approx E_f(\mathbf{k}),$$

the identity

$$\frac{E_f^2(\mathbf{k})}{\omega_\mu^2 - E_f^2(\mathbf{k})} \equiv \frac{\omega_\mu^2}{\omega_\mu^2 - E_f^2(\mathbf{k})} - 1,$$

and the sum rule

$$\sum_f \omega_f (\mathbf{d}_f \mathbf{u}_{\mathbf{k}\alpha})^2 = \frac{e^2}{2m} S,$$

which follows from (5.8), we can transform diagonal matrix elements (5.20) to the simple form

$$T_{\alpha\alpha} = -\omega_0^2 + \omega_0^2 \omega_\mu^2 S^{-1} \sum_f \frac{F_{f0}^\alpha}{\omega_\mu^2 - E_f^2(\mathbf{k})}, \tag{5.21}$$

where

$$F_{f0}^\alpha = 2m\omega_f (\mathbf{d}_f \mathbf{u}_{\mathbf{k}\alpha})^2 e^{-2}$$

is the oscillator strength of the quantum transition $0 \to f$.

If we know $\mathbf{d}_f$ and the laws of dispersion $E_f(\mathbf{k})$ in all of the exciton bands, from the condition of nontrivial solvability of system of homogeneous equations (5.19) we can determine the energies $\omega_\mu(\mathbf{k})$ of the photoexcitons for all values of the real wave vectors $\mathbf{k}$ from the first Brillouin zone.

Equations (5.19) have an especially simple form for optically isotropic crystals. Either transverse or longitudinal excitons are possible in optically isotropic crystals. Longitudinal excitons do not interact with the transverse electromagnetic field and form an independent branch of elementary excitations.

For transverse excitons, $(\mathbf{d}_f \mathbf{k}) = 0$. Therefore, if the polarization unit vector $\mathbf{u}_{\mathbf{k}1}$ of a photon is directed along $\mathbf{d}_f$, then $(\mathbf{u}_{\mathbf{k}2}\mathbf{d}_f) = 0$. Therefore, nondiagonal matrix elements (5.20) are equal to zero, and equation system (5.19), when expression (5.21) is taken into account, takes the simple form

$$\left[1 - S^{-1} \sum_f \frac{\omega_0^2 F_{f0}}{\omega_\mu^2 - E_f^2(\mathbf{k})} - \frac{c^2\mathbf{k}^2}{\omega_\mu^2}\right] u_{\mathbf{k}\alpha;\,\mu} = 0,$$

where $E_f(\mathbf{k})$ are the transverse-exciton dispersion laws. The conditions of nontrivial solvability of this equation comes down to

$$\frac{c^2\mathbf{k}^2}{\omega^2} = 1 - \sum_f \frac{\omega_p^2 F_{f0}}{\omega^2 - E_f^2(\mathbf{k})}, \tag{5.22}$$

where

$$\omega_p^2 = S^{-1}\omega_0^2 = \frac{4\pi e^2}{mv}.$$

Equation (5.22) is a dispersion equation that determines the frequencies of elementary excitations (photoexcitons) in a system of interacting excitons and photons as functions of the real wave vector **k**. The solutions of (5.22) have the form

$$\omega_\mu = \omega_\mu(\mathbf{k}), \tag{5.23}$$

where μ determines the type of elementary excitation.

If the lower electron band $E_1(\mathbf{k})$ is separated from all the others, then, since we are interested in excitations in the energy region $E_1(\mathbf{k})$, we can eliminate the dominant resonance term in (5.22) and transform it to

$$\frac{c^2\mathbf{k}^2}{\omega^2} = \varepsilon_0 + \frac{\omega_p^2 F_{10}}{E_1^2(\mathbf{k}) - \omega^2}, \tag{5.24}$$

where

$$\varepsilon_0 = 1 + \sum_{f>1} \frac{\omega_p^2 F_{f0}}{E_f^2(\mathbf{k}) - \omega^2}$$

is a function that is slightly dependent upon ω at $\omega \approx E_1(\mathbf{k})$. If we consider ε_0 as a constant, we can find two positive solutions of Eq. (5.24):

$$\omega_{1,2}(\mathbf{k}) = \frac{1}{\sqrt{2}}\left\{A \pm \sqrt{A^2 - 4E_1^2(\mathbf{k})\frac{c^2\mathbf{k}^2}{\varepsilon_0}}\right\}^{1/2}, \tag{5.25}$$

where

$$A = E_1^2(\mathbf{k}) + \frac{c^2\mathbf{k}^2 + \omega_p^2 F_{10}}{\varepsilon_0}.$$

The two dispersion curves $\omega_1(\mathbf{k})$ and $\omega_2(\mathbf{k})$ (in units of cm^{-1}) are shown on the right side of Fig. 18 for $0 \le ka \le 0.04$. In this range of ka, it is assumed that $E_1(\mathbf{k}) = E_1(0) = 25{,}300\ \mathrm{cm}^{-1}$. It is also assumed that $\omega_0^2 = 2.2 \times 10^8\ \mathrm{cm}^{-2}$, $F_{10} = 0.1$, and $\varepsilon_0 = 4$.

Each branch of steady elementary excitations is a "mixture" of photons and excitons. Returning to canonical transformations

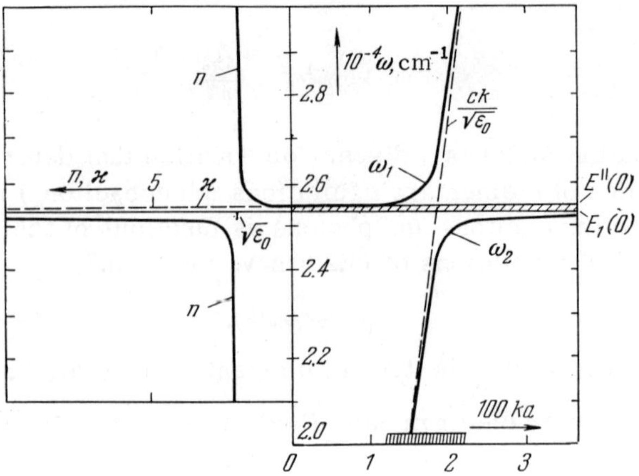

Fig. 18. The right side shows the energy of the two types of photoexcitons in isotropic crystals as a function of k for small ka. $E_1(0)$ is the limiting energy of transverse excitons, and $E^{\|}(0)$ is the energy of longitudinal excitons. The left side shows the refractive index n and the attenuation factor $\varkappa$ of the electromagnetic waves as functions of frequency.

(5.14), we see that the creation operator ξ_μ^+ for the new excitations (photoexcitons) is represented by a linear combination of the creation and annihilation operators for excitons and photons:

$$\xi_\mu^+(\mathbf{k}) = \sum_f \{B_{\mathbf{k}f}^+ u_{\mathbf{k}f;\,\mu} + B_{-\mathbf{k},\,f}^+ u_{-\mathbf{k},\,f;\,\mu} - B_{\mathbf{k}f} v_{\mathbf{k}f;\,\mu} - B_{-\mathbf{k},\,f} v_{-\mathbf{k},\,f;\,\mu}\} +$$

$$+ \sum_\alpha \{a_{\mathbf{k}\alpha}^+ u_{\mathbf{k}\alpha;\,\mu} + a_{-\mathbf{k},\alpha}^+ u_{-\mathbf{k},\,\alpha;\,\mu} - a_{\mathbf{k}\alpha} v_{\mathbf{k}\alpha,\,\mu} - a_{-\mathbf{k},\,\alpha} v_{-\mathbf{k},\,\alpha;\,\mu}\}.$$

The relative values of the coefficients $u_{\mathbf{k}f;\,\mu}$, $u_{\mathbf{k}\alpha;\mu}$,... determine the contributions of the excitons and photons to the total excitation. According to (5.18), the values of the coefficients $u_{\mathbf{k}f;\mu}$,... are independent of the sign of k. Further, it is easy to see that at k values for which $\omega_1(\mathbf{k}) \approx E_1(\mathbf{k})$, i.e., on the right of the ka region shaded in Fig. 18, the functions $v_{\mathbf{k}f;1}$ and $v_{\mathbf{k}\alpha;1}$ are close to zero and $u_{\mathbf{k}\alpha;1} \ll u_{\mathbf{k}f;1}$. At these k values, therefore, photoexcitons $\omega_1(\mathbf{k})$ coincide with excitons. Conversely, at k values for which $c^2k^2/\varepsilon_0 \approx E_1^2(0)$, the mixing of excitons with photons is especially great.

It also follows from Fig. 18 that the shaded energy region lying between the limiting energies of transverse $E_1(0)$ and longitudinal excitons is forbidden for photoexcitons

$$E^{\parallel}(0)=\sqrt{E_1^2(0)+\frac{\omega_p^2 F_{10}}{\varepsilon_0}}.$$

Evidently, Born and Huang ([9], see Section 8) first called attention to the "mixing" of electromagnetic waves with the polarization waves of the medium. They investigated (without allowance for absorption) the interaction of the optical lattice vibrations of ionic crystals with electromagnetic waves and showed that this interaction results in the creation in the system of new elementary excitations. In the resonance region (when $c^2\mathbf{k}^2 \approx E_0^2\varepsilon_0$, where E_0 is the energy of the optical vibrations), such elementary excitations are in mixture in comparable fractions of electromagnetic waves and optical lattice vibrations. These types of excitations become more and more completely separated with distance from the resonance region.

When $ka \ll 1$, the right side of (5.24) can be considered the real part of the dielectric constant [see (IV.7.33)]

$$\varepsilon(\omega)=\varepsilon_0+\frac{\omega_p^2 F_{10}}{E_1^2(0)-\omega^2}+i\pi\delta\left[E_1^2(0)-\omega^2\right], \tag{5.26}$$

which determines the refractive index $n(\omega)$ and the attenuation factor $\varkappa(\omega)$ of electromagnetic waves of frequency ω [not equal to $E_1(0)$] propagated in an optically isotropic dielectric medium. In this case, the wave vector **k** is considered a function of the real frequency ω.

If the refractive index n and attenuation factor $\varkappa$ are introduced by means of the equation $k = (\omega/c)(n + i\varkappa)$, then, using the equation $\varepsilon(\omega) = (n + i\varkappa)^2$, we obtain two equations:

$$n^2(\omega)-\varkappa^2(\omega)=\varepsilon_0+\frac{\omega_p^2 F_{10}}{E_1^2(0)-\omega^2},$$
$$2n(\omega)\varkappa(\omega)=\pi\delta\left[E_1^2(0)-\omega^2\right].$$

The solutions of these equations as functions of ω (in units of cm^{-1}) are shown on the left side of Fig. 18. It is apparent from the figure that, in the range of frequencies ω that fall within the shaded interval $E_1(0)$, $E^{\parallel}(0)$, only spatially inhomogeneous waves whose amplitude decreases as $\exp(-\varkappa z)$ can be propagated in the crystal. These waves do not coincide with photoexcitons. They characterize the forced "vibrations" of the system. In the transparency region, i.e., outside the shaded interval $E_1(0)$, $E^{\parallel}(0)$, $\varkappa = 0$ and the forced solutions coincide with elementary excitations – photoexcitons. Thus, in the transparency region, assignment of the dispersion law of photoexcitons, i.e., of the functions $\omega_\mu(\mathbf{k})$, allows us to determine the absolute values of the wave vectors of the corresponding excitations $|\mathbf{k}| = k(\omega_\mu)$ as functions of their frequencies. Then, using the equation

$$n^2(\omega) = \frac{c^2\mathbf{k}^2(\omega)}{\omega^2}, \tag{5.27}$$

we can also determine the refractive index of the corresponding waves. These waves are sometimes called *normal electromagnetic waves*. This name is justified in the transparency region of the crystal for photoexcitons that contain considerable admixtures of photons.

The curves in Fig. 18 correspond to small ka, for which in crystals (with not very wide exciton bands) we can let $E_1(\mathbf{k}) = E_1(0)$. The second important assumption used in plotting these curves was that true absorption was completely absent for all $\omega \neq E(0)$. This assumption is a rough idealization, based on the assumption that the molecules are rigidly fixed at the lattice points and that the natural width of the intramolecular electronic excitations is negligible. Let us consider, however, the results of a study of this idealized model in the case of systems with wide exciton bands.

The right sides of Figs. 19 and 20 show qualitative dependences of the photoexciton energy upon ka for optically isotropic systems with, respectively, positive and negative exciton effective masses m^*, i.e., when the exciton energy can be represented by

$$E(\mathbf{k}) = E(0) + \frac{\mathbf{k}^2}{2m^*}; \quad m^* > 0 \quad \text{or} \quad m^* < 0. \tag{5.28}$$

The left sides of these figures show the corresponding refractive indices n, which can be calculated using Eq. (5.27). A distinctive feature of the refractive-index curves in Figs. 19 and 20 is that

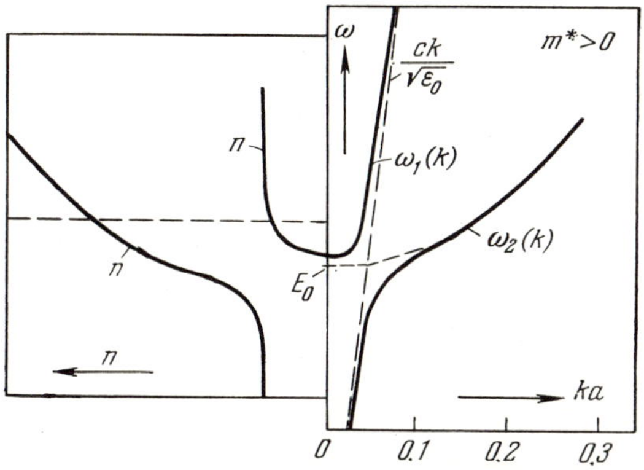

Fig. 19. The right side shows the photoexciton energy as a function of k*a* for a positive exciton effective mass. The left side shows n and ϰ of the corresponding normal electromagnetic waves.

in a certain frequency range (for example, in the region indicated by the horizontal broken line), each frequency ω corresponds to not one but two waves, which are propagated in the same direction and have the same polarization. This phenomenon in the theory of excitons was first pointed out by Pekar [21]. He showed that when the dependence of exciton energy upon the wave vector **k** is taken into account, the order of the equation determining the square of the refractive index is increased. For example, accord-

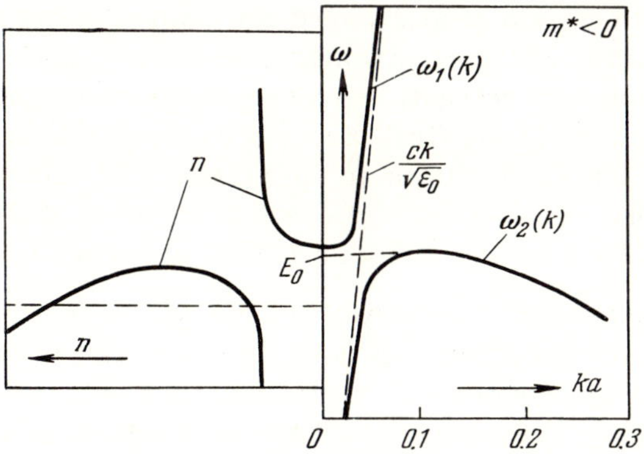

Fig. 20. The same as in Fig. 19, but for negative exciton effective mass.

ing to (5.28) and (5.27), providing

$$\beta \ll E(0), \quad \text{where } \beta \equiv \frac{\omega^2}{m^* c^2},$$

we can write

$$E^2(\mathbf{k}) \approx E^2(0) + \beta n^2 E_0.$$

Then Eq. (5.24) reduces to a second-order equation in n^2:

$$E_0\beta n^4 - [E_0^2 - \omega^2 - E_0\beta\varepsilon_0]\, n^2 - \omega_p^2 F - \varepsilon_0 [E_0 - \omega^2] = 0, \; E_0 = E(0). \quad (5.29)$$

This equation has one solution

$$n^2 = \varepsilon_0 + \frac{\omega_p^2 F}{E_0^2 - \omega^2} \quad (5.30)$$

when $\beta = 0$ and two solutions n_1^2 and n_2^2 when $\beta \neq 0$, one of which, n_1^2, differs comparatively little from (5.30) far from resonance. If both solutions are positive, two electromagnetic waves with the same frequency and polarization but with different wavelengths can be propagated in the crystal. The waves associated with the second refractive index are called new or anomalous waves.

Anomalous waves usually correspond to $ck > E_0\sqrt{\varepsilon_0}$. As we have already noted, such photoexcitons contain a small fraction of photons. Their energy almost coincides with the exciton energy. The properties of such excitations are very similar to the properties of excitons and differ from properties of electromagnetic waves. Their corresponding high refractive indices, which were obtained with the aid of (5.27), are not related to the macroscopic refractive index of the electromagnetic waves passing through the crystal. Such high refractive indices of electromagnetic waves are not realized, in the first place, because of attenuation of the exciton excitations due to their interaction with lattice vibrations and, secondly, because the macroscopic concept of the refractive index loses its meaning at high n (small wavelengths).

Agranovich and Konobeev [17] have shown that the mixing of exciton and photon states can play a certain part in explaining the very weak longwave edge of the absorption band in molecular crystals. They took into account the interactions of photoexcitons with low-frequency phonons. The virtual conversion of a photon to an exciton, which interacts with phonons, results in very weak absorption at photon energies that are less than the energy E_{min} of the lower edge of the exciton band, as calculated without allowance for the interaction of excitons with phonons and photons. Unfortunately, this effect is difficult to distinguish from other, more important effects. For example, as will be shown in the following chapter, allowance for the interaction of excitons with phonons results in displacement of the edge of the absorption band to the low-frequency side even at absolute zero. This effect increases considerably with an increase in temperature. Moreover, the presence of radiation attenuation leads to the appearance of a "Lorentz tail" of the absorption curve, which must be observed in the lower-frequency region from the maximum of the absorption curve.

Chapter IV

Interaction of Excitons with Phonons and Photons

1. The Exciton – Phonon Interaction Operator

In the preceding sections, we examined the excited states of the molecules of a crystal under the assumption that the molecules were rigidly fixed at the lattice points. This assumption was necessary to simplify the first stage of a theoretical study of the properties of molecular crystals. Since the masses of molecules are considerably greater than the masses of electrons and individual atoms, their velocities are lower than the velocities of electrons, and in the zero-th approximation of the theory the molecules can be considered fixed. Now let us move to the next approximations of the theory, in which the motion of the molecules is taken into account.

For simplicity, in this chapter we shall consider a model of an ideal crystal containing one neutral molecule in each unit cell and take into account only one lower exciton band corresponding to electronic excitation of the molecules. This model allows us to study without unnecessary complications the main characteristics of the interactions of excitons with phonons and photons.

Let us consider a molecular crystal with one molecule in each unit cell and take into account only the electronic ground and one excited state of the molecules. In the adiabatic approximation, i.e., when the molecules are fixed in the crystal, the energy operator of the crystal can be written as

$$H_0 = W(R) + H_{ex}(R), \tag{1.1}$$

where R characterizes the set of space coordinates and the orientation of the molecules; W(R) is the sum of electronic energies of all of the molecules that are in the ground state; and $H_{ex}(R)$ is the energy operator of the electronic excited states of the crystal for a particular spatial arrangement of the molecules.

If the centers of gravity of the molecules are displaced little from the points of the translation lattice of a crystal without excitons and the orientation of the molecules differs little from the equilibrium orientation, then the molecules can be denoted by the vectors **n**, **m**, ..., which characterize the position of the lattice points. In this case, in the occupation-number representation, the energy operator of the electronic excited states of the crystal has in the Heitler–London approximation (III.2.21) the following form:

$$H_{ex}(R) = \sum_{\mathbf{n}} [\Delta\varepsilon + D_{\mathbf{n}}(R)] B_{\mathbf{n}}^{+} B_{\mathbf{n}} + \sum_{\mathbf{n},\mathbf{m}}{}' M_{\mathbf{nm}}(R) B_{\mathbf{m}}^{+} B_{\mathbf{n}}, \tag{1.2}$$

$$D_{\mathbf{n}}(R) = \sum_{\mathbf{m}}{}' D_{\mathbf{nm}}(R). \tag{1.2a}$$

The second term in the brackets of operator (1.2) characterizes the change in the energy of van der Waals interaction of the molecule **n** with the surrounding molecules when it is excited. The greater this term, the greater is the tendency toward lattice deformation in the area of the excited molecule. On the other hand, the resonance-interaction matrix elements M_{nm} cause "blurring" of the excitation over a certain region of the crystal. If the resonance interaction is more considerable than the van der Waals interaction and the crystal lattice is "rigid," exciton states will be formed in the crystal without local deformation. Owing to translational invariance, $D_{\mathbf{n}}(R)$ is not a function of the location of the excited molecule and we can write

$$D_{\mathbf{n}}(R) = \sum_{\mathbf{m}(\neq 0)} D_{0\mathbf{m}}(R). \tag{1.2b}$$

In this case, the interaction of excitons with molecular vibrations amounts to the creation and absorption of phonons, i.e., of molecular vibration quanta relative to the equilibrium positions of the molecules in a crystal without excitons. This case we shall call weak coupling of excitons with phonons. With weak coupling, single-phonon processes play the leading role in the interaction of excitons and phonons.

In "soft" lattices, electronic excitations that are associated with a type of exciton state (distributed throughout the entire crystal and characterized by the wave vector k) are also possible, but they are accompanied by deformation of the entire lattice when its translational invariance is preserved. This case we shall call strong coupling of excitons with phonons. With strong coupling, many-phonon processes, along with single-phonon processes, play an important role in the interaction of excitons with phonons.

When the resonance interactions are small and the crystal is easily deformable, local lattice deformation can occur simultaneously with excitation of a molecule. This crystal state may be called a local electronic excitation. This state is no longer characterized by a specific value of the wave vector. It corresponds to the case of strong coupling of an electronic excitation with lattice vibrations.

Cases that are intermediate with respect to the above usually occur under real conditions. In some crystals, cases similar to some of the above-mentioned limiting cases can correspond to different intramolecular excitations. The nature of the coupling of excitons and electronic excitations with phonons is also a function of the temperature of the crystal. As a rule, the coupling force increases with an increase in temperature.

Only weak coupling of excitons with phonons will be examined in this chapter.

Strong coupling of an electronic excitation with molecular vibrations will be considered in Chapter V. Cases of strong coupling of excitons with phonons, just as cases of intermediate coupling, have so far not been studied.

In a crystal without excitons, the minimum energy W(R) corresponds to the equilibrium position of molecules of a definite orientation at the lattice points. We shall denote these equilibrium positions arbitrarily by R = 0. Then, let R_n^j determine the six deviations (j = 1, 2,..., 6) from the equilibrium positions and orientations of the molecule occupying location **n**. The values j = 1, 2, 3 refer to translations and the values j = 4, 5, 6 refer to changes of orientation.

The coupling between exciton excitations and molecular vibrations in the lattice is determined by the dependence of the matrix elements $M_{nm}(R)$ and $D_n(R)$ in operator (1.2) upon the displacements of the molecules from their equilibrium positions. If we expand these values in powers of the equilibrium-position deviations of the molecules in a crystal without excitons and retain only the terms not over the second order of deviations, we obtain

$$H_0 = H_{ex}(0) + W(R) + H^{(1)}_{int} + H^{(2)}_{int}, \tag{1.3}$$

where

$$H_{ex}(0) = \sum_{\mathbf{n}} [\Delta\varepsilon + D(0)] B^+_{\mathbf{n}} B_{\mathbf{n}} + \sum_{\mathbf{n},\mathbf{m}}{}' M_{\mathbf{nm}}(0) B^+_{\mathbf{n}} B_{\mathbf{m}} \tag{1.4}$$

is the operator of exciton excitations in an undeformed lattice; and

$$H^{(1)}_{int} = \sum_{\mathbf{n},\mathbf{m}}{}' B^+_{\mathbf{n}} B_{\mathbf{m}} \sum_{j=1}^{6} \left\{\left(R^j_{\mathbf{n}} \frac{\partial}{\partial R^j_{\mathbf{n}}} + R^j_{\mathbf{m}} \frac{\partial}{\partial R^j_{\mathbf{m}}}\right) M_{\mathbf{nm}}\right\}_0, \tag{1.5}$$

$$H^{(2)}_{int} = \sum_{\mathbf{n},\mathbf{m}}{}' B^+_{\mathbf{n}} B_{\mathbf{n}} \sum_{j=1}^{6} \left\{\left(R^j_0 \frac{\partial}{\partial R^j_0} + R^j_{\mathbf{m}} \frac{\partial}{\partial R^j_{\mathbf{m}}}\right) D_{0\mathbf{m}}(R)\right\}_0 \tag{1.6}$$

are the operators of interaction of exciton excitations with lattice vibrations. The zero subscript on the braces indicates that the derivatives with respect to the displacements are taken at zero displacements. The form of operator (1.6) is determined by our assumption that excitons, not electronic excitations, are generated in the crystal. Exciton production can lead to overall lattice deformation without loss of translational symmetry.

Operator (1.4) is diagonalized by a canonical transformation from the operators B_n to the operators $B(\mathbf{k})$,

$$B_{\mathbf{n}} = \frac{1}{\sqrt{N}} \sum_{\mathbf{k}} B(\mathbf{k}) \exp(i\mathbf{kn}), \tag{1.7}$$

to the form

$$H_{ex}(0) = \sum_{\mathbf{k}} E(\mathbf{k}) B^+(\mathbf{k}) B(\mathbf{k}), \tag{1.8}$$

where N is the number of unit cells in the crystal; **k** is the wave vector of an exciton;

$$E(\mathbf{k}) = \Delta\varepsilon + D(0) + \sum_{m}' M_{\mathbf{nm}}(0) \exp\{i\mathbf{k}(\mathbf{n} - \mathbf{m})\} \tag{1.9}$$

is the energy of one exciton; and $B^{+}(\mathbf{k})$ and $B(\mathbf{k})$ are the creation and annihilation operators of excitons with the wave vector **k**. The set of $E(\mathbf{k})$ values for all possible values of the wave vectors in the first Brillouin zone forms the exciton band of the crystal.

The operator W(R) in (1.3) determines the potential energy of interaction of the molecules in a crystal without excitons. In the approximation of small vibrations about the equilibrium positions, which correspond to R = 0, we can write

$$W(R) = W(0) + \frac{1}{2} \sum U^{ij}_{\mathbf{nm}} R^{i}_{\mathbf{n}} R^{j}_{\mathbf{m}}.$$

The classical total energy of molecular vibrations in the crystal [without the constant term W(0)] equals the sum

$$E_{\text{vib}} = T_R + W(R) - W(0), \tag{1.10}$$

where

$$T_R = \frac{1}{2} \sum_{\mathbf{n}, j} I_j (\dot{R}^{j}_{\mathbf{n}})^2$$

is the kinetic energy of small vibrations about the equilibrium positions; and I_j are the mass coefficients corresponding to the three translational (j = 1, 2, 3) and three rotational (j = 4, 5, 6) degrees of freedom of a molecule.

Total energy (1.10) is conveniently expressed in terms of the time-dependent 6N complex normal coordinates $\Phi_s(\mathbf{q})$, which are introduced by the transformation

$$R^{j}_{\mathbf{n}} = (I_j N)^{-1/2} \sum_{s, \mathbf{q}} e^{j}_{s}(\mathbf{q}) \Phi_s(\mathbf{q}) \exp(i\mathbf{qn}), \tag{1.11}$$

where the subscript s = 1, 2, ..., 6 numbers the normal-vibration branches of the rigid molecules in the crystal; **q** is the wave vector, which takes N discrete values equal to the number of unit cells; and $e_s^j(\mathbf{q})$ are the components of the unit polarization vectors of the s-th vibration branch with the wave vector $\mathbf{q}$.

The components of the unit polarization vectors $e_s^j(\mathbf{q})$ and the cyclic frequencies $\Omega_s(\mathbf{q})$ of normal vibrations are determined by the system of equations

$$\sum_{j=1}^{6} e_s^j(\mathbf{q})\, e_{s'}^j(\mathbf{q}') = \delta_{ss'}\delta_{\mathbf{q}\mathbf{q}'},$$

$$\sum_{j=1}^{6} G_{ij}(\mathbf{q})\, e_s^j(\mathbf{q}) = \Omega_s^2(\mathbf{q})\, e_s^i(\mathbf{q}),$$

where

$$G_{ij}(\mathbf{q}) = G_{ji}^*(\mathbf{q}) = G_{ij}^*(-\mathbf{q}) = \sum_{\mathbf{m}} U_{\mathbf{nm}}^{ij} \frac{\exp\{i\mathbf{q}(\mathbf{m}-\mathbf{n})\}}{\sqrt{I_i I_j}}. \tag{1.12}$$

In order to satisfy the conditions of realness of displacements (1.11), we must let

$$e_s^i(\mathbf{q}) = e_s^{*i}(-\mathbf{q}), \qquad \Phi_s(\mathbf{q}) = \Phi_s^*(-\mathbf{q}).$$

Since the matrix composed of elements (1.12) is Hermitian, it determines the real cyclic frequencies $\Omega_s(\mathbf{q})$ of the six modes or branches of lattice vibrations of a molecular crystal. The three branches with frequencies which approach zero as $\mathbf{q}$ approaches zero are called *acoustic branches*. We shall give these branches the values s = 1, 2, 3. The other three branches (s = 4, 5, 6) correspond to rotational molecular vibrations. They have nonzero limiting (when $\mathbf{q} \to 0$) frequencies and are called *optical branches*. In molecular crystals, the frequencies of the optical branches usually exceed by a factor of 5-10 the frequencies of the acoustic branches. Under these conditions, the optical and acoustic can with good approximation be considered independent.

If we substitute (1.11) into expression (1.10), we find

$$E_{\text{vib}} = \frac{1}{2} \sum_{s,\mathbf{q}} \{\dot{\Phi}_s^*(\mathbf{q})\, \dot{\Phi}_s(\mathbf{q}) + \Omega_s^2(\mathbf{q})\, \Phi_s^*(\mathbf{q})\, \Phi_s(\mathbf{q})\}. \tag{1.13}$$

It follows from expression (1.13) that the "coordinates" $\Phi_s(\mathbf{q})$ and $\Phi_s^*(\mathbf{q})$ correspond to the conjugate "momenta"

$$\Pi_s(\mathbf{q}) = \dot{\Phi}_s^*(\mathbf{q}), \qquad \Pi_s^*(\mathbf{q}) = \dot{\Phi}_s(\mathbf{q}).$$

We convert from classical vibration energy (1.13) to the Hamiltonian quantum operator by replacing Φ_s and Π_s by operators according to the following rule:

$$\left.\begin{aligned} \Phi_s(\mathbf{q}) &\Rightarrow \sqrt{\frac{\hbar}{2\Omega_s(\mathbf{q})}}\,(b_{\mathbf{q}s}+b^+_{-\mathbf{q},s}),\\ \Pi_s(\mathbf{q}) &\Rightarrow i\sqrt{\frac{\hbar\Omega_s(\mathbf{q})}{2}}\,(b^+_{\mathbf{q}s}-b_{-\mathbf{q},s}). \end{aligned}\right\} \tag{1.14}$$

The creation b^+_{qs} and annihilation b_{qs} operators for phonons sq satisfy the commutation relations

$$[b_{\mathbf{q}s}, b^+_{\mathbf{q}'s'}] = \delta_{ss'}\delta_{\mathbf{q}\mathbf{q}'}, \qquad [b_{\mathbf{q}s}, b_{\mathbf{q}'s'}] = 0. \tag{1.14a}$$

Having made transformations (1.14) in Eq. (1.13), we find the energy operator for molecular vibrations in a crystal without excitons:

$$H^0_{\text{ph}} = \sum_{s,\,\mathbf{q}} \hbar\Omega_s(\mathbf{q})\left[b^+_{\mathbf{q}s}b_{\mathbf{q}s}+\frac{1}{2}\right]. \tag{1.15}$$

If we make transformations (1.14) in expression (1.11), we obtain the operator of the l-th displacement component of the n-th molecule:

$$R^l_{\mathbf{n}} = \sum_{s,\,\mathbf{q}} \left[\frac{\hbar}{2I_l N\Omega_s(\mathbf{q})}\right]^{1/2} e^l_s(\mathbf{q})\,\varphi_{\mathbf{q}s}\exp(i\mathbf{q}\mathbf{n}), \tag{1.16}$$

where

$$\varphi_{\mathbf{q}s} = \varphi^+_{-\mathbf{q},s} \equiv (b_{\mathbf{q}s}+b^+_{-\mathbf{q},s}). \tag{1.17}$$

If we substitute expression (1.7) into exciton–phonon interaction operators (1.5) and (1.6), which were obtained in an adiabatic approximation, and replace the displacements R^l_n by operators (1.16), after simple transformations we find the exciton–phonon interaction operators in the occupation-number representation for excitons and phonons:

$$H^{(1)}_{\text{ex-ph}} = \frac{1}{\sqrt{N}} \sum_{s,\mathbf{k},\mathbf{q}} F_s(\mathbf{k},\mathbf{q})\,B^+(\mathbf{k}+\mathbf{q})\,B(\mathbf{k})\,\varphi_{\mathbf{q}s}, \tag{1.18}$$

$$H^{(2)}_{\text{ex-ph}} = \frac{1}{\sqrt{N}} \sum_{s,\mathbf{k},\mathbf{q}} \chi_s(\mathbf{q})\,B^+(\mathbf{k})\,B(\mathbf{k})\,\varphi_{\mathbf{q}s}. \tag{1.19}$$

Here, the coupling functions for excitons with phonons are determined by the expressions

$$F_s(\mathbf{k}, \mathbf{q}) = \sum_{l,\, m(\neq 0)} e_s^l(\mathbf{q}) \left[\frac{\hbar}{2I_l\Omega_s(\mathbf{q})}\right]^{1/2} \left\{\left(\frac{\partial}{\partial R_0^l} + e^{i\mathbf{q}\mathbf{m}} \frac{\partial}{\partial R_{\mathbf{m}}^l}\right) M_{0\mathbf{m}}\right\}_0 e^{i\mathbf{k}\mathbf{m}}, \quad (1.20)$$

$$\chi_s(\mathbf{q}) = \sum_{l,\, m(\neq 0)} e_s^l(\mathbf{q}) \left[\frac{\hbar}{2I_l\Omega_s(\mathbf{q})}\right]^{1/2} \left\{\left(\frac{\partial}{\partial R_0^l} + e^{i\mathbf{q}\mathbf{m}} \frac{\partial}{\partial R_{\mathbf{m}}^l}\right) D_{0\mathbf{m}}\right\}_0. \quad (1.21)$$

Exciton–phonon coupling functions (1.20) and (1.21) satisfy the relations

$$F_s(\mathbf{k}, \mathbf{q}) = F_s^+(\mathbf{k} + \mathbf{q}, -\mathbf{q}), \qquad \chi_s(\mathbf{q}) = \chi_s^+(-\mathbf{q}), \quad (1.22)$$

which ensure that the operators $H_{ex\text{-}ph}$ are Hermitian.

Thus, the Hamiltonian of a system of interacting excitons and phonons may be written as

$$H = H_{ex}(0) + H_{ph}^0 + H_{ex\text{-}ph}^{(1)} + H_{ex\text{-}ph}^{(2)}, \quad (1.23)$$

where the operators $H_{ex}(0)$, H_{ph}^0, $H_{ex\text{-}ph}^{(1)}$, and $H_{ex\text{-}ph}^{(2)}$ are defined by expressions (1.8), (1.15), (1.18), and (1.19), respectively. With weak exciton–phonon coupling, the inequality $H_{ex\text{-}ph}^{(1)} \gg H_{ex\text{-}ph}^{(2)}$ is satisfied. With strong coupling, the reverse inequality is satisfied or both operators have comparable values.

Exciton–phonon interaction operator (1.18) describes processes of elastic and inelastic scattering of excitons in which the number of excitons remains constant while the number of phonons is changed. These processes occur when the laws of conservation of quasi-momentum is satisfied. Operator (1.19) is proportional to the number of excitons in a given exciton band (which corresponds to a particular electronic excitation) regardless of the values of their wave vectors. This operator, as we shall see below, characterizes the displacement of the equilibrium positions of the molecular vibrations, i.e., the lattice deformation over the entire region in which the exciton excitation is distributed.

Since the coupling functions $F_s(\mathbf{k}, \mathbf{q})$ and $\chi_s(\mathbf{q})$ are determined by the different properties of molecular crystals, the relative role of operators (1.18) and (1.19) can differ considerably in different molecular crystals and even in the same crystals when the exciton states are associated with different intramolecular electronic excitations. In some cases, the inequality

$$F_s(\mathbf{k}, \mathbf{q}) \gg \chi_s(\mathbf{q}),$$

is satisfied, and in other cases the reverse inequality holds. Sometimes, along with the terms in operator (1.19) that are linear in the phonon creation and annihilation operators, we must take into account quadratic terms, which are obtained when allowance is made in (1.6) not only for the linear but also for the quadratic displacements of the molecules from their equilibrium positions.

In real crystals, processes of total conversion of excitons to phonons (conversion of an electronic excitation to heat) or the reverse processes (thermal excitation of excitons) are also possible. To describe these processes, nonadiabatic corrections must be made in the initial operator of adiabatic approximation (1.3).

In the following chapters, we shall assume that weak coupling of excitons with phonons is realized. Operator (1.19) can be omitted in this approximation. In the first approximation, therefore, molecular vibrations in a crystal with excitons occur about the same equilibrium values as in a crystal without excitons. Single-phonon processes play the leading role in this case.

In the case of strong coupling of excitons with phonons, exciton creation causes overall deformation of the crystal lattice. Molecular vibrations in the presence of excitons occur about new equilibrium positions. In this case, the lattice energy is changed by a value equal to the energy of tens and hundreds of phonons – lattice vibration quanta about the old equilibrium positions. In order to take into account such many-phonon processes, the effect of lattice deformation should be taken into account in the first stage of calculation. In this case, the new phonons in the crystal characterize the molecular vibrations about the new equilibrium positions. In other words, we must move from the creation and annihilation operators of phonons in an undeformed lattice to new operators such that this transformation includes the lattice deformation for exciton excitation.

Since operator (1.19) is independent of the exciton wave vectors and is a function only of their number, it is convenient to unite this operator with molecular-vibration operator (1.15) for a crystal without excitons:

$$H^{\theta}_{\text{ph}} + H^{(2)}_{\text{ex-ph}} = \sum_{s,\mathbf{q}} \hbar\Omega_s(\mathbf{q})\,[b^{+}_{s\mathbf{q}}b_{s\mathbf{q}} + \Lambda(s,\mathbf{q})(b_{s\mathbf{q}} + b_{s,-\mathbf{q}})\,\hat{n}], \tag{1.24}$$

where

$$\Lambda(s, \mathbf{q}) = \frac{\chi_s(\mathbf{q})}{\hbar\Omega_s(\mathbf{q})\sqrt{N}}, \qquad \hat{n} = \sum_{\mathbf{k}} B^+(\mathbf{k}) B(\mathbf{k}). \tag{1.25}$$

New operator (1.24) can be diagonalized by converting from the operators $b_{s\mathbf{q}}$, $B(\mathbf{k})$ to the new operators $\tilde{b}_{s\mathbf{q}}$, $\tilde{B}(\mathbf{k})$ by means of the unitary operator

$$S = \exp\left\{\hat{\sigma} \sum_{\mathbf{k}} \tilde{B}^+(\mathbf{k}) \tilde{B}(\mathbf{k})\right\}, \tag{1.26}$$

where

$$\hat{\sigma} = \sum_{s, \mathbf{q}} (\Lambda^*(s, \mathbf{q}) \tilde{b}^+_{s\mathbf{q}} - \Lambda(s, \mathbf{q}) \tilde{b}_{s\mathbf{q}}). \tag{1.27}$$

Thus, we obtain

$$\left.\begin{aligned} b_{s\mathbf{q}} &= S\tilde{b}_{s\mathbf{q}}S^+ = \tilde{b}_{s\mathbf{q}} - \sum_{\mathbf{k}} \tilde{B}^+(\mathbf{k}) \tilde{B}(\mathbf{k}) \Lambda^*(s, \mathbf{q}), \\ B(\mathbf{k}) &= S\tilde{B}(\mathbf{k}) S^+ = \exp(-\sigma) \tilde{B}(\mathbf{k}); \end{aligned}\right\} \tag{1.28}$$

$$\tilde{H}_{\text{ph}} = \tilde{H}^0_{\text{ph}} + \tilde{H}^{(2)}_{\text{ex-ph}} = \sum_{s,\mathbf{q}} \hbar\Omega_s(\mathbf{q}) \left[\tilde{b}^+_{s\mathbf{q}}\tilde{b}_{s\mathbf{q}} - |\Lambda(s, \mathbf{q})|^2 \hat{n} + \frac{1}{2}\right], \tag{1.29}$$

where

$$\hat{n} = \sum_{\mathbf{k}} \tilde{B}^+(\mathbf{k}) \tilde{B}(\mathbf{k})$$

is the exciton-number operator.

Further, using relations (1.28), we find that exciton operator (1.8) retains its form:

$$H_{\text{ex}} = \sum_{\mathbf{k}} E(\mathbf{k}) \tilde{B}^+(\mathbf{k}) \tilde{B}(\mathbf{k}). \tag{1.30}$$

Exciton–phonon interaction operator (1.18) is transformed to

$$\tilde{H}_{\text{ex-ph}} = \frac{1}{\sqrt{N}} \sum_{s, \mathbf{q}, \mathbf{k}} F_s(\mathbf{k}, \mathbf{q}) \tilde{B}^+(\mathbf{k} + \mathbf{q}) \tilde{B}(\mathbf{k}) [\tilde{\varphi}_{s\mathbf{q}} - 2\Lambda^*(s, \mathbf{q}) \hat{n}]. \tag{1.31}$$

Thus, complete Hamiltonian (1.23) of the system of interacting excitons and phonons, after conversion to the new creation and annihilation operators of the new elementary excitations, has the form

$$\widetilde{H} = \widetilde{H}_{ex} + \widetilde{H}_{ph} + \widetilde{H}_{ex\text{-}ph}, \tag{1.32}$$

where $\widetilde{H}_{ex}$ is the exciton operator, and $\widetilde{H}_{ph}$ is the operator of the new phonons, which characterize the molecular vibrations about the new equilibrium positions. This operator is diagonal in the new phonon creation and annihilation operators. It take into account the lattice deformation, i.e., the principal part of the interaction of excitons with the original phonons. The operator $\widetilde{H}_{ex\text{-}ph}$ takes into account the interaction of excitons with the new phonons.

To find the definite form of the displacements of the molecules to the new equilibrium positions, we must, using transformations (1.28), find an expression for operator (1.16) of the l-th component of the displacement of the **n**-th molecule in the new operators. The part of this operator

$$\Delta\widetilde{R}_{\mathbf{n}}^{l} = -\hat{n}\sum_{s,\mathbf{q}}\left(\frac{\hbar}{2I_l N\Omega_s(\mathbf{q})}\right)^{1/2} e_s^l(\mathbf{q})[e^{-i\mathbf{q}\mathbf{n}}\Lambda(s,\mathbf{q}) + \text{complex conj.}],$$

that does not contain the creation and annihilation operators of the new phonons characterizes such static displacements.

When the exciton–phonon coupling force is sufficiently great, not only the equilibrium positions of the molecules but also the frequencies of the normal vibrations are changed. To study the effect of the change in the normal-vibration frequencies, the quadratic displacement terms must be taken into account in expansion (1.6).

At the conclusion of this section, we shall calculate the explicit form of the exciton–phonon coupling functions for some elementary cases.

Let us consider functions (1.20), which appear in exciton–phonon interaction operator (1.18), for a model of a one-dimensional crystal consisting of N ($\gg 1$) identical anisotropic molecules situated along the z axis at a distance **a** from one another. Let the dipole moments **d** of the intramolecular electron transitions be directed along the equilibrium positions, which lie in the plane xz and make an angle ϑ with the z axis. The equilibrium positions of the centers of gravity of the molecules (in the absence of exciton excitation) are determined by the coordinates $\mathbf{n}_l = \mathbf{a}l$, where $l = 0, \pm 1, \pm 2, \ldots$. The instantaneous displacements of a molecule l from the equilibrium positions are denoted by x_l, y_l, z_l for translations, and by the angle α_l for rotations in the plane xz, and by the angle β_l, which determines the deviation of the axis of the dipole moment **d** from the plane xz. The matrix elements for resonance interaction between the molecules 0 and l in this model have the form

$$M_{0l} = \mathrm{d}^2 r_{0l}^{-3} \{\sin\beta_0 \sin\beta_l + \sin(\vartheta + \beta_0)\sin(\vartheta + \beta_l) - \\ - 2\cos(\vartheta + \alpha_0)\cos(\vartheta + \alpha_l)\}, \tag{1.33}$$

where

$$r_{0l} = [(al + z_l - z_0)^2 + (y_l - y_0)^2 + (x_l - x_0)^2]^{1/2}.$$

The nonzero derivatives of matrix elements (1.33) with respect to the six possible displacements of each molecule (for zero displacements) are

$$\left.\begin{aligned} \left\{\frac{\partial M_{0l}}{\partial z_l}\right\}_0 &= -\left\{\frac{\partial M_{0l}}{\partial z_0}\right\}_0 = -\frac{3lL}{4a\,|\,l\,|^5}, \\ \left\{\frac{\partial M_{0l}}{\partial \alpha_l}\right\}_0 &= \left\{\frac{\partial M_{0l}}{\partial \alpha_0}\right\}_0 = \frac{3L\sin(2\vartheta)}{4(1-3\cos^2\vartheta)}, \end{aligned}\right\} \tag{1.34}$$

where

$$L = \frac{4\mathrm{d}^2}{a^3}(1 - 3\cos^2\vartheta)$$

is equal in absolute value to the width of the exciton band. Thus, excitons in a one-dimensional crystal interact with only two (of the six) molecular-vibration branches: with the longitudinal acoustic vibrations and with the rotational vibrations lying in the plane that passes through the dipole moment of the quantum transition and through the axis of the crystal.

Let us assume that the dispersions of the acoustic and optical lattice-vibration branches are determined by the formulas

$$\Omega_{\mathrm{ac}}(\mathbf{q}) = v_0 \left|\sin\left(\frac{1}{2}\mathbf{qa}\right)\right|, \quad \Omega_{\mathrm{op}}(\mathbf{q}) = \Omega_0\left(1 + \frac{1}{2}\xi\cos\mathbf{qa}\right), \quad \mathbf{q}\parallel\mathbf{a}.$$

If we substitute (1.34) into (1.20), we obtain

$$F_{\mathrm{ac}}(\mathbf{k}, \mathbf{q}) = -i3L\frac{\beta}{a}\sum_{l\geqslant 1}\frac{\cos\left[\left(\mathbf{k} + \frac{1}{2}\mathbf{q}\right)al\right]\sin\left(\frac{1}{2}\mathbf{q}al\right)}{l^4\sqrt{\left|\sin\left(\frac{1}{2}\mathbf{qa}\right)\right|}}, \tag{1.35}$$

$$F_{\text{op}}(\mathbf{k},\mathbf{q}) = 3L\gamma \frac{\sin 2\vartheta}{1-3\cos^2\vartheta} \sum_{l\geqslant 1} \frac{\cos\left(\frac{1}{2}\,\mathbf{q}al\right)\cos\left[\left(\mathbf{k}+\frac{1}{2}\,\mathbf{q}\right)al\right]}{l^3\left[1+\frac{1}{2}\,\xi\cos(\mathbf{q}a)\right]^{1/2}}, \quad (1.36)$$

where $\beta^2 = \hbar(2I_z\nu_0)^{-1}$ is the mean square amplitude of the zero-point vibrations of a harmonic oscillator with mass coefficient I_z and frequency ν_0; and $\gamma^2 = \hbar(2I\Omega_0)^{-1}$ is the mean square amplitude of the zero-point angular vibrations of a pendulum with moment of inertia I and frequency Ω_0.

It follows from (1.35) that

$$\lim_{q\to 0} F_{\text{ac}}(\mathbf{k},\mathbf{q}) = 0.$$

If interaction only with the nearest neighbors is taken into account in (1.34) and (1.36), then

$$F_{\text{ac}}(\mathbf{k},\mathbf{q}) = -3iL\frac{\beta}{a}\,\frac{\cos\left[\left(\mathbf{k}+\frac{1}{2}\,\mathbf{q}\right)a\right]\sin\left(\frac{1}{2}\,\mathbf{q}a\right)}{\sqrt{\left|\sin\frac{1}{2}\,\mathbf{q}a\right|}}, \quad (1.37)$$

$$F_{\text{op}}(\mathbf{k},\mathbf{q}) = 3\gamma L\frac{\sin 2\vartheta}{1-3\cos^2\vartheta}\,\frac{\cos\left[\left(\mathbf{k}+\frac{\mathbf{q}}{2}\right)a\right]\cos\left(\frac{1}{2}\,\mathbf{q}a\right)}{\sqrt{1+\frac{1}{2}\,\xi\cos\mathbf{q}a}}. \quad (1.38)$$

The first calculations of functions (1.37) and (1.38) were made by the author [1] (see also [2]).

Let us calculate in explicit form the functions $\chi_s(\mathbf{q})$ (1.21) for the one-dimensional model. The values D_{0m} in (II.2.20) determine the change in the energy of van der Waals interaction of the molecules 0 and m when the former is excited. For our linear model ($m = al$) we can write

$$D_{0l} = -\mathcal{P}r_{0l}^{-6}, \quad (1.39)$$

where $\mathcal{P}$ is a positive value, which is a function of the orientation of the molecules and is proportional to the difference between their polarizabilities in the excited and ground states. If we substitute (1.39) into (1.21), we find, considering only the interaction of adjacent molecules,

$$\chi_{\rm ac}(\mathbf{q}) = i\,\frac{4\mathcal{P}}{a^6}\,\frac{\beta q}{a\,|\mathbf{q}|}\cos\left(\frac{1}{2}\,\mathbf{qa}\right)\sqrt{\left|\sin\left(\frac{1}{2}\,\mathbf{qa}\right)\right|}\,, \tag{1.40}$$

$$\chi_{\rm op}(\mathbf{q}) = 4C\gamma\,\frac{\cos^2\left(\frac{1}{2}\,\mathbf{qa}\right)}{\sqrt{1+\frac{1}{2}\,\xi\cos\mathbf{qa}}}\,, \tag{1.41}$$

where

$$C = -\frac{1}{a^6}\left\{\frac{\partial\mathcal{P}}{\partial\alpha_0}\right\}_0.$$

Calculation of coupling functions (1.20) in the general case of three-dimensional crystals involves great difficulties. We must, therefore, consider particular cases and introduce a number of simplifications. Let us consider orthorhombic, tetragonal, and cubic crystals with one molecule in a unit cell [2].

Let the dipole moment $\mathbf{d}$ of the exciton states be directed along one of the basis vectors $\mathbf{a}_j$. If this vector is $\mathbf{a}_2$, then, according to (II.2.3) and (7.27), the exciton energy is expressed by the formula

$$E(\mathbf{k}) = E_0 + \frac{4\pi\mathbf{d}^2}{\varepsilon_0 v}\cos^2\varphi + \frac{1}{2}\sum_{j=1}^{3} L_j\cos(\mathbf{ka}_j); \tag{1.42}$$

where ε_0 is the dielectric constant of the crystal (in the frequency range $\sim E_0$), which is determined by all of the electronic states except the state responsible for the exciton band $E(\mathbf{k})$; φ is the angle between the vectors $\mathbf{k}$ and $\mathbf{a}_2$; v is the volume of a unit cell; and

$$L_j \approx \frac{2\mathbf{d}^2}{\varepsilon_0 a_j^3}\{1 - 3\cos^2(\widehat{\mathbf{a}_2\mathbf{a}_j})\}$$

are coefficients that determine the widths of the exciton bands for the senses of $\mathbf{k}$ along the basis vectors $\mathbf{a}_j$.

Considering the comparatively fast convergence of the sums in expression (1.20), the function $F_s(\mathbf{k}, \mathbf{q})$ can be calculated with allowance for only the nearest-neighbor molecules in the crystal. Then, for the acoustic branches for the wave-vector senses $\mathbf{k}\,\|\,\mathbf{q}\,\|\,\mathbf{a}_j$ we have

$$F_{\rm ac}^{j}(\mathbf{k}, \mathbf{q}) = -3i\,\frac{\beta_j}{a_j}\,L_j\,\frac{\cos\left[\left(\mathbf{k}+\frac{1}{2}\,\mathbf{q}\right)\mathbf{a}_j\right]\sin\left(\frac{1}{2}\,\mathbf{qa}_j\right)}{\sqrt{\left|\sin\left(\frac{1}{2}\,\mathbf{qa}_j\right)\right|}}\,,$$

where

$$\beta_j^2 = \frac{\hbar}{2I_j\nu_{0j}}$$

is the mean square amplitude of zero-point molecular vibrations along the axis $\mathbf{a}_j$. When $\mathbf{k}\mathbf{a}_j \ll 1$, function (1.43) can be approximated by the simpler expression

$$F^j_{\text{ac}}(0, \mathbf{q}) = \mp 3iL_j \frac{\beta_i}{a_j}\left[1-\left(\frac{\mathbf{q}\mathbf{a}_j}{\pi}\right)^2\right]\sqrt{|\mathbf{q}\mathbf{a}_j|}. \tag{1.44}$$

Function (1.44) is equal to zero at the center and at the boundaries ($\mathbf{q}\mathbf{a}_j = \pm\pi$) of the first Brillouin zone for the phonon wave vectors. Owing to the vector nature of dipole excitons, their interaction with low-frequency phonons is not isotropic even in simple cubic lattices, because L_1, L_2, and L_3 in expressions (1.43) and (1.44) have different absolute values. This distinguishes the interaction of molecular excitons from the interaction of free electrons with low-frequency phonons, which is isotropic in simple cubic lattices. When the values of the vectors $\mathbf{k}$ and $\mathbf{q}$ are small, the interaction of electrons with low-frequency phonons is determined by the function (see, for example, Section 6 in [3])

$$F_{\text{ac}} = ig\sqrt{|\mathbf{q}|}.$$

Sometimes such an isotropic expression is used in approximate calculations (see, for example, [4]) and in studies of the interaction of excitons with low-frequency phonons.

Above, we stipulated that the dipole moment of molecular transitions in a unit cell formed with the basis vectors $\mathbf{a}_1$, $\mathbf{a}_2$, $\mathbf{a}_3$ the equilibrium angles $\vartheta_1 = \pi/2$, $\vartheta_2 = 0$, and $\vartheta_3 = \pi/2$, respectively. The three branches of molecular rotational vibrations in the molecular crystals in question are determined by the small variations ($\Delta\vartheta_j$) in these angles. The interaction of excitons of band (1.42) with the j-th rotational-vibration branch when the vectors $\mathbf{k}$ and $\mathbf{q}$ are directed along the basis vector $\mathbf{a}_j$ is expressed by the function

$$F^j_{\text{op}}(\mathbf{k}, \mathbf{q}) = 3\gamma_j B_j \frac{\cos\left[\left(\mathbf{k}+\frac{1}{2}\mathbf{q}\right)\mathbf{a}_j\right]\cos\left(\frac{1}{2}\mathbf{q}\mathbf{a}_j\right)}{\sqrt{1+\frac{1}{2}\xi_j\cos(\mathbf{q}\mathbf{a}_j)}}, \tag{1.45}$$

where

$$B_2 = 0, \quad B_j = \frac{16d^2}{\varepsilon_0 (a_j^2 + a_2^2)^{3/2}}, \qquad \text{if } j = 1, 3;$$

$\gamma^2_i = \hbar[2I_j\Omega_{0j}]^{-1}$ is the mean square value of the rotational deviations ϑ_j from the equilibrium positions; and ξ_j determine the widths (in limiting-frequency units) of the corresponding optical branches of molecular vibrations.

Usually, the dispersion of high-frequency phonons is small ($\xi_j \ll 1$). Under this condition and when $ka_j \ll 1$, function (1.45) can be approximated by the expression

$$F^j_{op}(0, \mathbf{q}) = 3\gamma_j B_j \left[1 - \left(\frac{qa_j}{\pi}\right)^2\right]^2. \tag{1.46}$$

Function (1.46) has its maximum value at the center of and is equal to zero at the boundaries ($\mathbf{q}a_j = \pm\pi$) of the first Brillouin zone. Therefore, molecular excitons mainly interact with limiting ($\mathbf{q} \approx 0$) high-frequency phonons.

It is useful to compare the obtained coupling functions for molecular excitons with phonons with the corresponding functions for electrons and Wannier excitons. The interaction of electrons with high-frequency phonons in isotropic crystals at small $\mathbf{q}$ is determined by the function

$$F_{op}(\mathbf{q}) = g\left(\frac{1}{\varepsilon_\infty} - \frac{1}{\varepsilon_0}\right)\sqrt{\frac{\Omega(\mathbf{q})}{|\mathbf{q}|}}, \tag{1.47}$$

where ε_∞ and ε_0 are the dielectric constants of the crystal at high and low frequencies.

For Wannier excitons, the exciton–phonon interaction operator is determined on condition that an electron and a hole interact with lattice vibrations independently. Under this assumption, the coupling functions for excitons with low- and high-frequency phonons have the form

$$F_{op}(\mathbf{q}) = g_{op}[w_e(\mathbf{q}) - w_h(\mathbf{q})]\left(\frac{1}{\varepsilon_\infty} - \frac{1}{\varepsilon_0}\right)\frac{1}{|\mathbf{q}|}, \tag{1.48}$$

$$F_{ac}(\mathbf{q}) = [c_e w_e(\mathbf{q}) - c_h w_h(\mathbf{q})]\sqrt{|\mathbf{q}|}, \tag{1.49}$$

where the functions $w_e(\mathbf{q})$ and $w_h(\mathbf{q})$, which determine the contributions to the interaction from electrons and holes, are Fourier transforms of their charge distribution in the corresponding state of internal motion; and c_e and c_h are constants, which, generally speaking, are not different from one another. At small $\mathbf{q}$ in the state 1s, the functions w have the form

$$w_i(\mathbf{q}) = (1 + \alpha_i \mathbf{q}^2)^{-2}, \quad i = e, h.$$

At small $\mathbf{q}$, therefore, for Wannier excitons

$$\left.\begin{aligned} F_{op}(\mathbf{q}) &\sim |\mathbf{q}|, \\ F_{ac}(\mathbf{q}) &\sim \sqrt{|\mathbf{q}|}. \end{aligned}\right\} \tag{1.50}$$

Functions (1.48) and (1.49) have been used by Ansel'm and Firsov [5] in a calculation of the free path of Wannier excitons and by Toyozawa [6] in a study of the form of the exciton absorption bands in ionic crystals.

The coupling functions of Wannier excitons with high-frequency phonons (1.50) approaches zero as $\mathbf{q} \to 0$ while the interaction of molecular excitons (1.45) remains different from zero. The substantial difference in the behavior of the coupling functions for high-frequency phonons with molecular excitons (1.40) and with Wannier excitons (1.50) is due to the difference in the nature of the optical vibration branches in molecular and ionic crystals. In molecular crystals, high-frequency phonons are associated with rotational vibrations of neutral anisotropic molecules. In ionic crystals, high-frequency phonons correspond to displacements of the positive ions relative to the negative.

2. The Green Function Method in the Theory of Excitons

In the previous section, it was shown that with weak coupling the Hamiltonian of a system of interacting excitons and phonons can be written as

$$H = H_{ex} + H_{ph} + H_{ex\text{-}ph}. \tag{2.1}$$

In this expression, the operators H_{ex}, H_{ph}, and $H_{ex\text{-}ph} = H^{(1)}_{ex\text{-}ph} + H^{(2)}_{ex\text{-}ph}$ coincide with operators (1.8), (1.15), (1.18), and (1.19), respectively. In the absence of exciton–phonon interaction, the elementary excitations of the crystal are characterized by a certain number of independent excitons and phonons. The exciton steady states are determined by the number of the exciton band μ, the energy $E_\mu(\mathbf{k})$, the wave vector $\mathbf{k}$, and the specific dipole moment of the transition $p_\mu(\mathbf{k})$. The phonons, in turn, are characterized by the number s of its vibration branch, the wave vector $\mathbf{q}$, the energy $\hbar\Omega_s(\mathbf{q})$, and the components of the polarization vector, i.e., the vector that indicates the direction of displacement of the molecules in the lattice.

Below, we shall consider only the case of weak coupling, when the inequality

$$H^{(1)}_{\text{ex-ph}} \gg H^{(2)}_{\text{ex-ph}},$$

is satisfied, and ignore the operator $H^{(2)}_{\text{ex-ph}}$. In the presence of exciton–phonon interaction described by the operator $H^{(1)}_{\text{ex-ph}}$, the exciton state $\{\mathbf{k}, \mathbf{E}_\mu(\mathbf{k}), p_\mu(\mathbf{k})\}$ is no longer steady. Although this operator does not change the number of excitons in the system, it causes exciton transitions to other states.

Below, for simplicity, we shall consider crystals that contain one neutral molecule in each unit cell and only the one lowest exciton band corresponding to electronic excitation of the molecules. In this case, the operator $H^{(1)}_{\text{ex-ph}}$ causes transitions only between the sublevels of this exciton band. Each exciton state is quasi-steady with a definite lifetime τ_Φ with respect to transition to other exciton states in the same band.

Along with the operator $H^{(1)}_{\text{ex-ph}}$, which does not change the number of excitons in the system, there are two more operators $V_{\text{ex-}\gamma}$ and T_{int}, which characterize exciton creation and annihilation in the crystal. The operator $V_{\text{ex-}\gamma}$ determines the interaction of excitons with photons – with the quanta of the electromagnetic field of the light waves. When light is absorbed, exciton states are produced in the crystal. The reverse transition corresponds to the emission of a photon with the disappearance of an exciton. The lifetime of an exciton (τ_V) with respect to the "luminescence" process we shall call the radiation lifetime. The second operator T_{int} determines the interactions that result in nonradiative conversions of excitons into a large number of phonons, i.e., transitions of the entire electronic excitation of the molecules into thermal vibrational energy in the lattice. If we let τ_T be the lifetime with respect to nonradiative transitions, then the total lifetime of an exciton (τ) is given by the relation

$$\frac{1}{\tau}=\frac{1}{\tau_\Phi}+\frac{1}{\tau_V}+\frac{1}{\tau_T}. \qquad (2.2)$$

If $\tau_T \ll \tau_V$ and τ_Φ the crystal does not luminesce and the energy absorbed by the crystal is converted to heat. A number of molecular crystals formed from aromatic molecules luminesce with a quantum yield close to unity. It has been found that in such crystals, during the exciton lifetime with respect to luminescence a quasi-thermal equilibrium is established, i.e., the excitons are statistically distributed over the sublevels of the lowest exciton band. Therefore, the partial exciton lifetimes satisfy the inequalities

$$\tau_T \gg \tau_V \gg \tau_\Phi.$$

When these inequalities are satisfied, we can ignore the operator T_{int} and consider the operator $V_{ex\text{-}ph}$ as a perturbation that determines the quantum transitions between the quasi-steady states of the crystal corresponding to the operator H. In this case, the excited states of a crystal in a thermostat with a particular temperature which contains n excitons (in the first exciton band) should be described not by the wave function but by the density matrix

$$\rho_0=\exp\{\beta[\Omega-\mathcal{H}]\}, \qquad (2.3)$$

where $1/\beta$ = kT is the temperature in energy units;

$$\mathcal{H}=H-\mu\hat{n}; \qquad (2.4)$$

H is the Hamiltonian operator (2.1), which is independent of time;

$$\hat{n}=\sum_{\mathbf{k}} B^+(\mathbf{k})\,B(\mathbf{k}) \qquad (2.5)$$

is the exciton-number operator, which commutes with the operator H; and μ is the exciton chemical potential, which is determined from the condition that the number of excitons is equal to their average number

$$n=\mathrm{Sp}\{\rho_0\hat{n}\}.$$

Here and below, the operation Sp indicates taking the sum of all diagonal matrix elements. Summation is over all possible states of the crystal. In Eq. (2.3), Ω denotes the thermodynamic potential of the crystal. It is determined from the normalization condition of density matrix (2.3) and equals

$$\Omega = -\frac{1}{\beta} \ln \mathrm{Sp} \{\mu \hat{n} - H\}.$$

It is easy to see that the phonon-number operator $\sum_{s,\mathbf{q}} b^+_{\mathbf{q}s} b_{\mathbf{q}s}$ does not commute with H. Hence, the number of phonons is not preserved and their chemical potential is zero.

The mean value of any operator F in the state defined by the density matrix ρ_0 we shall denote briefly by the symbol $\langle\langle F \rangle\rangle$, i.e.,

$$\langle\langle F \rangle\rangle \equiv \mathrm{Sp} \{\rho_0 F\}.$$

The interaction of a crystal with an external transverse "homogeneous" plane electromagnetic wave whose electric-field strength has the form

$$\left.\begin{aligned} &\mathbf{E}(\mathbf{r}, t) = \frac{1}{2} \mathbf{E}_0 \exp(i[\mathbf{Q}\mathbf{r} - \omega t]) + \text{complex conj.} \\ &(\mathbf{E}_0 \mathbf{Q}) = 0, \end{aligned}\right\} \quad (2.6)$$

is characterized by the complex auxiliary transverse tensor $\varepsilon^{\perp}(\omega, \mathbf{Q})$. As was shown in Section 3 of Chapter I, this tensor is defined by means of the equation

$$\langle \mathbf{P}(\mathbf{r}, t) \rangle = \frac{1}{2} \left\{ \frac{\varepsilon^{\perp}(\omega, \mathbf{Q}) - 1}{4\pi} E_0 \exp(i[\mathbf{Q}\mathbf{r} - \omega t]) + \text{Hermitian conj.} \right\} \quad (2.7)$$

the mean value of the specific electric moment produced in the crystal by transverse field (2.6).

The operator for interaction of a crystal with field (2.6) (in a linear approximation of the field) has in the coordinate representation the form

$$V(t) = -\frac{e}{mc} \sum_{\mathbf{n}} \{\mathbf{A}_0 \exp\{i[\mathbf{Q}(\mathbf{n} + \mathbf{r}_\mathbf{n}) - \omega t]\} \hat{\mathbf{p}}_\mathbf{n} + \text{Hermitian conj.}\},$$

where $\mathbf{A}_0 = -\frac{ic}{2\omega} \mathbf{E}_0$ is the amplitude of the vector potential in the Coulomb calibration; $\mathbf{r}_\mathbf{n}$ and $\hat{\mathbf{p}}_\mathbf{n}$ are the total operators of the coordinates and momenta of all electrons of the molecule n; and m and e are the electron mass and charge.

Using rule (III.1.10), let us transform operator (2.8) to the exciton occupation-number representation. Considering the ground $|0\rangle$ and only the first electronic excited $|f\rangle$ states and Eqs. (III.2.13) and (III.2.22), we find

$$V(t) = w \exp(-i\omega t + \eta t) + \text{Hermitian conj.} \tag{2.9}$$

where

$$w = \frac{ie\sqrt{N}}{2m\omega} \mathbf{E}_0 \{\langle f | e^{i\mathbf{Q}\mathbf{r}_n} \hat{\mathbf{p}}_n | 0\rangle B^+(\mathbf{Q}) + \langle 0 | e^{i\mathbf{Q}\mathbf{r}_n} \hat{\mathbf{p}}_n | f\rangle B(-\mathbf{Q})\}. \tag{2.9a}$$

In order to avoid the effect of the unsteady processes produced at the moment of interaction, we shall assume that interaction (2.9) was begun adiabatically in the infinite past. For this, into expression (2.9) we introduce the factor $\exp(\eta t)$, where η is a small positive quantity, which is assumed to be zero in the final formulas.

In the longwave approximation ($\mathbf{Qa} \ll 1$) (see [7], p. 333), we can write

$$\langle f | e^{i\mathbf{Q}\mathbf{r}_n} \hat{\mathbf{p}}_n | 0\rangle = im\omega_f \langle f | \mathbf{r}_n | 0\rangle,$$

where ω_f is the frequency of the intramolecular electron transition. In this case, operator (2.9a) is simplified:

$$w = -\frac{\omega_f \sqrt{N}}{2\omega} (\mathbf{E}_0 \mathbf{d}_f) [B^+(\mathbf{Q}) - B(-\mathbf{Q})], \tag{2.10}$$

where

$$\mathbf{d}_f = e \langle f | \mathbf{r}_n | 0\rangle$$

is the electric dipole moment of the molecular quantum transition. Interaction (2.9) changes density matrix (2.3). In the interaction representation, this change is characterized by the equation[1]

$$i \frac{\partial \tilde{\rho}(t)}{\partial t} = [\tilde{V}(t), \tilde{\rho}(t)] \tag{2.11}$$

under the initial conditions

$$[\tilde{\rho}(t)]_{t=-\infty} = \rho_0.$$

[1]Here and below in this section, we shall use a system of units in which $\hbar = 1$.

On the right side of Eq. (2.11), the operators are in the Heisenberg representation

$$\left.\begin{aligned}\tilde{\rho}(t) &= e^{iHt}\rho(t)e^{-iHt},\\ \tilde{V}(t) &= w(t)\exp\{-i\omega t+\eta t\} + \text{Hermitian conj.}\end{aligned}\right\} \quad (2.12)$$

where

$$w(t) = -\frac{\omega_f\sqrt{N}}{2\omega}(\mathbf{E}_0\mathbf{d}_f)\left[B^+(\mathbf{Q}, t) - B(-\mathbf{Q}, t)\right], \quad (2.13)$$

$$B(\mathbf{Q}, t) = e^{iHt}B(\mathbf{Q})e^{-iHt}. \quad (2.14)$$

If we solve Eq. (2.11) in a linear approximation of the perturbation operator, we find (see [8]) a density matrix that determines the exciton states of the crystal with allowance for the interaction of excitons with the electromagnetic wave:

$$\tilde{\rho}(t) = \rho_0 - i\int_{-\infty}^{t}[\tilde{V}(\tau), \rho_0]\,d\tau. \quad (2.15)$$

With the aid of density matrix (2.15), the mean value of the specific moment at point $\mathbf{n}$ in the crystal is expressed by the formula

$$\langle\mathbf{P}(\mathbf{n}, t)\rangle_f = \mathrm{Sp}\{\tilde{\rho}(t)\mathbf{P}(\mathbf{n}, t)\}, \quad (2.16)$$

where

$$\mathbf{P}(\mathbf{n}, t) = \frac{\mathbf{d}}{v\sqrt{N}}\sum_{\mathbf{k}}\{B^+(-\mathbf{k}, t) + B(\mathbf{k}, t)\}e^{i\mathbf{kn}} \quad (2.17)$$

is the operator of the specific electric moment of the crystal in the exciton occupation-number representation and in the interaction representation. Operator (2.17) is obtained by general rules from the operator of the specific electric moment, which in the coordinate representation has the form

$$\hat{\mathbf{P}}_{\mathbf{n}} = \frac{1}{v}e\mathbf{r}_{\mathbf{n}},$$

where v is the volume of a unit cell.

If we substitute expressions (2.15) and (2.17) into Eq. (2.16) and assume that the mean specific electric dipole moment of the crystal without an external field, which is determined by the expression $\mathrm{Sp}\{\rho_0 \mathbf{P}(\mathbf{n}, t)\}$, is equal to zero, we obtain after averaging about the point $\mathbf{n}$

$$\langle \mathbf{P}(\mathbf{n}, t)\rangle_f = \frac{i\mathbf{d}(\mathbf{E}_0\mathbf{d})\,\omega_f}{2v\omega}\int_{-\infty}^{t}\{\langle\!\langle [B(\mathbf{Q}, t), B^+(\mathbf{Q}, \tau)]\rangle\!\rangle -$$

$$-\langle\!\langle [B^+(-\mathbf{Q}, t), B(-\mathbf{Q}, \tau)]\rangle\!\rangle\}\exp\{i[\mathbf{Q}\mathbf{n} - \omega\tau] + \eta\tau\}\,d\tau + \text{Hermitian conj.}, \tag{2.18}$$

where

$$\langle\!\langle [B(\mathbf{Q}, t), B^+(\mathbf{Q}, \tau)]\rangle\!\rangle \equiv \mathrm{Sp}\{\rho_0 [B(\mathbf{Q}, t), B^+(\mathbf{Q}, \tau)]\}.$$

If we introduce under the integral in this expression the step function

$$\Theta(t-\tau) = \begin{cases} 1, & \text{if } t > \tau, \\ 0, & \text{if } t < \tau, \end{cases}$$

the upper limit of integration can be replaced by infinity. In this case, under the integral will be expressions that reduce to two-time (temperature) retarded Green functions for excitons that interact with phonons. In fact, using the definitions[2] of two-time retarded Green functions (see [8])

$$\left.\begin{aligned} G_r(\mathbf{Q}, t-\tau) &= -i\Theta(t-\tau)\langle\!\langle [B(\mathbf{Q}, t), B^+(\mathbf{Q}, \tau)]\rangle\!\rangle, \\ G_r^+(\mathbf{Q}, t-\tau) &= -i\Theta(t-\tau)\langle\!\langle [B^+(\mathbf{Q}, t), B(\mathbf{Q}, \tau)]\rangle\!\rangle, \end{aligned}\right\} \tag{2.19}$$

we can rewrite expression (2.18) as

$$\langle \mathbf{P}(\mathbf{n}, t)\rangle_f = -\frac{\mathbf{d}(\mathbf{d}\mathbf{E}_0)\,\omega_f}{2v\omega}\exp(i[\mathbf{Q}\mathbf{n} - \omega t])\times$$

$$\times\int_{-\infty}^{\infty}\{G_r(\mathbf{Q}, \tau) - G_r^+(-\mathbf{Q}, \tau)\}\,e^{i\omega\tau}\,d\tau + \text{Hermitian conj.}$$

Then, introducing the Fourier components of the time retarded Green functions by means of the relation

$$G_r(\mathbf{Q}, \omega) = \int_{-\infty}^{\infty} G_r(\mathbf{Q}, \tau)\,e^{i\omega\tau}\,d\tau,$$

[2]The Green functions are defined by expressions (2.19) only when $t \neq \tau$, since the latter contain the discontinuous function $\Theta(t-\tau)$. When $t = \tau$, the Green function is defined by an additional condition, which will be considered below in this section.

we obtain a final expression for the mean value of the specific electric dipole moment due to interaction of the field of a transverse electromagnetic wave with excitons:

$$\langle \mathbf{P}(\mathbf{n}, t)\rangle_f = -\frac{\mathbf{d}(\mathbf{E}_0\mathbf{d})\,\omega_f}{2v\omega}[G_r(\mathbf{Q}, \omega) - G_r^+(-\mathbf{Q}, -\omega)]\,e^{i(\mathbf{Qn}-\omega t)} + $$
$$+ \text{ Hermitian conj.} \tag{2.20}$$

Formula (2.20) takes into account only one (f-th) excited state of a molecule. The mean total dipole moment of the crystal can be written as

$$\langle \mathbf{P}(\mathbf{n}, t)\rangle = \langle \mathbf{P}(\mathbf{n}, t)\rangle_f + \frac{1}{2}\beta_0^{\perp}\mathbf{E}_0\,e^{i(\mathbf{Qn}-\omega t)} + \text{ Hermitian conj.},$$

where $\beta_0^{\perp}$ is a tensor that takes into account the effect of all the remaining excited states of the molecules.

If we compare this expression with (2.7), taking into account (2.20), we find in the coordinate system of the wave vector (see Chapter I, Section 3) the relationship between the components of the two-dimensional tensor $\varepsilon^{\perp}(\omega, \mathbf{Q})$ and the Fourier components of the time retarded Green functions for excitons that interact with phonons:

$$\varepsilon_{xy}^{\perp} - \delta_{xy} = 4\pi\beta_{xy,0}^{\perp} - \frac{4\pi d^x d^y \omega_f}{v\omega}\{G_r(\mathbf{Q}, \omega) - G_r^+(-\mathbf{Q}, -\omega)\}. \tag{2.21}$$

Methods of calculating retarded exciton Green functions (2.19) will be discussed in the following sections. We shall see that the exciton Green function $G_r(\mathbf{Q}, \omega)$ has a resonance nature in the frequency range associated with exciton excitations. In this frequency range, the function $G_r^+(-\mathbf{Q}, -\omega)$ is a very small in absolute value smooth function of ω. In this same frequency range, $\beta_{xy,0}^{\perp}$ is also a smooth function of frequency if the energy of the molecular excitation ω_f differs from the energies of the other excited states by a value that exceeds the width of the exciton band.

Now let us calculate the probabilities of transitions, induced by light wave (2.6), per unit time that result in a change in the number of excitons in the crystal by one with the simultaneous emission and absorption of photons. Let $|l, n_k\rangle$ be the eigenfunctions of operators (2.1) and (2.5), i.e.,

$$\{H - E(l, n_{\mathbf{k}})\}\,|l, n_{\mathbf{k}}\rangle = 0; \quad \{\hat{n} - n_{\mathbf{k}}\}\,|l, n_{\mathbf{k}}\rangle = 0.$$

Here, n_k is the number of excitons with the wave vector **k** in the first exciton band; and l are all of the remaining quantum numbers that characterize the states of the crystal, particularly the lattice-vibration states.

Let us consider the part of operator (2.9) with the form

$$-\frac{\omega_f}{2\omega}(E_0 d)\sqrt{N}\,B^+(Q)\,e^{-i\omega t}. \tag{2.22}$$

Operator (2.22) causes processes of exciton creation at the expense of photon energy. In this case, the energy of the final state $E(l', n+1)$ of a crystal containing n + 1 excitons with momenta **Q** must be equal to the sum of the photon energy ω and the initial energy of the crystal $E(l, n)$, i.e.,

$$E(l', n+1) = E(l, n) + \omega. \tag{2.23}$$

The probability of such a process per unit time averaged over the initial states $|\,l, n\,\rangle$ and summed over the possible final states $|\,l', n+1\,\rangle$ has the form

$$\frac{dw_{+1}^{-\gamma}}{dt} = 2\pi\left(\frac{\omega_f}{\omega}\right)^2\left(\frac{E_0 d}{2}\right)^2 N\times$$

$$\times\sum_{l,l',n}\rho_0(l,n)\left|\langle l', n+1\,|\,B^+(Q)\,|\,l, n\rangle\right|^2\delta\,[E(l', n+1)-E(l, n)-\omega], \tag{2.24}$$

where

$$\rho_0(l, n) = \exp\{\beta\,[\Omega - E(l, n) + n\mu]\}.$$

The reverse transitions are caused by the part of operator (2.9) that has the form

$$-\frac{\omega_f}{2\omega}(E_0 d)\sqrt{N}\,B(Q)\,e^{i\omega t}. \tag{2.25}$$

The probability of such transitions (emission of a photon and annihilation of an exciton) per unit time is determined by

$$\frac{dw_{-1}^{+\gamma}}{dt} = 2\pi N\left(\frac{\omega_f E_0 d}{2\omega}\right)^2\times$$

$$\times\sum_{l,l',n}\rho_0(l', n+1)\left|\langle\, l, n\,|\,B(Q)\,|\,l', n+1\,\rangle\right|^2\times$$

$$\times\,\delta\,[E(l', n+1) - E(l, n) - \omega], \tag{2.26}$$

where

$$\rho_0(l', n+1) = \exp\{\beta[\Omega - E(l', n+1) + \mu(n+1)]\}.$$

If we compare expressions (2.24) and (2.26), taking into account (2.23), we obtain the simple relation

$$\frac{dw_{+1}^{-\gamma}}{dt} = \frac{dw_{-1}^{+\gamma}}{dt} \exp\{\beta[\omega - \mu]\}. \qquad (2.27)$$

Besides operators (2.22) and (2.25), operator (2.9) contains two other operators:

$$\frac{\omega_f}{2\omega}(\mathbf{E}_0\mathbf{d}) B^+(-\mathbf{Q}) e^{i\omega t} + \text{Hermitian conj.} \qquad (2.28)$$

The first of these causes the simultaneous creation of an exciton and a photon at the expense of the thermal energy of the crystal. By comparing this operator with operator (2.22), we obtain the probability of such processes per unit time $\frac{dw_{+1}^{+\gamma}(\mathbf{Q}, \omega)}{dt}$ by simple transformation of expression (2.24):

$$\frac{dw_{+1}^{+\gamma}(\mathbf{Q}, \omega)}{dt} = \frac{dw_{+1}^{-\gamma}(-\mathbf{Q}, -\omega)}{dt}. \qquad (2.29)$$

The operator corresponding to the second term in (2.28) causes the simultaneous conversion of an exciton and a photon to a large number of phonons. The probability of such processes $\frac{dw_{-1}^{-\gamma}(\mathbf{Q}, \omega)}{dt}$ per unit time is linked with expression (2.26) by the simple relation

$$\frac{dw_{-1}^{-\gamma}(\mathbf{Q}, \omega)}{dt} = \frac{dw_{-1}^{+\gamma}(-\mathbf{Q}, -\omega)}{dt}. \qquad (2.30)$$

The obtained probabilities of quantum transitions per unit time (2.24), (2.26), (2.29), and (2.30) are expressed in terms of the Fourier components of the time correlation functions of the exciton operators. The *time correlation functions of the exciton operators* are the mean values over a Gibbs grand canonical ensemble of the products of the exciton operators. Let us consider two types of exciton-operator time correlation functions

$$G_<(\mathbf{Q}, t-\tau) = \langle\langle B^+(\mathbf{Q}, \tau) B(\mathbf{Q}, t)\rangle\rangle, \tag{2.31}$$

$$G_>(\mathbf{Q}, t-\tau) = \langle\langle B(\mathbf{Q}, t) B^+(\mathbf{Q}, \tau)\rangle\rangle. \tag{2.32}$$

When $t = \tau$, time correlation function (2.31) determines the average number of excitons in the state with momentum **Q**:

$$\bar{n}_{\mathbf{Q}} = \langle\langle B^+(\mathbf{Q}, t) B(\mathbf{Q}, t)\rangle\rangle = \{G_<(\mathbf{Q}, t-\tau)\}_{\tau=t}.$$

Using the total set of eigenfunctions $|\, l, n\rangle$ of operators (2.1) and (2.5) and expression (2.14), function (2.31) can be written as

$$G_<(\mathbf{Q}, t) = \sum_{l, l', n} \rho_0(l', n+1)\, |\langle l, n | B(\mathbf{Q}) | l', n+1\rangle|^2 \times$$
$$\times \exp\{i[E(l', n+1) - E(l, n)]\, t\}.$$

Moving to the Fourier components $G_<(\mathbf{Q}, \omega)$ by means of the transformation

$$G_<(\mathbf{Q}, t) = \frac{1}{2\pi}\int_{-\infty}^{\infty} e^{-i\omega t} G_<(\mathbf{Q}, \omega)\, d\omega, \tag{2.33}$$

we find

$$G_<(\mathbf{Q}, \omega) = 2\pi \sum_{l, l', n} \rho_0(l', n+1)\, \Big|\langle l, n | B(\mathbf{Q}) | l', n+1\rangle\Big|^2 \times$$
$$\times \delta[E(l', n+1) - E(l, n) - \omega].$$

If we compare this expression with induced-transition probability (2.26), we obtain

$$\frac{dw_{-1}^{+\gamma}(\mathbf{Q}, \omega)}{dt} = \frac{1}{4} N \left(\frac{\omega_f E_0 d}{\omega}\right)^2 G_<(\mathbf{Q}, \omega). \tag{2.34}$$

If we make similar transformations with correlation function (2.32) and compare it with probability (2.24), we have

$$\frac{dw_{+1}^{-\gamma}(\mathbf{Q}, \omega)}{dt} = \frac{N}{4} \left(\frac{\omega_f E_0 d}{\omega}\right)^2 G_>(\mathbf{Q}, \omega). \tag{2.35}$$

Further, as a consequence of Eqs. (2.29), (2.30), (2.34), and (2.35), we have

$$\frac{dw_{+1}^{+\gamma}(\mathbf{Q},\omega)}{dt}=\frac{N}{4}\left(\frac{\omega_f E_0 d}{\omega}\right)^2 G_>(-\mathbf{Q},-\omega), \tag{2.36}$$

$$\frac{dw_{-1}^{-\gamma}(\mathbf{Q},\omega)}{dt}=\frac{N}{4}\left(\frac{\omega_f E_0 d}{\omega}\right)^2 G_<(-\mathbf{Q},-\omega). \tag{2.37}$$

From the definitions of functions (2.19), (2.31), and (2.32) follows a relationship between the retarded Green functions and the time correlation functions:

$$G_r(\mathbf{Q},t)=-i\Theta(t)\{G_>(\mathbf{Q},t)-G_<(\mathbf{Q},t)\}. \tag{2.38}$$

Expression (2.38) contains the discontinuous function $\Theta(t)$. Therefore, it defines the retarded Green function only when $t \neq 0$ (accordingly, Eqs. (2.19) define the retarded functions when $t \neq \tau$). The Green function's definition must be supplemented at the point $t = 0$. Such a supplementary definition of a Green function is usually made by indicating a rule for calculating the time integrals that contain the Green functions. We shall show how this is done by the example of conversion to the energy representation by means of the equation

$$G_r(\mathbf{Q},E)=\int_{-\infty}^{\infty} G_r(\mathbf{Q},t)e^{iEt}\,dt. \tag{2.39}$$

Using relation (2.33), we convert on the right of (2.38) to the Fourier components of the correlation function and substitute the obtained expression $G_r(\mathbf{Q}, t)$ into integral (2.39). Then we have

$$G_r(\mathbf{Q},E)=-\frac{i}{2\pi}\int_{-\infty}^{\infty} d\omega\int_{-\infty}^{\infty} dt\, e^{i(E-\omega)t}\Theta(t)\{G_>(\mathbf{Q},\omega)-G_<(\mathbf{Q},\omega)\}.$$

As a supplementary definition of the Green function at the point $t = 0$ let us take the following rule for calculating the time integral:

$$\int_{-\infty}^{\infty}\Theta(t)e^{i(E-\omega)t}\,dt=\lim_{\eta\to+0}\int_{-\infty}^{\infty}\Theta(t)e^{i(E-\omega)t-\eta t}\,dt=\lim_{\eta\to+0}\left(\frac{i}{E-\omega+i\eta}\right).$$

Using this rule,[3] we find

$$G_r(\mathbf{Q}, E) = \lim_{\eta \to +0} \frac{1}{2\pi} \int_{-\infty}^{\infty} \frac{\{G_>(\mathbf{Q}, \omega) - G_<(\mathbf{Q}, \omega)\}}{E - \omega + i\eta} d\omega. \tag{2.39a}$$

The small positive value η in integral (2.39a) determines the rule of circumvention of the pole $E = \omega$. After calculating the integral, we should pass to the limit $\eta \to +0$. The sign of passage to the limit is usually not written in explicit form.

From Eq. (2.27) and expressions (2.34) and (2.35) follows a simple relationship between the Fourier components of the exciton-operator correlation functions:

$$G_>(\mathbf{Q}, \omega) = G_<(\mathbf{Q}, \omega) e^{\beta(\omega-\mu)}. \tag{2.40}$$

Therefore, expression (2.39a) can also be written as

$$G_r(\mathbf{Q}, E) = \frac{1}{2\pi} \int_{-\infty}^{\infty} \frac{\{e^{\beta(\omega-\mu)} - 1\} G_<(\mathbf{Q}, \omega)\, d\omega}{E - \omega + i\eta}. \tag{2.41}$$

Since $G_<(\mathbf{Q}, \omega)$ is a real function of ω, it follows from (2.41) that $G_r(\mathbf{Q}, E)$ as a function of the complex variable $Q_0 = E + i\gamma$ does not have singularities in the upper half-plane of this variable for any $\gamma \geq 0$. Therefore, the retarded Green function $G_r(\mathbf{Q}, E)$ can be analytically continued from the real axis of the variable E to all of the complex values $Q_0 = E + i\gamma$ $(\gamma \geq 0)$ lying in the upper

[3]This supplementary definition of the retarded Green function, introduced by converting from the real frequency ω to the complex frequency $\omega - i\eta$, reflects the objective properties of physical systems. With the aid of the retarded Green function, we calculate the steady-state reaction of a macroscopic system to an external influence of given frequency ω. In particular, dielectric constant (2.21), which determines the forced steady-state reaction of a macroscopic body to an external macroscopic electromagnetic field, is calculated in this way. In ideal systems without attenuation, it is impossible in principle to operate with the concept of a forced steady-state reaction. Attenuation always exists in real systems. Even with extremely weak attenuation, after a sufficiently long period of time the natural excitations in a system are attenuated and only forced excitations remain. Replacing ω by $\omega - i\eta$ takes into account the attenuation that always exists in any real system. Thus, the small value $-i\eta$ must be included in the frequency ω in intermediate calculations even when studying "ideal systems" without attenuation.

half-plane. From this [and also directly from definition (2.19)] it follows that the time Green function $G_r(\mathbf{Q}, t)$ itself is identically equal to zero for all negative t.

Expressions such as (2.41) were first obtained by Lehmann [9] for the Green functions of quantum electrodynamics: they are called spectral representations of the retarded Green function. Here,

$$\rho(\mathbf{Q}, \omega) = G_>(\mathbf{Q}, \omega) - G_<(\mathbf{Q}, \omega) = (e^{\beta(\omega-\mu)} - 1)\, G_<(\mathbf{Q}, \omega) \qquad (2.42)$$

may be called the density (intensity) of the spectral representation. The spectral representations permit calculation of the correlation functions if the delayed functions are known, and vice versa. In particular, using the realness of the function $G_<(\mathbf{Q}, E)$ and the symbolic identity

$$(x + i\eta)^{-1} = \mathcal{P}x^{-1} - i\pi\delta(x), \qquad \eta \to +0, \qquad (2.43)$$

from Eq. (2.41) we find the imaginary and real parts of the retarded Green function:

$$\left.\begin{aligned} \operatorname{Im} G_r(\mathbf{Q}, E) &= -\frac{1}{2}\{e^{\beta(E-\mu)} - 1\}\, G_<(\mathbf{Q}, E) = -\frac{1}{2}\rho(\mathbf{Q}, E), \\ \operatorname{Im} G_r^+(\mathbf{Q}, -E) &= \frac{1}{2}\{e^{-\beta(E+\mu)} - 1\}\, G_<(\mathbf{Q}, -E) = \frac{1}{2}\rho(\mathbf{Q}, -E), \end{aligned}\right\} \qquad (2.44a)$$

$$\operatorname{Re} G_r(\mathbf{Q}, E) = \frac{\mathcal{P}}{2\pi}\int_{-\infty}^{\infty} \frac{\{e^{\beta(\omega-\mu)} - 1\}\, G_<(\mathbf{Q}, \omega)\, d\omega}{E - \omega}, \qquad (2.44b)$$

where $\mathcal{P}$ indicates that the integral is calculated in the sense of the principal value.

If we substitute into the integrand of Eq. (2.44b) the density of the spectral representation $\rho(\mathbf{Q}, \omega)$ from (2.44a), we find a relationship between the imaginary and real parts of the retarded Green function:

$$\operatorname{Re} G_r(\mathbf{Q}, E) = \frac{\mathcal{P}}{\pi}\int_{-\infty}^{\infty} \frac{\operatorname{Im} G_r(\mathbf{Q}, \omega)\, d\omega}{\omega - E}. \qquad (2.45)$$

Spectral-representation density (2.42) of the retarded Green functions satisfies the very important integral relation

$$\frac{1}{2\pi}\int_{-\infty}^{\infty} \rho(\mathbf{Q}, E)\, dE = 1. \qquad (2.46)$$

To prove Eq. (2.46), let us consider the transformation inverse to (2.39)

$$G_r(\mathbf{Q}, t) = \frac{1}{2\pi} \int_{-\infty}^{\infty} G_r(\mathbf{Q}, E) e^{-iEt} dE. \qquad (2.47)$$

According to definition (2.19)

$$\lim_{t \to +0} G_r(\mathbf{Q}, t) = -i. \qquad (2.48)$$

Therefore, passage to the limit on both sides of (2.47) at $t \to +0$ gives

$$i = -\lim_{t \to +0} \frac{1}{2\pi} \int_{-\infty}^{\infty} G_r(\mathbf{Q}, E) e^{-iEt} dE.$$

After substituting (2.41) into the right side of this equation, taking (2.42) into account, we find

$$i = \lim_{\eta, t \to +0} \frac{1}{(2\pi)^2} \int_{-\infty}^{\infty} \frac{\rho(\mathbf{Q}, \omega) e^{-iEt} dE\, d\omega}{E - \omega + i\eta}.$$

Then, using the equation

$$\lim_{\eta \to +0} \frac{1}{2\pi} \left(\frac{1}{x + i\eta} - \frac{1}{x - i\eta} \right) = -i\delta(x),$$

we obtain integral relation (2.46).

With passage of light wave (2.6) in a crystal, the energy absorbed in a unit volume per unit time is, according to (2.34)-(2.37), determined by the expression

$$u(\mathbf{Q}, \omega) = \frac{\omega}{vN} \left\{ \frac{dw_{+1}^{-\gamma}}{dt} - \frac{dw_{-1}^{+\gamma}}{dt} + \frac{dw_{-1}^{-\gamma}}{dt} - \frac{dw_{+1}^{+\gamma}}{dt} \right\} =$$

$$= \frac{(\omega_f \mathbf{E}_0 \mathbf{d})^2}{4v\omega} \{ [e^{\beta(\omega-\mu)} - 1] G_<(\mathbf{Q}, \omega) + [e^{-\beta(\omega+\mu)} - 1] G_<(-\mathbf{Q}, -\omega) \}.$$

With the aid of (2.44a), this expression can be transformed to

$$u(\mathbf{Q}, \omega) = -\frac{(\omega_f \mathbf{E}_0 \cdot \mathbf{d})^2}{2v\omega} \operatorname{Im} \{ G_r(\mathbf{Q}, \omega) - G_r^+(-\mathbf{Q}, -\omega) \}.$$

Let us assume that **E** is directed along the x axis in the coordinate system of the wave vector **Q**, then $\mathbf{E}_0 \cdot \mathbf{d} = E_0 d^x$. In this case, if we compare (2.49) and (2.21) and assume that Im $\beta_0^{\perp} = 0$ in the frequency range $\omega \sim \omega_f$, we find

$$u(\mathbf{Q}, \omega) = \frac{\omega E_0^2}{8\pi} \operatorname{Im} \varepsilon_{xx}^{\perp}. \qquad (2.50)$$

Therefore, the energy absorption by the crystal is determined by the imaginary part of the tensor $\varepsilon_{xx}^{\perp}$.

The mean value of the Poynting vector, i.e., the energy flux density oflight wave (2.6) along the wave vector, is determined by the expression (see [10], Sections 61-63)

$$S_z = \frac{cE_0^2 \sqrt{\operatorname{Re} \varepsilon_{xx}^{\perp}}}{8\pi} \quad \text{at} \quad \operatorname{Re} \varepsilon_{xx}^{\perp} > 0.$$

The energy absorption per unit volume per one second must satisfy the equation $u = -dS_z/dz$. Considering (2.50), we can write

$$-\frac{dS_z}{dz} = \frac{\omega}{c\sqrt{\operatorname{Re} \varepsilon_{xx}^{\perp}}} S_z \operatorname{Im} \varepsilon_{xx}^{\perp}.$$

Solution of this equation gives the energy flux density of the light wave for passage of distance z:

$$S_z(z) = S_z(0) \exp\left\{-\frac{\omega}{c\sqrt{\operatorname{Re} \varepsilon_{xx}^{\perp}}} z \operatorname{Im} \varepsilon_{xx}^{\perp}\right\}. \qquad (2.51)$$

The coefficient $\dfrac{\omega \operatorname{Im} \varepsilon_{xx}^{\perp}}{c\sqrt{\operatorname{Re} \varepsilon_{xx}^{\perp}}}$, which has the dimensionality cm^{-1}, determines the relative decrease in energy flux per unit distance in the direction of propagation. Considering (I.3.5), we can write

$$\frac{\omega \operatorname{Im} \varepsilon_{xx}^{\perp}}{c\sqrt{\operatorname{Re} \varepsilon_{xx}^{\perp}}} = \frac{2\omega \varkappa_x}{c},$$

where $\varkappa_x$ is the dimensionless attenuation factor of a plane electromagnetic wave that is propagated along the z axis and whose electric-field strength is parallel to the x axis.

Below, in Section 4, it will be shown that in the frequency range $\omega = E(\mathbf{Q})$ the inequality

$$G_<(\mathbf{Q}, \omega) \gg G_<(\pm \mathbf{Q}, -\omega)$$

is satisfied. For light-wave induced transitions, therefore, we can take into account only transitions (2.34) and (2.35). Then we obtain the approximate expression

$$u(\mathbf{Q}, \omega) \approx \frac{\omega_f^2 (\mathbf{E}_0 \cdot \mathbf{d})^2}{4v\omega} [e^{\beta(\omega-\mu)} - 1] G_<(\mathbf{Q}, \omega). \tag{2.52}$$

The function $G_<(\mathbf{Q}, \omega)$ is always positive; hence, the sign of (2.52) is a function of the sign of the difference $\omega - \mu$. When $\omega > \mu$, $u(\mathbf{Q}, \omega)$ is positive, i.e., the crystal absorbs the energy of the light wave. When $\omega < \mu$, $u(\mathbf{Q}, \omega)$ is negative and the energy of the crystal is imparted to the light wave.

In concluding this section, let us estimate the chemical potential μ for a cubic crystal containing one molecule in each unit cell with edge a. With weak exciton-phonon coupling, the average number of excitons with energy $E(\mathbf{k})$ and momentum $\mathbf{k}$ is

$$\bar{n}_{\mathbf{k}} = \{\exp[\beta(E(\mathbf{k}) - \mu] - 1\}^{-1}.$$

Let n_{ex} be the density of all excitons. Then their chemical potential is determined by

$$n_{ex} = \frac{1}{V} \sum_{\mathbf{k}} \bar{n}_{\mathbf{k}}.$$

Moving from the sum to the integral, we have

$$n_{ex} = \frac{1}{(2\pi)^2} \int_0^{\pi/a} \frac{k^2 dk}{\exp\{\beta[E(\mathbf{k}) - \mu]\} - 1} \cdot \tag{2.53}$$

If the minimum of $E(\mathbf{k})$ corresponds to $E(0)$, then in the effective-mass approximation (see Chapter II, Section 2)

$$E(\mathbf{k}) = E(0) + \frac{\hbar^2 \mathbf{k}^2}{2m^*} \quad \text{at} \quad ka \ll 1. \tag{2.54}$$

If we substitute this value into (2.53), we see that the principal contribution to the integral is made by the range

$$\frac{\hbar^2 k^2 \beta}{2m^*} \leqslant 1.$$

If we use approximation formula (II.2.20), which expresses the exciton effective mass in terms of the width ΔL of the exciton band, the latter inequality takes the form

$$k^2a^2 \leqslant \frac{4}{\beta\Delta L}. \qquad (2.55)$$

Thus, when $\beta\Delta L \gg 1$, we can use expansion (2.54) in calculating integral (2.53). If we convert to the dimensionless variable

$$x = \frac{\hbar k\sqrt{\beta}}{\sqrt{2m^*}}$$

and extend the upper limit in (2.53) to infinity, we have

$$2\pi^2\left(\frac{\hbar^2\beta}{2m^*}\right)^{3/2} n_{ex} = \int_0^\infty \frac{x^2\,dx}{Ae^x - 1}, \qquad (2.56)$$

where

$$A = \exp\{\beta[E(0) - \mu]\}.$$

By solving transcendental equation (2.56) we can calculate A [or $E(0) - \mu$] as a function of β, m^*, and n_{ex}. Equation (2.56) has solutions only for $A \geq 1$, i.e., if

$$E(0) - \mu \geqslant 0.$$

Consequently, for a positive exciton effective mass, the exciton chemical potential is always below the bottom of the exciton band. The limiting value $A = 1$ corresponds to the exciton concentrations

$$n_{ex}^0 \approx 0.081\left(\frac{2m^*}{\pi\hbar^2\beta}\right)^{3/2}. \qquad (2.57)$$

If we express the exciton effective mass in terms of the width of the exciton band, we can transform this expression to

$$n_{ex}^0 \approx \frac{0.12}{a^3}(\beta\Delta L)^{-3/2}. \qquad (2.58)$$

It follows from (2.58) that if $a = 10^{-7}$ cm and $\beta = 100/L$, then $n_{ex}^0 \approx 10^{17}$ cm^{-3}.

When $n_{ex} > n_{ex}^0$, the formulas obtained above lose their meaning, since Bose condensation must set in at these exciton densities [11-13]. In particular, our formulas are inapplicable (moving from summation to integration in (2.53) is invalid) at absolute zero, since when $1/\beta = 0$, $n_{ex}^0 = 0$ and the inequality $n_{ex} > n_{ex}^0$ is satisfied for any number of excitons.

The value A and the difference $E(0) - \mu$ increase monotonically as the exciton density n_{ex} is decreased relative to n_{ex}^0. In particular, at very small exciton concentrations ($n_{ex} \ll n_{ex}^0$), $A \gg 1$. In this limiting case, integral (2.56) is easily calculated and we obtain

$$E(0) - \mu = \frac{1}{\beta} \ln\left\{\frac{1}{n_{ex}}\left(\frac{2m^*}{\pi^{4/3}\beta\hbar^2}\right)^{3/2}\right\} \approx \frac{1}{\beta} \ln\left\{\frac{(1/4\beta\Delta L)^{-3/2}}{\pi^2 n_{ex} a^3}\right\}.$$

If the exciton energy is read from E(0), then (in the effective-mass approximation) the excitons in a crystal can be considered an ideal Bose gas. As must be the case, the chemical potential will be negative ($1/\beta \neq 0$),

$$\mu = -\frac{1}{\beta} \ln\left\{\frac{(1/4\beta\,\Delta L)^{-3/2}}{n_{ex} a^3}\right\}. \qquad (2.59)$$

At low exciton densities, chemical potential (2.59) has high negative values and the exciton distribution function becomes the Boltzmann distribution function:

$$\bar{n}_{\mathbf{k}} = \exp\left\{\beta\left[\mu - \frac{\hbar^2 \mathbf{k}^2}{2m^*}\right]\right\}. \qquad (2.60)$$

The exciton gas in a crystal exerts the pressure

$$p = \frac{2}{3}\frac{E}{V},$$

where V is the volume of the crystal; and

$$E = \sum_{\mathbf{k}} \bar{n}_{\mathbf{k}} [E(\mathbf{k}) - E(0)] = \sum_{\mathbf{k}} \frac{\hbar^2 \mathbf{k}^2}{2m^*} \bar{n}_{\mathbf{k}}$$

is the total kinetic energy of the excitons in the crystal. At high exciton densities and a low exciton mass (great exciton-band widths), this pressure can damage the crystal.

3. Relationship between the Dielectric Constant and the Retarded Green Function for Photons

In this section, we shall find a relationship between the dielectric constant and the retarded Green function for photons (see also [14], Section 28). For this, according to (I.3.6), we must find a relation for the mean value (over the canonical ensemble, since the photon chemical potential is zero) of the operator of the vector potential created by the extrinsic current

$$\mathbf{j}^{\text{ст}}(\mathbf{r}, t) = \mathbf{j}(\mathbf{Q}) \exp\{i(\mathbf{Q}\mathbf{r} - \omega t) + \eta t\}, \tag{3.1}$$

which was connected adiabatically in the infinite past ($\eta \to +0$).

The vector-potential operator in the Coulomb calibration is, according to (III.5.4), expressed by the formulas

$$\mathbf{A}(\mathbf{r}) = \sum_{\mathbf{k},\alpha} \sqrt{\frac{2\pi c}{V|\mathbf{k}|}}\, \mathbf{u}_{\mathbf{k}\alpha} \gamma_{\mathbf{k}\alpha} \exp(i\mathbf{k}\mathbf{r}), \qquad \alpha = 1, 2; \tag{3.2}$$

$$\gamma_{\mathbf{k}\alpha} = \overset{+}{\gamma}_{-\mathbf{k},\alpha} \equiv a_{\mathbf{k}\alpha} + \overset{+}{a}_{-\mathbf{k},\alpha} \tag{3.3}$$

in terms of the photon creation and annihilation operators ($\mathbf{k}$, α). Here, V is the volume of the crystal. The interaction operator for extrinsic currents with photons at time t, as is known [7], has the form

$$w(t) = -\frac{1}{c}\int \mathbf{A}(\mathbf{r})\, \mathbf{j}^{\text{ст}}(\mathbf{r}, t)\, d^3r.$$

If we substitute (3.1) and (3.2) into this expression, we find

$$w(t) = -\sqrt{\frac{2\pi V}{c|\mathbf{Q}|}} \sum_{\beta=1}^{2} \overset{+}{\gamma}_{\mathbf{Q},\beta} j_\beta(\mathbf{Q}) \exp\{-i\omega t + \eta t\}. \tag{3.4}$$

Let us assume that when $t = -\infty$ the system was described by the density matrix

$$\rho_0 = \exp\{\beta(F - H)\}, \quad e^{-\beta F} = \mathrm{Sp}\,(e^{-\beta H}),$$

where $\beta = 1/kT$; and H is the Hamiltonian of the interacting excitons, phonons, and photons. Then, when extrinsic currents (3.1) are switched on, the density matrix $\rho(t)$ at time t is defined in the interaction representation by the equation

$$i\frac{\partial\widetilde{\rho}(t)}{\partial t}=[\widetilde{w}(t),\widetilde{\rho}(t)],\tag{3.5}$$

where

$$\left.\begin{aligned}\widetilde{\rho}(t)=\rho_0+e^{iHt}\Delta\rho(t)e^{-iHt},\quad \Delta\rho(t)=\rho(t)-\rho_0,\\ \Delta\rho(-\infty)=0,\quad \widetilde{w}(t)=e^{iHt}w(t)e^{-iHt}.\end{aligned}\right\}.\tag{3.6}$$

Therefore, in a linear approximation of the interaction operator we have

$$\widetilde{\Delta\rho}(t)=\frac{1}{i}\int\limits_{-\infty}^{t}[\widetilde{w}(\tau),\rho_0]\,d\tau.\tag{3.7}$$

The mean value of the α-th component (in the wave-vector coordinate system) of the operator of the vector potential produced by time t under the influence of an extrinsic current is determined by the expression

$$\langle A_{\mathbf{k}\alpha}(\mathbf{r},t)\rangle=\mathrm{Sp}\{\widetilde{A}_{\mathbf{k}\alpha}(\mathbf{r},t)\,\widetilde{\Delta\rho}(t)\},\tag{3.8}$$

where

$$\left.\begin{aligned}\widetilde{A}_{\mathbf{k}\alpha}(\mathbf{r},t)=\sqrt{\frac{2\pi c}{V|\mathbf{k}|}}\,\hat{\gamma}_{\mathbf{k}\alpha}(t)\,e^{i\mathbf{kr}},\\ \hat{\gamma}_{\mathbf{k}\alpha}(t)=e^{iHt}\gamma_{\mathbf{k}\alpha}e^{-iHt}.\end{aligned}\right\}\tag{3.9}$$

If we substitute (3.7) into (3.8), we obtain

$$\langle A_{\mathbf{k}\alpha}(\mathbf{r},t)\rangle=\frac{1}{i}\int\limits_{-\infty}^{t}\mathrm{Sp}\{\rho_0[\widetilde{A}_{\mathbf{k}\alpha}(\mathbf{r},t),\widetilde{w}(\tau)]\,d\tau.$$

With the aid of (3.4), (3.6), and (3.9), we can transform this equation to

$$\langle A_{\mathbf{k}\alpha}(\mathbf{r},t)\rangle=\delta_{\mathbf{kQ}}\frac{2i\pi}{|\mathbf{Q}|}\sum_{\beta}j_{\beta}(\mathbf{Q})\exp\{i[\mathbf{Qr}-\omega t]+\eta t\}\times$$

$$\times\int\limits_{0}^{\infty}\langle\!\langle[\hat{\gamma}_{\mathbf{Q}\alpha}(\tau),\hat{\gamma}^{+}_{\mathbf{Q}\beta}(0)]\rangle\!\rangle\exp[i\omega\tau-\eta\tau]\,d\tau.\tag{3.10}$$

In Eq. (3.10), we used the short notation

$$\langle\langle[\hat{\gamma}_{Q\alpha}(\tau), \hat{\gamma}_{Q\beta}(0)]\rangle\rangle \equiv \mathrm{Sp}\,\{\rho_0[\hat{\gamma}_{Q\alpha}(\tau), \hat{\gamma}^{+}_{Q\beta}(0)]\}.$$

Now we introduce the retarded function for photons by means of the equation

$$\Gamma^{\mathrm{ret}}_{\alpha\beta}(\mathbf{Q}, t) = -i\Theta(t)\langle\langle[\hat{\gamma}_{Q\alpha}(t), \hat{\gamma}^{+}_{Q\beta}(0)]\rangle\rangle, \tag{3.11}$$

where

$$\Theta(t) = \begin{cases} 1, & \text{if } t \geqslant 0; \\ 0, & \text{if } t < 0. \end{cases}$$

When (3.11) is taken into account, the integral in (3.10) is transformed to

$$\int_0^\infty e^{i\omega\tau-\eta\tau}\langle\langle[\hat{\gamma}_{Q\alpha}(\tau), \gamma^{+}_{Q\beta}(0)]\rangle\rangle\, d\tau =$$

$$= i\int_{-\infty}^{\infty} e^{i\omega\tau-\eta\tau}\,\Gamma^{\mathrm{ret}}_{\alpha\beta}(\mathbf{Q}\tau)\, d\tau = i\Gamma^{\mathrm{ret}}_{\alpha\beta}(\mathbf{Q}, \omega), \tag{3.12}$$

where $\Gamma^{\mathrm{ret}}_{\alpha\beta}(\mathbf{Q}, \omega)$ is the Fourier transform of the retarded Green function for photons.

If we substitute (3.12) into (3.10), we find

$$\langle A_{Q\alpha}(\mathbf{r}, t)\rangle = -\frac{2\pi}{|\mathbf{Q}|}\sum_{\beta}\Gamma^{\mathrm{ret}}_{\alpha\beta}(\mathbf{Q}, \omega)\, j_\beta(\mathbf{Q})\, e^{i(\mathbf{Qr}-\omega t)+\eta t}.$$

The obtained mean value is related to the amplitude $A_\alpha(\mathbf{Q})$ of the macroscopic vector potential produced by extrinsic currents by the equation

$$\langle A_{Q\alpha}(\mathbf{r}, t)\rangle = A_\alpha(\mathbf{Q})\, e^{i(\mathbf{Qr}-\omega t)+\eta t}.$$

Therefore,

$$A_\alpha(\mathbf{Q}) = -\frac{2\pi}{|\mathbf{Q}|}\sum_{\beta}\Gamma^{\mathrm{ret}}_{\alpha\beta}(\mathbf{Q}, \omega)\, j_\beta(\mathbf{Q}).$$

If we compare this equation with Eq. (I.4.9) of macroscopic electrodynamics, we find a relationship between the transverse dielectric constant and the photon retarded Green function:

$$\varepsilon^{\perp}_{\alpha\beta}(\mathbf{Q}, \omega) = \frac{c^2 Q^2}{\omega^2}\delta_{\alpha\beta} + \frac{2|\mathbf{Q}|c}{\omega^2}\{\Gamma^{\mathrm{ret}}(\mathbf{Q}, \omega)\}^{-1}_{\alpha\beta}. \tag{3.13}$$

As a simple illustration of the use of this formula, let us calculate the dielectric constant of a vacuum. The vacuum is isotropic, the photons are free, and the averaging for calculation of Green function (3.11) must be over the vacuum state. For free photons,

$$\Gamma^0_{\alpha\beta}(\mathbf{Q}, t) = -i\Theta(t)\langle 0 | [\hat{\gamma}_{\mathbf{Q}\alpha}(t), \hat{\gamma}^+_{\mathbf{Q}\beta}(0)] | 0\rangle,$$

where

$$\hat{\gamma}_{\mathbf{Q}\alpha}(t) = e^{iH_\gamma t}\gamma_{\mathbf{Q}\alpha}e^{-iH_\gamma t}, \qquad H_\gamma = \sum_{\mathbf{k},\,\alpha} c|\mathbf{k}|\, a^+_{\mathbf{k}\alpha}a_{\mathbf{k}\alpha}.$$

Using (3.3), we immediately find

$$\Gamma^0_{\alpha\beta}(\mathbf{Q}, t) = -i\delta_{\alpha\beta}\Theta(t)\,[e^{-i|\mathbf{Q}|ct} - e^{i|\mathbf{Q}|ct}].$$

If we substitute this value into (3.12), we find the Fourier transform of the retarded Green function for free photons in the form of the scalar function

$$\Gamma^0_{\alpha\beta}(\mathbf{Q}, \omega) = \frac{2|\mathbf{Q}|c}{\omega^2 - Q^2c^2}\delta_{\alpha\beta}. \tag{3.14}$$

In this case, inversion of the Fourier transform of the Green function does not involve great difficulty:

$$[\Gamma^0(\mathbf{Q}, \omega)]^{-1}_{\alpha\beta} = \frac{\omega^2 - Q^2c^2}{2|\mathbf{Q}|c}\delta_{\alpha\beta}.$$

Therefore, if we substitute the obtained value into (3.13), we find the explicit form of the dielectric constant of a vacuum:

$$\varepsilon^0_{\alpha\beta}(\mathbf{Q}, \omega) = \delta_{\alpha\beta}.$$

For anisotropic crystals, inversion of the Fourier transform of the retarded Green function is a complicated task. In practical calculations of the dielectric-constant tensor, therefore, it is convenient to use not relation (3.13) but formula (2.21), in which the transverse-dielectric-constant tensor is directly expressed in terms of the Fourier transform of the retarded Green function for excitons.

4. Green Functions for Excitons at Absolute Zero

In the preceding sections, it was shown that the optical properties of solids, just as of many other systems of interacting particles, can be studied with the aid of the mathematical apparatus of quantum field theory. In this case, many physical results are expressed in terms of one-particle Green functions.

Calculations using Green functions at absolute zero are depicted by Feynman diagrams, which give simple and graphic rules for writing any term of the perturbation-theory series. In some cases, the dominant terms of this series can be isolated and summed.

To solve practical problems, one must use variants of the Green-function method, which differ in the choice of the Green functions (causal, retarded, etc.) that are calculated first of all. The various Green functions are interconnected by simple relations (see below). Therefore, it is sufficient to calculate one of them and from it to determine all the others, in terms of which the various properties of the system are expressed. The use of Green functions in solid-state problems was developed in papers by Galitskii and Migdal [15], Martin and Schwinger [16], and in a number of other papers which are cited in review articles and monographs (for example,[3, 8, 14]).

In this section, we shall study the state of a system at absolute zero. For absolute zero, it is advisable to calculate the causal Green functions, since this can be done using the graphical methods of field theory.

First of all, let us consider free excitons and phonons. The Hamiltonian for free excitons and phonons has the form

$$H_0 = \sum_{\mathbf{k}} E(\mathbf{k}) B^+(\mathbf{k}) B(\mathbf{k}) + \sum_{s,\mathbf{q}} \Omega_s(\mathbf{q}) b^+_{\mathbf{q}s} b_{\mathbf{q}s}. \tag{4.1}$$

The one-particle causal functions for free excitons at absolute zero are defined as the averages over the ground (vacuum) state $|0\rangle$ (when there are no excitons or phonons in the system) of the time-ordered product of the exciton creation and annihilation operators in the Heisenberg representation

$$G^0(\mathbf{k}, t-\tau) = -i\langle 0|T\{B(\mathbf{k}, t) B^+(\mathbf{k}, \tau)\}|0\rangle, \tag{4.2}$$

where

$$T\{B(\mathbf{k}, t) B^+(\mathbf{k}, \tau)\} = \begin{cases} B(\mathbf{k}, t) B^+(\mathbf{k}, \tau), & \text{if } t > \tau; \\ B^+(\mathbf{k}, \tau) B(\mathbf{k}, t), & \text{if } t < \tau, \end{cases}$$

is the time-ordered product of the operators; and

$$B(\mathbf{k}, t) = e^{iH_0t} B(\mathbf{k}) e^{-iH_0t} \tag{4.3}$$

are the operators in the Heisenberg representation.

The causal Green function for free phonons is defined by the expression

$$D^{0}(s\mathbf{q}, t-\tau) = -i\langle 0 | T \{\varphi_{\mathbf{q}s}(t) \varphi^{+}_{s\mathbf{q}}(\tau)\} | 0 \rangle, \tag{4.4}$$

where

$$\varphi_{\mathbf{q}s}(t) = b_{\mathbf{q}s}(t) + b^{+}_{-\mathbf{q},s}(t), \tag{4.5}$$

$$b_{\mathbf{q}s}(t) = e^{iH_0t} b_{\mathbf{q}s} e^{-iH_0t} \tag{4.6}$$

are the operators in the Heisenberg representation.

With the aid of definitions (4.2) and (4.4), it is easy to calculate in explicit form the Green functions for free excitons

$$G^0(\mathbf{k}, t) = \begin{cases} -i \exp\{-iE(\mathbf{k})t\}, & \text{if } t > 0; \\ 0, & \text{if } t < 0, \end{cases} \tag{4.7}$$

and free phonons

$$D^0(s\mathbf{q}; t) = \begin{cases} -i \exp\{-i\Omega_s(\mathbf{q})t\}, & \text{if } t > 0; \\ -i \exp\{i\Omega_s(-\mathbf{q})t\}, & \text{if } t < 0. \end{cases} \tag{4.8}$$

In crystals with a center of symmetry $\Omega_s(\mathbf{q}) = \Omega_s(-\mathbf{q})$; hence, the simple equation

$$D^0(s\mathbf{q}; t) = D^0(s, -\mathbf{q}; t)$$

follows from (4.8).

Causal Green functions (4.2) and (4.4), just as the retarded Green functions of Section 2, are defined only for time $t \neq \tau$. Their definition for $t = \tau$ is supplemented also by indicating a rule for calculating the time integrals. With such a supplemented definition, in the energy representation the causal function for free excitons $G^0(\mathbf{k}, \omega)$ is obtained from (4.7) by the transformation

$$G^0(\mathbf{k}, \omega) = \lim_{\eta \to +0} \int_{-\infty}^{\infty} G^0(\mathbf{k}, t) \exp\{i\omega t - \eta t\}\, dt =$$

$$= \lim_{\eta \to +0} [\omega - E(\mathbf{k}) + i\eta]^{-1}.$$

Usually, the limit sign in this expression is omitted and the small positive value η is left,[4] in order that the rule for circumvention of the pole be clearly defined in future calculations of the integrals with respect to ω with the Green functions $G^0(\mathbf{k}, \omega)$.

Thus, the causal Green function for free excitons in the energy representation has the form

$$G^0(\mathbf{k}, \omega) = [\omega - E(\mathbf{k}) + i\eta]^{-1}. \tag{4.9}$$

If we make similar transformations with the causal Green function for free phonons, we obtain in the energy representation

$$D^0(s\mathbf{q}; \omega) = [\omega - \Omega_s(\mathbf{q}) + i\eta]^{-1} - [\omega + \Omega_s(\mathbf{q}) - i\eta]^{-1}. \tag{4.10}$$

If we replace in expression (4.9) the real variable ω by the complex variable $k_0 = \omega + i\gamma$, we see that the real part of the pole of the function $G^0(\mathbf{k}, k_0)$ in the lower half-plane of the variable k_0 determines the free-exciton energy $E(\mathbf{k})$. Similarly, it follows from expression (4.10) that the real part of the pole of the Green function $D^0(s\mathbf{q}, q_0)$ in the lower half-plane of the complex variable $q_0 = \omega + i\gamma$ determines the energy $\Omega_s(\mathbf{q})$ of a free phonon of branch s and wave vector $\mathbf{q}$.

It is easy to see that at zero temperature the exciton causal Green functions coincide with the retarded Green functions. In fact, according to definition (2.19) the exciton retarded Green functions at zero temperature have the form

$$G_r(\mathbf{k}, t) = -i\Theta(t)\langle 0|[B(\mathbf{k}, t), B^+(\mathbf{k}, 0)]|0\rangle = \begin{cases} -i\langle 0|B(\mathbf{k}, t)B^+(\mathbf{k}, 0)|0\rangle, & \text{if } t>0; \\ 0, & \text{if } t<0. \end{cases}$$

On the other hand, according to definition (4.2),

$$G(\mathbf{k}, t) = \begin{cases} -i\langle 0|B(\mathbf{k}, t)B^+(\mathbf{k}, 0)|0\rangle, & \text{if } t>0; \\ -i\langle 0|B^+(\mathbf{k}, 0)B(\mathbf{k}, t)|0\rangle = 0, & \text{if } t<0. \end{cases}$$

Therefore,

$$G(\mathbf{k}, t) = G_r(\mathbf{k}, t). \tag{4.11}$$

[4]The small positive value η, as was indicated in the previous footnote, takes into account the attenuation that is always present in any physical system.

Time correlation functions (2.31) and (2.32) at zero temperature for the case of free excitons are written as

$$\left.\begin{aligned} G^0_>(\mathbf{k}, t) &= \langle 0 | B(\mathbf{k}, t) B^+(\mathbf{k}, 0) | 0 \rangle, \\ G^0_<(\mathbf{k}, t) &= 0. \end{aligned}\right\} \tag{4.12}$$

Considering (4.3), we find the explicit expression

$$G^0_>(\mathbf{k}, t) = \exp\{-iE(\mathbf{k})t\}. \tag{4.13}$$

Moving to the energy representation, we obtain

$$\left.\begin{aligned} G^0_>(\mathbf{k}, \omega) &= 2\pi\delta[\omega - E(\mathbf{k})], \\ G^0_<(\mathbf{k}, \omega) &= 0. \end{aligned}\right\} \tag{4.14}$$

According to (2.39a) and (4.14), the spectral representation of the free-exciton retarded Green function at zero temperature has the form

$$G^0_r(\mathbf{k}, \omega) = \frac{1}{2\pi} \int\limits_{-\infty}^{\infty} \frac{G^0_>(\mathbf{k}, E)\, dE}{\omega - E + i\eta} = [\omega - E(\mathbf{k}) + i\eta]^{-1}. \tag{4.15}$$

Therefore, the free-exciton spectral-representation density at zero temperature is given by

$$\rho^0(\mathbf{k}, E) = G^0_>(\mathbf{k}, E) = 2\pi\delta[E - E(\mathbf{k})]. \tag{4.16}$$

Thus, the free excitons in a crystal at zero temperature are characterized by Green function (4.9), which has the pole $k_0 = E(\mathbf{k}) - i\eta$, which is immediately below the real axis ($\eta \to +\infty$), and by spectral density (4.16), which has a deltoid peak at the same value $E = E(\mathbf{k})$, which corresponds to the energy of free excitons with the wave vector **k**. Time correlation function (4.13) oscillates with a frequency corresponding to the exciton energy.

Considering symbolic identity (2.43), we can write Green function (4.15) as

$$G^0_r(\mathbf{k}, \omega) = \mathcal{P}[\omega - E(\mathbf{k})]^{-1} - i\pi\delta[\omega - E(\mathbf{k})].$$

Then, using (4.16), we find

$$\rho(\mathbf{k}, E) = G^0_>(\mathbf{k}, E) = -2\,\mathrm{Im}\, G^0_r(\mathbf{k}, E), \tag{4.17}$$

which is a particular case of the more general Eq. (2.44).

According to (2.35), the probability of induced light absorption per unit time is proportional to $G_>(\mathbf{Q}, \omega)$. If we substitute (4.16) into Eq. (2.35), we find

$$\frac{dw_{+1}^{-\gamma}(\mathrm{Q}, \omega)}{dt} = \frac{\pi}{4}\left(\frac{\omega_f}{\omega}\right)^2 (\mathbf{E}_0\mathbf{d})^2 N\boldsymbol{\delta}[\omega - E(\mathbf{k})]\, \delta_{\mathbf{Qk}}. \tag{4.18}$$

It follows from (4.18) that in the absence of exciton–phonon interaction, a light wave excites exciton states in a crystal when the following conservation laws are satisfied:

a) $\mathbf{Q} = \mathbf{k}$, i.e., the exciton and photon quasi-momenta must coincide;

b) $\omega = E(\mathbf{Q})$, i.e., the photon energy must equal the exciton energy;

c) $(\mathbf{E}_0\mathbf{d}) \neq 0$, i.e., the projection of the electric vector of the light wave onto the direction of the dipole moment of the transition is not zero.

According to (4.18), the absorption spectrum must have a deltoid peak at $\omega = E(\mathbf{Q})$. If in (4.15) η does not tend toward zero but equals $\gamma_0/2$, where γ_0 is the intrinsic width of the molecular excited state determined by the spontaneous emission of a free molecule, then, according to (4.17), the spectral density takes the form

$$G_>(\mathrm{Q}, \omega) = \frac{\gamma_0}{[\omega - E(\mathrm{Q})]^2 + \frac{1}{4}\gamma_0^2}.$$

In accordance with this, the absorption curve will have the Lorentz form with half-width γ_0.

Using (2.21) and (4.15), we can obtain an expression for the components of the two-dimensional auxiliary tensor $\varepsilon_{xy}^{\perp}$ which determines (see Section 3, Chapter I) the refractive index and absorption coefficient of light waves in the crystal:

$$\varepsilon_{xy}^{\perp} - \delta_{xy} = 4\pi\beta_{xy,0}^{\perp} - \frac{4\pi d^x d^y \omega_f}{v\omega}\left\{\frac{1}{\omega - E(\mathrm{Q}) + i\eta} + \frac{1}{\omega + E(\mathrm{Q}) + i\eta}\right\}. \tag{4.19}$$

The second term inside the braces does not contain a pole; therefore, in the range $\omega \approx E(\mathbf{Q})$ we can write

$$\varepsilon_{xy}^{\perp} - \delta_{xy} = 4\pi\beta_{xy,0}^{\perp} + \frac{4\pi d^x d^y \left(\frac{\omega_f}{v\omega}\right)}{E(\mathrm{Q}) - \omega - i\eta}. \tag{4.19a}$$

To obtain the intrinsic width of the spectral line, we must substitute $\gamma_0/2$ for η.

In an optically isotropic crystal, tensor (4.19) reduces to a scalar. If the square of the dipole moment of the transition is expressed by (II.5.5) in terms of the oscillator strength F (at $\hbar = 1$), then the scalar dielectric constant of an isotropic crystal is represented by the formula

$$\varepsilon = \varepsilon_0 + \frac{\omega_p^2 F}{E^2(\mathbf{Q}) - \omega^2 - i\eta\omega},$$

$$G(\mathbf{k}; t-\tau) = -i\langle 0|\hat{T}\{\tilde{B}(\mathbf{k}, t)\tilde{B}^+(\mathbf{k}, \tau)\}|0\rangle, \tag{4.19b}$$

where $\omega_p^2 \equiv 4\pi e^2/vm$ is the square of the plasma frequency; and $\varepsilon_0 = 1 + 4\pi\beta_0$ is the dielectric constant of the crystal as determined by all of its excited states except the exciton states with energy E(Q).

Let us see how the above results are changed when exciton–phonon interaction is taken into account. The causal Green function for excitons that interact with phonons at zero temperature is defined by

$$\tilde{B}(\mathbf{k}, t) = e^{iHt} B(\mathbf{k}) e^{-iHt} \tag{4.20}$$

which differs from (4.2) in that now the operators

(4.21)

are the exciton operators in the Heisenberg representation with the total Hamiltonian

$$H = H_0 + H_{\text{ex-ph}}, \tag{4.22}$$

where

$$H_{\text{ex-ph}} = \frac{1}{\sqrt{N}} \sum_{\mathbf{k},\mathbf{q},s} F_s(\mathbf{k}, \mathbf{q}) B^+(\mathbf{k}+\mathbf{q}) B(\mathbf{k}) \varphi_{\mathbf{q}s}. \tag{4.23}$$

It can be shown (see, for example, [14]) that expression (4.20) can be brought to a form that contains free-exciton and free-phonon operators (4.3) and (4.5). Namely,

$$G(\mathbf{k}; t-\tau) = -i\langle 0|T\{B(\mathbf{k}, t)B^+(\mathbf{k}, \tau)\}S(\infty)|0\rangle_c, \tag{4.24}$$

where the subscript c indicates that the average over the vacuum state is calculated in a definite way, about which we shall speak below;

$$S(\infty) = T\exp\left\{-i\int_{-\infty}^{\infty} H_{\text{ex-ph}}(t)\,dt\right\} \equiv 1 + \frac{1}{i}\int_{-\infty}^{\infty} H_{\text{ex-ph}}(t)\,dt +$$

$$+\frac{1}{2i^2}\int_{-\infty}^{\infty} T\{H_{\text{ex-ph}}(t_1)\,H_{\text{ex-ph}}(t_2)\}\,dt_1\,dt_2 + \ldots$$

Substituting this expression into Eq. (4.24), let us represent the causal Green function for excitons that interact with phonons as the sum

$$G(\mathbf{k}; t-\tau) = \sum_{n=0}^{\infty} G^{(n)}(\mathbf{k}; t-\tau), \tag{4.25}$$

where n takes only even values (the terms for odd n are zero); and

$$i\,G^{(n)}(\mathbf{k}; t-\tau) = \frac{1}{n!\,i^n}\int\ldots\int_{-\infty}^{\infty} dt_1\ldots dt_n \,\langle 0\,|\,T\{B(\mathbf{k}, t)\,H_{\text{ex-ph}}(t_1)\ldots$$

$$\ldots H_{\text{ex-ph}}(t_n)\,B^+(\mathbf{k}, \tau)\}\,|\,0\rangle_c \tag{4.26}$$

is the exciton Green function in the n-th order of perturbation theory, which contains n interaction operators.

Considering the explicit form of interaction operators (4.23) we see that the integrand in (4.26) contains the average value over the vacuum (ground) state of the product of the exciton and phonon operators. In the theory of Green functions (see, for example, [3, 14]) it has been demonstrated that this average value decomposes (Wick theorem) to the sum of all possible products of the averages over the ground state for individual pairs of exciton operators and pairs of phonon operators, i.e., to a sum of products of factors of the form

$$\langle 0\,|\,T\{B(\mathbf{k}, t)\,B^+(\mathbf{k}, \tau)\}\,|\,0\rangle, \quad \langle 0\,|\,T\{\varphi_{\mathbf{q}s}(t)\,\overset{+}{\varphi}_{\mathbf{q}s}(\tau)\}\,|\,0\rangle.$$

Such averages of the products of two operators are called pairings. The subscript c in (4.26) indicates that only pairings that result in "coupled" exciton processes should be taken into account. In other words, each term of the sum by which the integrand in (4.26) is represented must contain all possible pairings of the phonon operators $\varphi_{\mathbf{q}s}(t_i)$ which pertain to the different operators $H_{\text{ex-ph}}(t_i)$, and the pairing of exciton operator $B(\mathbf{k}, t)$ with the operator $B^+(k, t)$ from $H_{ex\text{-}ph}(t_1)$, the pairing of $B(k, t_1)$ from $H_{ex\text{-}ph}(t_1)$ with the operator $B^+(k, t_2)$ from $H_{ex\text{-}ph}(t_2)$, and so forth, until $B^+(k, \tau)$ is reached without having omitted one operator $H_{ex\text{-}ph}$.

If we recall the definitions of causal Green functions for free excitons (4.2) and phonons (4.4), we can see that the pairings are expressed in terms of the free-particle Green functions by means of the equations

$$\left.\begin{aligned} \langle 0 | T \{B(\mathbf{k}, t) B^+(\mathbf{k}', \tau)\} | 0 \rangle &= i\delta_{\mathbf{kk}'} G^0(\mathbf{k}; t-\tau), \\ \langle 0 | T \{\varphi_{\mathbf{q}s}(t) \varphi^+_{\mathbf{q}s'}(\tau)\} | 0 \rangle &= i\delta_{ss'}\delta_{\mathbf{qq}'} D^0(s\mathbf{q}; t-\tau). \end{aligned}\right\} \tag{4.27}$$

In (4.26), integration is with respect to the times $t_1, \ldots, t_n$. Therefore, the products of the pairings, which differ in the permutations of $t_1, t_2, \ldots, t_n$, are equivalent. The number of such permutations is n!, so in (4.26) we can select a particular sequence and omit n! in the denominator. In this case, the integrand is represented by a sum of terms that differ in the method of pairing of the phonon operators.

Let us consider some simple examples. When $n = 0$, Green function (4.26) reduces to the Green function for a free exciton. When $n = 2$, the integrand of (4.26) contains only the two operators $H_{ex\text{-}ph}(t_1)$ and $H_{ex\text{-}ph}(t_2)$. Thus, there is only one way of pairing the phonon operators $\langle 0 | T \{\varphi_{s\mathbf{q}}(t_1) \varphi_{s\mathbf{q}}(t_2)\} | 0 \rangle$ and $G^{(2)}$ is expressed by a single term.

When $n = 4$, the integrand of (4.26) contains four interaction operators. Hence, three pairings of the phonon operators are possible:

$$\langle 0 | T \{\varphi_{\mathbf{q}s}(t_1) \varphi^+_{\mathbf{q}s}(t_2)\} | 0 \rangle \langle 0 | T \{\varphi_{\mathbf{q}s}(t_3) \varphi^+_{s\mathbf{q}}(t_4)\} | 0 \rangle,$$
$$\langle 0 | T \{\varphi_{\mathbf{q}s}(t_1) \varphi^+_{\mathbf{q}s}(t_3)\} | 0 \rangle \langle 0 | T \{\varphi_{\mathbf{q}s}(t_2) \varphi^+_{s\mathbf{q}}(t_4)\} | 0 \rangle,$$
$$\langle 0 | T \{\varphi_{\mathbf{q}s}(t_1) \varphi^+_{\mathbf{q}s}(t_4)\} | 0 \rangle \langle 0 | T \{\varphi_{\mathbf{q}s}(t_2) \varphi^+_{\mathbf{q}s}(t_3)\} | 0 \rangle$$

and $G^{(4)}$ is expressed by three terms.

Thus, considering the explicit form of operator (4.23), when $n = 2$ we have

$$\begin{aligned} &\langle 0 | T \{B(\mathbf{k}, t) H_{\text{ex-ph}}(t_1) H_{\text{ex-ph}}(t_2) B^+(\mathbf{k}, \tau)\} | 0 \rangle = \\ &= \frac{1}{N} \sum F_{s_1}(\mathbf{k}_1, \mathbf{q}_1) F_{s_2}(\mathbf{k}_2, \mathbf{q}_2) \langle 0 | T \{B(\mathbf{k}, t) B^+(\mathbf{k}_1 + \mathbf{q}_1; t_1)\} | 0 \rangle \times \\ &\times \langle 0 | T \{B(\mathbf{k}_1, t_1) B^+(\mathbf{k}_2 + \mathbf{q}_2; t_2)\} | 0 \rangle \langle 0 | T \{B(\mathbf{k}_2, t_2) B^+(\mathbf{k}, \tau)\} | 0 \rangle \times \\ &\times \langle 0 | T \{\varphi_{\mathbf{q}_1 s_1}(t_1) \varphi^+_{\mathbf{q}_2 s_2}(t_2)\} | 0 \rangle. \end{aligned}$$

If we substitute this expression into (4.26), we find, taking (4.27) into account, the exciton causal Green function in the second approximation of perturbation theory

$$iG^{(2)}(\mathbf{k}; t-\tau) = \frac{1}{Ni^2}\sum_{s,\mathbf{q}}\int dt_1 \int dt_2\, F_s(\mathbf{k}-\mathbf{q}, \mathbf{q})\, F_s(\mathbf{k}, -\mathbf{q}) \times$$
$$\times\, iG^0(\mathbf{k}; t-t_1)\, iG^0(\mathbf{k}-\mathbf{q}; t_1-t_2)\, iG^0(\mathbf{k}; t_2-\tau)\, iD^0_{\mathbf{q}s}(t_1-t_2). \quad (4.28)$$

Green functions (4.28) can be associated with the diagram in Fig. 21. In Fig. 21, the two nodes (vertices) t_1 and t_2 correspond to the second order of perturbation theory. The solid lines represent the exciton and the broken line the phonon Green functions for free particles. The momenta of the corresponding particles are arranged on the lines such that the law of conservation of momentum is satisfied at each vertex. The rules for associating analytic expressions with the diagram in Fig. 21 boil down to the following.

1) Each solid line

is associated with the free-exciton Green function $iG^0(\mathbf{k}; t_1 - t_2)$; each broken line

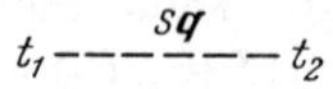

is associated with the free-phonon Green function $iD^0(s\mathbf{q}; t_1 - t_2)$; and each vertex

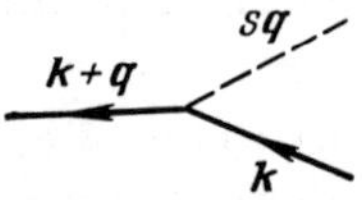

is associated with the function $(i/\sqrt{N})F_s(\mathbf{k}, \mathbf{q})$.

2) To calculate $iG^{(2)}(\mathbf{k}; t-\tau)$, the contributions from each line and vertex are multiplied and the result is summed over all possible s and **q** and integrated with respect to the intermediate times t_1 and t_2.

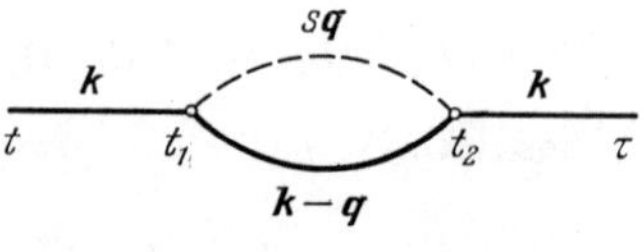

Fig. 21. Feynman diagram for calculating exciton Green function in second order of perturbation theory.

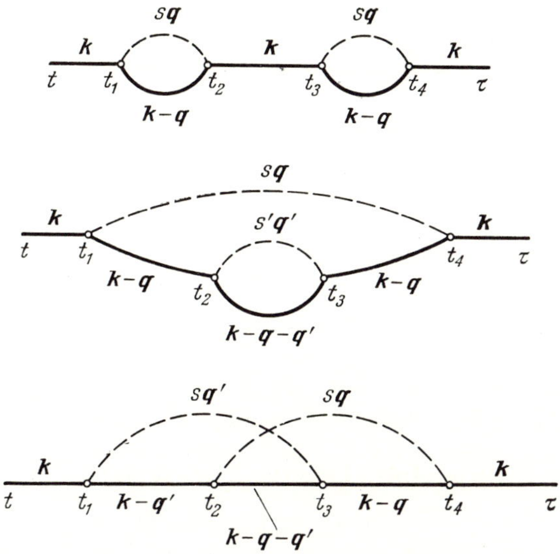

Fig. 22. Three Feynman diagrams for calculating exciton Green function in fourth order of perturbation theory.

The Green function in the fourth approximation of perturbation theory is expressed by a sum of three terms, which correspond to the three diagrams in Fig. 22.

Since the free-exciton and free-phonon Green functions are discontinuous functions of time, it is convenient to calculate them in the energy representation. To illustrate the transformation to the energy representation, let us consider the Green function of the second approximation of perturbation theory. Conversion from expression (4.28) to the energy representation is accomplished by the transformation

$$iG^{(2)}(\mathbf{k};\omega) = \int_{-\infty}^{\infty} iG^{(2)}(\mathbf{k};t)\exp(i\omega t)\,dt.$$

If we substitute into the right side of this equation expression (4.28) and

$$G^{0}(\mathbf{k};t) = \frac{1}{2\pi}\int_{-\infty}^{\infty} G^{0}(\mathbf{k};\omega)\exp(-i\omega t)\,d\omega,$$

$$D^{0}(s,\mathbf{q};t) = \frac{1}{2\pi}\int_{-\infty}^{\infty} D^{0}(s,\mathbf{q};\Omega)\exp(-i\Omega t)\,d\Omega,$$

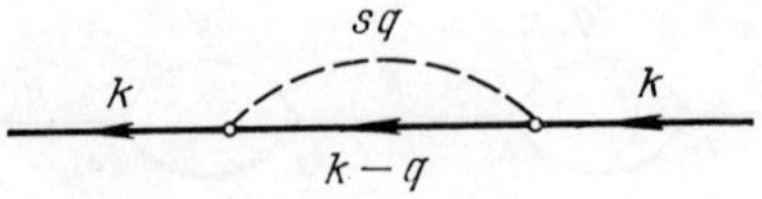

Fig. 23. Feynman diagram for calculating exciton Green function in second approximation in energy representation.

we find

$$\frac{i}{2\pi} G^{(2)}(k) = \frac{i}{2\pi} G^0(k) \sum_{s,\mathbf{q}} \frac{2\pi}{i\sqrt{N}} F_s(\mathbf{k}-\mathbf{q}, \mathbf{q}) \times$$

$$\times \frac{2\pi}{i\sqrt{N}} F_s(\mathbf{k}, -\mathbf{q}) \frac{i}{2\pi} G^0(k) \int_{-\infty}^{\infty} \frac{i}{2\pi} G^0(k-q) \frac{i}{2\pi} D^0(s, q)\, d\Omega, \quad (4.29)$$

where for brevity we introduce the four-vectors

$$k \equiv (\mathbf{k}, \omega) \text{ and } q \equiv (\mathbf{q}, \Omega). \quad (4.30)$$

Formula (4.29) can be associated with the Feynman diagram in Fig. 23. This diagram, which has two vertices, gives the value $(i/2\pi)G^{(2)}(k)$ if: 1) the four-vectors k, q, k − q are arranged on the diagram such that the laws of conservation of the space k, q,... and time ω, Ω,... parts are satisfied at each vertex; 2) each solid line is associated with the function $i(2\pi)^{-1}G^0(k)$, each broken line is associated with the function $i(2\pi)^{-1}D^0(s, q)$, and each vertex

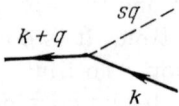

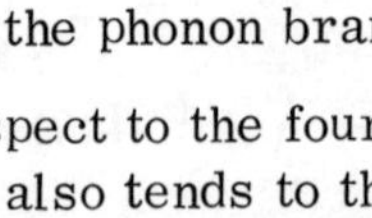

is associated with the function $(2\pi/i)F_s(\mathbf{k}, \mathbf{q})$; and 3) after multiplication of the contributions from all vertices and lines, summation $\frac{1}{N}\sum_{s,q}$ over all of the phonon branches s and wave vectors **q** and integration with respect to the fourth (time) component are accomplished. This rule also tends to the energy-representation calculation of the exciton Green functions $(i/2\pi)G^{(n)}(k)$, which correspond to the n-th approximation of perturbation theory. It must be taken into account only that the function is represented as a sum of diagrams that contain n vertices and differ from one another in the arrangement of the broken lines connecting the vertices. Summation over all s and **q** and integration with respect to Ω are necessary for each broken line when calculating the contributions from each diagram.

If we substitute $G^0(k-q)$ and $D^0(s, q)$ into (4.29) and calculate the integral, we can transform the exciton Green function in the second order of perturbation theory to

$$G^{(2)}(k) = G^0(k) M^{(2)}(k) G^0(k), \tag{4.31}$$

where

$$M^{(2)}(k) = \frac{i}{2\pi N} \sum_{s,\mathbf{q}} |F_s(\mathbf{k}-\mathbf{q}, \mathbf{q})|^2 \int_{-\infty}^{\infty} G^0(k-q) D^0(s, q)\, d\Omega =$$

$$= \frac{1}{N} \sum_{s,\mathbf{q}} \frac{|F_s(\mathbf{k}-\mathbf{q}, \mathbf{q})|^2}{\omega - \Omega_s(\mathbf{q}) - E(\mathbf{k}-\mathbf{q}) + i\eta} \tag{4.32}$$

is the exciton mass operator in the second order of perturbation theory. This mass operator takes into account all single-phonon interactions.

The total exciton Green function is given by the expression

$$G(k) = \sum_{n=0}^{\infty} G^{(n)}(k) = G^0(k) + G^0(k) M(k) G^0(k) +$$

$$+ G^0(k) M(k) G^0(k) M(k) G^0(k) + \ldots = G^0(k) \sum_{n=0}^{\infty} [M(k) G^0(k)]^n, \tag{4.33}$$

where M(k) is the total exciton mass operator. The total exciton mass operator (Fig. 24) corresponds to the sum of all compact diagrams, i.e., the sum of all diagrams none of which can be divided into two unconnected parts by breaking one solid line.

Equation (4.33) corresponds to the diagram in Fig. 25, where the heavy line represents the total exciton Green function. Let us rewrite (4.33) as follows:

$$G(k) = G^0(k) + G^0(k) M(k) \left\{ G^0(k) \sum_{n=0}^{\infty} [M(k) G^0(k)]^n \right\}. \tag{4.33a}$$

According to (4.33), the value inside the braces in (4.33a) coincides with G(k). Therefore, we have

$$G(k) = G^0(k) + G^0(k) M(k) G(k). \tag{4.34}$$

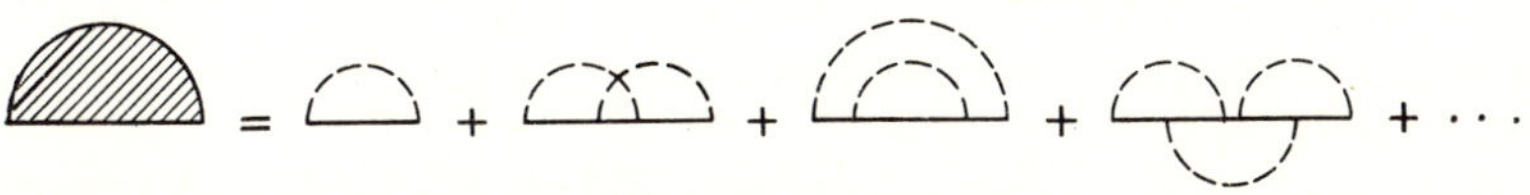

Fig. 24. Graphical representation of mass operator for excitons that interact with phonons.

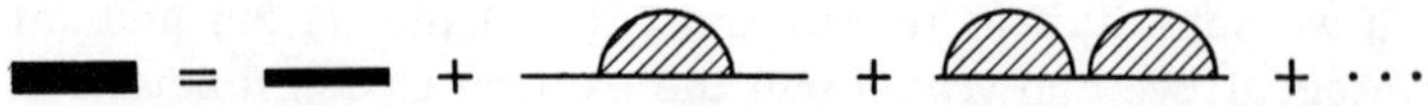

Fig. 25. Graphical representation of total Green function for excitons that interact with phonons.

This equation is called the **Dyson equation.** It determines the total exciton Green function when the free-exciton Green function and the mass operator that takes into account exciton–phonon interaction are known.

If we solve Eq. (4.34) for G(k) and substitute $G^0(k) = [\omega - E(k) + i\eta]^{-1}$, we find the Green function for excitons that interact with phonons in explicit form:

$$G(\mathbf{k}, \omega) = \{\omega - E(\mathbf{k}) - M(\mathbf{k}, \omega) + i\eta\}^{-1}. \quad (4.35)$$

In general, the mass operator $M(\mathbf{k}, \omega)$ when $\eta \to +0$ is a complex function of the real variables ω and $\mathbf{k}$:

$$M(k, \omega) = \Delta(k, \omega) - \frac{i}{2}\gamma(k, \omega). \quad (4.36)$$

In particular, if the mass operator is defined in second approximation (4.32), then, using identity (2.43), we find

$$\Delta^{(2)}(\mathbf{k}, \omega) = \frac{\mathscr{P}}{N} \sum_{s, \mathbf{q}} \frac{|F_s(\mathbf{k} - \mathbf{q}, \mathbf{q})|^2}{\omega - \Omega_s(\mathbf{q}) - E(\mathbf{k} - \mathbf{q})}, \quad (4.37)$$

$$\gamma^{(2)}(\mathbf{k}, \omega) = \frac{2\pi}{N} \sum_{s, \mathbf{q}} |F_s(\mathbf{k} - \mathbf{q}, \mathbf{q})|^2 \delta[\omega - E(\mathbf{k} - \mathbf{q}) - \Omega_s(\mathbf{q})]. \quad (4.38)$$

Substituting (4.36) into (4.35) and taking (4.11) into account, we find the retarded Green function for excitons that interact with phonons at zero temperature:

$$G_r(\mathbf{k}, \omega) = G(\mathbf{k}, \omega) = \left[\omega - E(\mathbf{k}) - \Delta(\mathbf{k}, \omega) + \frac{i}{2}\gamma(\mathbf{k}, \omega)\right]^{-1}. \quad (4.39)$$

Therefore, function (4.39) is analytic in the upper half-plane of the complex variable $k_0 = \omega + i\eta$ and defines the retarded Green function.

Here, ω and $\mathbf{k}$ are real variables and $\gamma(\mathbf{k}, \omega) \geq 0$ for all $\mathbf{k}$ and ω. With weak exciton-phonon interaction,

$$\Delta(\mathbf{k}, \omega) \ll E(\mathbf{k}), \quad \gamma(\mathbf{k}, \omega) \ll E(\mathbf{k}). \quad (4.40)$$

Therefore, Green function (4.39) can be replaced by the approximate function

$$G_r(\mathbf{k}, \omega) = \left\{\omega - E(\mathbf{k}) - \Delta^{(2)}(\mathbf{k}, \omega) + \frac{i}{2}\gamma^{(2)}(\mathbf{k}, \omega)\right\}^{-1}, \tag{4.41}$$

in which $\Delta^{(2)}(\mathbf{k}, \omega)$ and $\gamma^{(2)}(\mathbf{k}, \omega)$ are calculated by formulas (4.37) and (4.38). Approximate Green function (4.41) corresponds to the infinite sequence of Feynman diagrams shown in Fig. 26. It takes into account all single-phonon processes.

If we compare (4.39) with the free-exciton Green function, we see that when weak exciton–phonon interaction is induced, the deltoid [with its peak at $\omega = E(\mathbf{k})$] spectral density function (4.16) is replaced by the expression

$$\rho(\mathbf{k}, \omega) = -2\,\mathrm{Im}\, G_r(\mathbf{k}, \omega) = \frac{\gamma(\mathbf{k}, \omega)}{[\omega - E(\mathbf{k}) - \Delta(\mathbf{k}, \omega)]^2 + \frac{1}{4}\gamma^2(\mathbf{k}, \omega)}. \tag{4.42}$$

At fixed $\mathbf{k}$, the principal maximum of the spectral density function corresponds to $\omega = \mathscr{E}(\mathbf{k})$, which is defined by the condition

$$\{\omega - E(\mathbf{k}) - \Delta(\mathbf{k}, \omega)\}_{\omega=\mathscr{E}(\mathbf{k})} = 0. \tag{4.43}$$

Under conditions (4.40), Eq. (4.43) can be solved by the method of successive approximations. Then,

$$\mathscr{E}(\mathbf{k}) \approx E(\mathbf{k}) + \Delta(\mathbf{k}, E(\mathbf{k})). \tag{4.44}$$

The value $\mathscr{E}(\mathbf{k})$ determines the dispersion law for excitons that interact with phonons. The states of excitons with a definite $\mathbf{k}$ in the presence of interaction with phonons are no longer steady. The half-width $\gamma(\mathbf{k}, \omega)$ of distribution function (4.42) at $\omega = \mathscr{E}(\mathbf{k})$ characterizes the indeterminate form of the energy values, and $\{\gamma(\mathbf{k}, \mathscr{E}(\mathbf{k}))\}^{-1}$ determines the mean lifetime with respect to transition to exciton states with other values $\mathbf{k}' \neq \mathbf{k}$.

Fig. 26. Exciton Green function taking into account all single-phonon interaction processes.

5. Temperature Matsubara Green Functions for Interacting Excitons and Phonons

As was shown in the preceding section, diagram techniques can be used to calculate the causal Green functions for absolute zero. If the causal Green function is known, it is easy to determine the spectral density and the exciton retarded Green function in terms of which the light-crystal interaction phenomena that interest us are expressed. But the diagram methods do not allow direct extension to the case of nonzero temperatures. At nonzero temperatures, as Matsubara has shown ([17]; see also Chapter III of [14]), it is convenient to calculate beforehand auxiliary functions that depend not upon time but upon a parameter that varies in a finite range determined by the temperature of the system. These functions are called *temperature Matsubara functions*.

The temperature Matsubara functions for excitons and phonons are defined, respectively, by the equations

$$\left.\begin{aligned} \mathfrak{G}(\mathbf{k};\ \tau_1-\tau_2) &= -\langle\!\langle T_\tau\{B(\mathbf{k},\ \tau_1)\,B^+(\mathbf{k},\ \tau_2)\}\rangle\!\rangle, \\ \mathfrak{D}(s\mathbf{q};\ \tau_1-\tau_2) &= -\langle\!\langle T_\tau\{\varphi_{\mathbf{q}s}(\tau_1)\,\varphi^+_{\mathbf{q}s}(\tau_2)\rangle\!\rangle. \end{aligned}\right\} \tag{5.1}$$

In these expressions, the double brackets $\langle\!\langle\ldots\rangle\!\rangle$ indicate averaging over the states defined by density matrix (2.3). The operators in definition (5.1) are expressed in terms of the exciton operators B(k) and B$^+$(k) and the phonon operators $\varphi_{\mathbf{q}s}$ and $\varphi^+_{\mathbf{q}s}$ by means of the equations

$$\left.\begin{aligned} B(\mathbf{k},\ \tau) &= e^{\mathcal{H}\tau}B(\mathbf{k})\,e^{-\mathcal{H}\tau}, \\ B^+(\mathbf{k},\ \tau) &= e^{\mathcal{H}\tau}B^+(\mathbf{k})\,e^{-\mathcal{H}\tau}, \\ \varphi_{\mathbf{q}s}(\tau) &= e^{\mathcal{H}\tau}\varphi_{\mathbf{q}s}e^{-\mathcal{H}\tau}, \\ &\ldots\ldots\ldots\ldots\ldots \end{aligned}\right\} \tag{5.2}$$

where

$$\mathcal{H} = H - \mu\hat{n};$$

τ is a continuous parameter that varies in the finite range from $-\beta$ to 0; and T_τ is the ordering operator for the parameter τ. It is defined by the equation

$$T_\tau\{B(\mathbf{k},\ \tau_1)\,B^+(\mathbf{k},\ \tau_2)\} = \begin{cases} B(\mathbf{k},\ \tau_1)\,B^+(\mathbf{k},\ \tau_2), & \text{if } \tau_1 > \tau_2; \\ B^+(\mathbf{k},\ \tau_2)\,B(\mathbf{k},\ \tau_1), & \text{if } \tau_2 > \tau_1. \end{cases}$$

It follows directly from definitions (5.1) that the temperature Matsubara functions $\mathfrak{G}(\mathbf{k}, \tau)$ and $\mathfrak{D}(\mathbf{k}, \tau)$ are functions of the variable $\tau = \tau_1 - \tau_2$, which varies within the limits $-\beta \leq \tau \leq \beta$. It can be shown that at negative τ we have

$$\mathfrak{G}(\mathbf{k}, \tau) = \mathfrak{G}(\mathbf{k}, \tau + \beta), \tag{5.3}$$

$$\mathfrak{D}(s\mathbf{q}, \tau) = \mathfrak{D}(s\mathbf{q}, \tau + \beta). \tag{5.4}$$

These equations relate the Matsubara functions for negative and positive τ. Therefore, we need consider only positive τ.

It follows from definition (5.1) that when $\tau > 0$ the exciton Matsubara function can be transformed to

$$\mathfrak{G}(\mathbf{k}, \tau) = -\langle\langle B(\mathbf{k}, \tau) B^+(\mathbf{k}, 0)\rangle\rangle =$$
$$= -\sum_{n, l, l'} \rho_0(l, n) |\langle l', n+1 | B^+(\mathbf{k}, 0) | l, n\rangle|^2 \exp\{[E(l, n) -$$
$$- E(l', n+1) + \mu]\tau\} = -\frac{1}{2\pi}\int_{-\infty}^{\infty} e^{(\mu-E)\tau} \times$$
$$\times \left\{2\pi \sum_{l, l', n} \rho_0(l, n) |\langle l', n+1 | B^+_{\mathbf{k}}(0) | l, n\rangle|^2 \delta[E(l', n+1) -\right.$$
$$\left. - E(l, n) - E]\right\} dE.$$

The expression inside the braces under the integral in this equation coincides with the Fourier component of correlation function (2.32) for excitons. Therefore,

$$\mathfrak{G}(\mathbf{k}, \tau) = -\frac{1}{2\pi}\int_{-\infty}^{\infty} e^{(\mu-E)\tau} G_>(\mathbf{k}, E)\, dE, \quad \tau > 0. \tag{5.5}$$

This equation relates the correlation functions to the Matsubara functions for excitons.

It is convenient (see [18, 19]) to convert from Matsubara functions (5.1), which are functions of the variable τ, to their Fourier components. The Fourier components of the Matsubara functions are determined, taking (5.3) into account, by the Fourier series

$$\mathfrak{G}(\mathbf{k}, \tau) = \frac{1}{\beta}\sum_n e^{-i\omega_n\tau}\, \mathfrak{G}(\mathbf{k}, \omega_n), \tag{5.6}$$

where $\omega_n = 2\pi n/\beta$, $n = 0, \pm 1, \pm 2, \ldots, \infty$;

$$\mathfrak{G}(\mathbf{k}, \omega_n) = \int_0^\beta e^{i\omega_n\tau}\, \mathfrak{G}(\mathbf{k}, \tau)\, d\tau. \tag{5.7}$$

If we substitute expression (5.5) into (5.7), after integration with respect to τ with allowance for the equation $\exp(i\beta\omega_n) = 1$, we find the spectral representation of the Fourier component of the temperature Matsubara function:

$$\mathfrak{G}(\mathbf{k}, \omega_n) = \frac{1}{2\pi} \int_{-\infty}^{\infty} \frac{[1 - e^{\beta(\mu - E)}]\, G_>(\mathbf{k}, E)}{i\omega_n + \mu - E}\, dE.$$

This spectral representation is transformed by means of (2.40) to

$$\mathfrak{G}(\mathbf{k}, \omega_n) = \frac{1}{2\pi} \int_{-\infty}^{\infty} \frac{[e^{\beta(E - \mu)} - 1]\, G_<(\mathbf{k}, E)\, dE}{i\omega_n + \mu - E}. \tag{5.8}$$

Taking the real correlation function $G_<$ into account, the equation

$$\mathfrak{G}(\mathbf{k}, \omega_n) = \mathfrak{G}^*(\mathbf{k}, -\omega_n).$$

follows directly from expression (5.8). If on the right side of (5.8) we make the formal transformation

$$i\omega_n + \mu \Rightarrow \omega + i\eta = k_0, \tag{5.9}$$

then, taking (2.41) into account, we obtain the very important equation

$$\mathfrak{G}(\mathbf{k}, \omega_n) = \{G_r(\mathbf{k}, k_0)\}_{k_0 = i\omega_n + \mu}. \tag{5.10}$$

Equation (5.10) allows us, with the aid of the retarded Green function $G_r(\mathbf{k}, k_0)$, which is known and analytic in the upper half-plane of the variable k_0, to calculate the temperature Matsubara function $\mathfrak{G}(\mathbf{k}, \omega_n)$ for all positive ω_n. The inverse problem of finding the retarded Green function $G_r(\mathbf{k}, \omega)$ from the known (for all discrete positive ω_n) Matsubara function $\mathfrak{G}(\mathbf{k}, \omega_n)$ reduces to the analytic continuation of this function, with the discrete set of points ω_n, to the entire upper half-plane [18, 19]. If the function $\mathfrak{S}(\mathbf{k}, \omega_n)$ is known for the infinite sequence ω_n ($n = 0, 1, 2, \ldots, \infty$) and the function $F(\mathbf{k}, k_0)$, which is analytic in the upper half-plane of the complex variable, is such that

$$\{F(\mathbf{k},\ k_0)\}_{k_0=i\omega_n+\mu} = \mathfrak{G}(\mathbf{k},\ \omega_n),$$

then it follows from the uniqueness of the analytic function defined in the infinite sequence of points (with the limiting point ω_∞ in the range of analyticity) and from Eq. (5.10) that the functions $F(\mathbf{k},\ k_0)$ and $G_r(\mathbf{k},\ k_0)$ coincide everywhere in the upper half-plane k_0.

The possibility of transformation of the Matsubara function to the retarded Green function allows us to replace, when $1/\beta \neq 0$, calculation of the retarded Green function by the more simple calculation of the Matsubara function. Calculation of the Matsubara function is more convenient, since it can be done by diagram methods, by summation of the contributions for the successive approximations of perturbation theory. When we calculate the various perturbation-theory approximations of the function $\mathfrak{G}$ we can (see [14]) use the same Feynman diagrams that we used in calculating the causal Green function G at $1/\beta = 0$. In this case, the explicit analytic expressions for the individual terms of the perturbation-theory series can be obtained from the expressions for the causal Green function at absolute zero that correspond to the same diagrams. For this, in the corresponding expressions for the causal Green function at absolute zero we must replace the frequency $\omega + i\eta$ by the discrete values $i\omega_n + \mu$, and the integrals with respect to ω should be replaced by discrete sums according to the rule

$$\frac{1}{2\pi}\int \ldots d\omega \Rightarrow \frac{i}{\beta}\sum_n \ldots . \tag{5.11}$$

Although diagram methods allow us to obtain the Matsubara functions in any approximation, their practical calculation is very complicated and is often done only with considerable simplifications. It is sometimes convenient, therefore, to find the approximate values of the retarded temperature Green functions directly. In the next section, we shall consider methods proposed by Bogolyubov and Tyablikov [20] for direct approximate calculation of the temperature two-time retarded Green functions.

Now let us calculate in explicit form the temperature Matsubara functions for excitons and phonons.

According to definitions (5.1)-(5.3), when $\tau \geq 0$ the Matsubara function for free excitons has the form

$$\mathfrak{G}^0(\mathbf{k}, \tau) = -\langle\langle B(\mathbf{k}, \tau) B^+(\mathbf{k}, 0)\rangle\rangle = -\langle\langle B(\mathbf{k}, 0)^+ B(\mathbf{k}, 0)\rangle\rangle e^{\tau[\mu - E(\mathbf{k})]}. \quad (5.12)$$

If we consider that the average number of free excitons of type **k** is determined by the expression

$$\bar{n}_{\mathbf{k}} = \langle\langle B^+(\mathbf{k}, 0) B(\mathbf{k}, 0)\rangle\rangle = [e^{\beta[E(\mathbf{k}) - \mu]} - 1]^{-1},$$

then

$$1 + n_k = \langle\langle B(\mathbf{k}, 0) B^+(\mathbf{k}, 0)\rangle\rangle = [1 - e^{\beta(\mu - E(\mathbf{k}))}]^{-1}, \quad (5.13)$$

and Matsubara function (5.12) can be written as

$$\mathfrak{G}^0(\mathbf{k}, \tau) = -(\bar{n}_k + 1)\exp\{\tau[\mu - E(\mathbf{k})]\}.$$

Let us substitute this value into (5.7) and convert to the Fourier components of the Matsubara functions

$$\mathfrak{G}^0(\mathbf{k}, \omega_n) = -(\bar{n}_{\mathbf{k}} + 1)\int_0^\beta \exp\{i[\omega_n - E(\mathbf{k}) + \mu]\tau\}\,d\tau =$$

$$= \frac{(\bar{n}_{\mathbf{k}} + 1)\{1 - e^{i[\omega_n - E(\mathbf{k}) + \mu]\beta}\}}{i\omega_n - E + \mu}.$$

Considering the equation $\exp(i\beta\omega_n) = 1$ and (5.13), we find finally

$$\mathfrak{G}^0(\mathbf{k}, \omega_n) = \{i\omega_n - E(\mathbf{k}) + \mu\}^{-1}. \quad (5.14)$$

The Fourier component of the Matsubara function (5.14) for free excitons is transformed by formal transformation (5.9) to Green-function Fourier component (4.9).

The Matsubara functions for free phonons are obtained similarly.

In fact, when $\tau \geq 0$

$$\mathfrak{D}^0(s\mathbf{q}, \tau) = -\langle\langle \varphi_{\mathbf{q}s}(\tau)\varphi^+_{\mathbf{q}s}(0)\rangle\rangle = -(1 + \bar{\nu}_{\mathbf{q}s})e^{-\tau\Omega_s(\mathbf{q})} - \bar{\nu}_{\mathbf{q}s}e^{\tau\Omega_s(\mathbf{q})}, \quad (5.15)$$

where

$$\bar{\nu}_{\mathbf{q}s} = \langle\langle b^+_{\mathbf{q}s}(0)\,b_{\mathbf{q}s}(0)\rangle\rangle = [e^{\beta\Omega_s(\mathbf{q})} - 1]^{-1}.$$

Converting to Fourier components using (5.7), we find

$$\mathfrak{D}^0(s\mathbf{q}, \omega_n) = -\frac{2\Omega_s(\mathbf{q})}{\omega_n^2 + \Omega_s^2(\mathbf{q})}. \quad (5.16)$$

Let us use, finally, the above rule for formal transformation (5.11) from the energy representations of Green functions at absolute zero to Matsubara functions at finite temperatures. According to this rule, the equation that defines the Matsubara function at nonzero temperature is derived from Eq. (4.35) for the exciton Green functions at absolute zero if in this equation we replace $\mathbf{k} \equiv (\mathbf{k};\, \omega + i\eta)$ by $\mathbf{k} \equiv (\mathbf{k};\, i\omega_n + \mu)$. Thus, we obtain

$$\mathfrak{G}(\mathbf{k},\ \omega_n) = \{i\omega_n + \mu - E(\mathbf{k}) - \Xi(\mathbf{k},\ \omega_n)\}^{-1}, \tag{5.17}$$

where $\Xi(\mathbf{k},\ \omega_n)$ is the mass operator for the exciton Matsubara function. Let us calculate it in the second approximation of perturbation theory. For this, we make the inverse of transformation (5.9) in expression (4.32), which determines in the second approximation the exciton mass operator for zero temperature. This transforms the Green functions for free excitons and phonons to the Matsubara functions for free excitons and phonons. Further, using transformation (5.11), we replace integration with respect to ω by summation over all $\omega_n = 2\pi n/\beta$. Thus, we obtain

$$\Xi^{(2)}(\mathbf{k},\ \omega_n) = \frac{1}{N}\sum_{s,\,\mathbf{q}} |F_s(\mathbf{k}-\mathbf{q},\ \mathbf{q})|^2 \sum_{m\geqslant-\infty}^{\infty} \varphi(\omega_m), \tag{5.18}$$

where

$$\varphi(\omega_m) \equiv -\frac{1}{\beta}\mathfrak{G}^0(\mathbf{k}-\mathbf{q},\ \omega_n-\omega_m)\,\mathfrak{D}^0(s\mathbf{q},\ \omega_m).$$

Substituting the explicit form of Matsubara functions (5.14) and (5.16), we have

$$\varphi(\omega_m) = \frac{2i\Omega_s(\mathbf{q})}{\beta\,[\omega_m - \omega_n - i\,(E(\mathbf{k}-\mathbf{q}) - \mu)]\,[\omega_m^2 + \Omega_s^2(\mathbf{q})]}\,. \tag{5.19}$$

Summation of functions (5.19) in Eq. (5.18) can be accomplished using the following theorem of functions of a complex variable. If the function $\varphi(z)$ of the complex variable z has a finite number of simple poles at $z = a_j$ ($j = 1, 2, \ldots, l$) not equal to $2\pi m/\beta$ ($m = 0, \pm 1, \ldots$) and $\lim|\varphi(z)| \to 0$ when $|z| \to \infty$, then

$$\sum_{m\geqslant-\infty}^{\infty} \varphi(\omega_m) = -i\beta\sum_{j=1}^{l}\left\{\operatorname{Res}\left(\frac{\varphi(z)}{e^{i\beta z}-1}\right)\right\}_{z=a_j}, \tag{5.20}$$

where Res F(z) denotes the residue of the function F(z). The function

$$\varphi(z) = \frac{2i\Omega_s(\mathbf{q})}{\beta\,\{z - \omega_n - i\,[E(\mathbf{k}-\mathbf{q}) - \mu]\}\,[z^2 + \Omega_s^2(\mathbf{q})]} \tag{5.21}$$

satisfies the above conditions. It has three simple poles at z equal to

$$a_1 = \omega_n + i\,[E(\mathbf{k}-\mathbf{q}) - \mu], \quad a_2 = i\Omega_s(\mathbf{q}), \quad a_3 = -i\Omega_s(\mathbf{q}). \tag{5.22}$$

Considering the equation $\exp(i\beta\omega_n) = 1$ and definitions (5.13) and (5.15), we find

$$\left.\begin{aligned} &[\exp(i\beta a_1) - 1]^{-1} = -(1 + \bar{n}_{\mathbf{k}-\mathbf{q}}); \ [\exp(i\beta a_2) - 1]^{-1} = \bar{\nu}_{qs}; \\ &[\exp(i\beta a_3) - 1]^{-1} = -(1 + \bar{\nu}_{\mathbf{q}s}). \end{aligned}\right\} \tag{5.23}$$

Further, applying (5.20) to function (5.21), we obtain, taking (5.23) into account,

$$\sum_{m \geqslant -\infty}^{\infty} \varphi(\omega_m) = \frac{1 + \bar{n}_{\mathbf{k}-\mathbf{q}} + \bar{\nu}_{\mathbf{q}s}}{i\omega_n - E(\mathbf{k}-\mathbf{q}) - \Omega_s(q) + \mu} + \frac{\bar{\nu}_{\mathbf{q}s} - \bar{n}_{\mathbf{k}-\mathbf{q}}}{i\omega_n - E(\mathbf{k}-\mathbf{q}) + \Omega_s(\mathbf{q}) + \mu}.$$

If we substitute this value into Eq. (5.18), we find a final expression for the mass operator of the Matsubara Green function in the second order of perturbation theory:

$$\Xi^{(2)}(\mathbf{k}, \omega_n) = \frac{1}{N} \sum_{s,\,\mathbf{q}} |F_s(\mathbf{k}-\mathbf{q}, \mathbf{q})|^2 \times$$

$$\times \left\{ \frac{1 + \bar{n}_{\mathbf{k}-\mathbf{q}} + \bar{\nu}_{\mathbf{q}s}}{i\omega_n + \mu - E(\mathbf{k}-\mathbf{q}) - \Omega_s(\mathbf{q})} + \frac{\bar{\nu}_{\mathbf{q}s} - \bar{n}_{\mathbf{k}-\mathbf{q}}}{i\omega_n + \mu - E(\mathbf{k}-\mathbf{q}) + \Omega_s(\mathbf{q})} \right\}. \tag{5.24}$$

Applying formal transformation (5.9) to (5.17) and (5.24), we transform the temperature Matsubara function and its mass operator to the Green function for excitons that interact with phonons at nonzero temperature, and we find the corresponding exciton mass operator (in the second approximation of perturbation theory). Namely,

$$G_r(\mathbf{k}, \omega) = \{\omega - E(\mathbf{k}) - M^{(2)}(\mathbf{k}, \omega) + i\eta\}^{-1}, \tag{5.25}$$

where

$$M^{(2)}(\mathbf{k}, \omega) = \frac{1}{N} \sum_{s,\,\mathbf{q}} |F_s(\mathbf{k}-\mathbf{q}, \mathbf{q})|^2 \times$$

$$\times \left\{ \frac{1 + \bar{n}_{\mathbf{k}-\mathbf{q}} + \bar{\nu}_{\mathbf{q}s}}{\omega - E(\mathbf{k}-\mathbf{q}) - \Omega_s(\mathbf{q}) + i\eta} + \frac{\bar{\nu}_{\mathbf{q}s} - \bar{n}_{\mathbf{k}-\mathbf{q}}}{\omega - E(\mathbf{k}-\mathbf{q}) + \Omega_s(\mathbf{q}) + i\eta} \right\}. \tag{5.26}$$

Function (5.25) is analytic in the upper half-plane of the complex variable $k_0 = \omega + i\eta$ (at $\eta \geq 0$); therefore, it is the exciton retarded Green function. At absolute zero

$$\bar{n}_{\mathbf{k}} = \bar{\nu}_{\mathbf{q}s} = 0$$

and expression (5.26) coincides with mass operator (4.32).

6. Retarded Two-Time Green Functions for Excitons at Nonzero Temperatures

It was noted in Section 4 that at absolute zero the causal Green functions could be calculated using the graphical methods of field theory. This advantage of causal Green functions is lost in transition to nonzero temperatures.

At nonzero temperatures, therefore, it is more convenient to use retarded (or advanced) Green functions, which permit analytic continuation to the complex plane. As Bogolyubov and Tyablikov have shown [20], the retarded Green functions satisfy an infinite set of linked equations. Such an infinite set of equations is studied by approximately terminating it at a certain term and then solving the finite set of equations.

Below, we shall apply the Bogolyubov–Tyablikov method to the system described by the Hamiltonian operator

$$H = \sum_{\mathbf{k}} E(\mathbf{k}) B^{+}(\mathbf{k}) B(\mathbf{k}) + \sum_{s,\mathbf{q}} \Omega_s(\mathbf{q}) b^{+}_{\mathbf{q}s} b_{\mathbf{q}s} + \frac{1}{\sqrt{N}} \sum_{s,\mathbf{q},\mathbf{k}} F_s(\mathbf{k},\mathbf{q}) B^{+}(\mathbf{k}+\mathbf{q}) B(\mathbf{k}) \varphi_{\mathbf{q}s}, \tag{6.1}$$

where

$$\left.\begin{aligned} \varphi_{\mathbf{q}s} = \varphi^{+}_{-\mathbf{q},s} &= b_{\mathbf{q}s} + b^{+}_{-\mathbf{q},s}, \\ F_s(\mathbf{k},\mathbf{q}) = F^{+}_s(\mathbf{k}+\mathbf{q}, -\mathbf{q}), &\quad \Omega_s(\mathbf{q}) = \Omega_s(-\mathbf{q}). \end{aligned}\right\} \tag{6.2}$$

The number of excitons is preserved in the system described by operator (6.1). Thus, operator (6.1) characterizes the states of the excitons during a period that does not exceed their lifetime with respect to their complete conversion to photons or phonons. It is also assumed that during this time a statistical distribution of the excitons over the sublevels of the first exciton band is established.

In the Heisenberg representation, the time variation of any operator F is determined by the equation ($\hbar = 1$)

$$i\frac{dF}{dt} = [F,\ H].$$

For the system described by operator (6.1), therefore, the exciton and phonon operators satisfy the equations

$$\left.\begin{aligned} i\frac{dB(\mathbf{k})}{dt} &= E(\mathbf{k})\,B(\mathbf{k}) + \frac{1}{\sqrt{N}}\sum_{s,\,\mathbf{q}} F_s^{+}(\mathbf{k},\ \mathbf{q})\,B(\mathbf{k}+\mathbf{q})\,\varphi_{\mathbf{q}s}^{+}, \\ i\frac{db_{\mathbf{q}s}}{dt} &= \Omega_s(\mathbf{q}).b_{\mathbf{q}s} + \frac{1}{\sqrt{N}}\sum_{\mathbf{k}} F_s(\mathbf{k},\ -\mathbf{q})\,B^{+}(\mathbf{k}-\mathbf{q})\,B(\mathbf{k}), \\ -i\frac{db_{\mathbf{q}s}^{+}}{dt} &= \Omega_s(\mathbf{q})\,b_{\mathbf{q}s}^{+} + \frac{1}{\sqrt{N}}\sum_{\mathbf{k}} F_s(\mathbf{k},\mathbf{q})\,B^{+}(\mathbf{k}+\mathbf{q})\,B(\mathbf{k}). \end{aligned}\right\} \quad (6.3)$$

Let us introduce the exciton and phonon retarded Green functions

$$\left.\begin{aligned} G_r(\mathbf{k},\ t) &= -i\Theta(t)\langle\langle[B(\mathbf{k},\ t),\ B^{+}(\mathbf{k},\ 0)]\rangle\rangle, \\ \mathscr{D}_r(s\mathbf{q},\ t) &= -i\Theta(t)\langle\langle[b_{\mathbf{q}s}(t),\ b_{\mathbf{q}s}^{+}(0)]\rangle\rangle. \end{aligned}\right\} \quad (6.4)$$

If we differentiate these equations with respect to t and take into account Eqs. (6.3) and the equation $d\Theta(t)/dt = \delta(t)$, we obtain for the Green functions

$$i\frac{dG_r(\mathbf{k},\ t)}{dt} = \delta(t) + E(\mathbf{k})\,G_r(\mathbf{k},\ t) +$$

$$+\frac{1}{\sqrt{N}}\sum_{s,\,\mathbf{q}} F_s^{+}(\mathbf{k},\ \mathbf{q})\,\{P_1(\mathbf{k}+\mathbf{q},\ s\mathbf{q};\ \mathbf{k}\,|\,t) + P_2(\mathbf{k}+\mathbf{q},\ s\mathbf{q};\ \mathbf{k}\,|\,t\}, \quad (6.5)$$

$$i\frac{d\mathscr{D}_r(s\mathbf{q},\ t)}{dt} = \delta(t) + \Omega_s(\mathbf{q})\,\mathscr{D}_r(s\mathbf{q},\ t) +$$

$$+\frac{1}{\sqrt{N}}\sum_{\mathbf{k}} F_s(\mathbf{k},-\mathbf{q})\,P_3(\mathbf{k}-\mathbf{q},\ \mathbf{k};\ s\mathbf{q}\,|\,t). \quad (6.6)$$

The right sides of Eqs. (6.5) and (6.6) contain Green functions of a higher order than the initial functions, i.e.,

$$\left.\begin{aligned} P_1(\mathbf{k}+\mathbf{q},s\mathbf{q};\mathbf{k}\,|\,t) &= -i\Theta(t)\langle\langle[B(\mathbf{k}+\mathbf{q};\,t)\,b_{\mathbf{q}s}^{+}(t),\ B^{+}(\mathbf{k};\ 0)]\rangle\rangle, \\ P_2(\mathbf{k}+\mathbf{q},s\mathbf{q};\mathbf{k}\,|\,t) &= -i\Theta(t)\langle\langle[B(\mathbf{k}+\mathbf{q};\,t)\,b_{-\mathbf{q},s}(t),\ B^{+}(\mathbf{k},\ 0]\rangle\rangle, \\ P_3(\mathbf{k}-\mathbf{q},\mathbf{k};s\mathbf{q}\,|\,t) &= -i\Theta(t)\langle\langle[B^{+}(\mathbf{k}-\mathbf{q};\ t)\,B(\mathbf{k};\ t),\ b_{\mathbf{q}s}^{+}(0)]\rangle\rangle. \end{aligned}\right\} \quad (6.7)$$

If we differentiate Eqs. (6.7) with respect to time and use motion equations (6.3), we obtain equations for functions (6.7).

For example,

$$i\frac{dP_1(t)}{dt} = \delta(t)\,\delta_{\mathbf{q}0}\,\langle\langle b_{s0}^{+}\rangle\rangle + [E(\mathbf{k}+\mathbf{q}) - \Omega_s(\mathbf{q})]\,P_1 + \frac{1}{\sqrt{N}}\sum_{s',\mathbf{q}'} F_{s'}^{+}(\mathbf{k}+\mathbf{q};\,\mathbf{q}')\,\Phi_1(\mathbf{k},\,s\mathbf{q};\,s'\mathbf{q}'|t) - \frac{1}{\sqrt{N}}\sum_{\mathbf{k}'} F_s(\mathbf{k}';\,\mathbf{q})\,\Phi_2(\mathbf{k},\,s\mathbf{q};\,s'\mathbf{q}'\,|\,t), \tag{6.8}$$

where

$$\left.\begin{aligned} &\Phi_1(\mathbf{k},\,s\mathbf{q};\,s'\mathbf{q}'\,|\,t) \equiv \\ &\equiv -i\Theta(t)\,\langle\langle[B(\mathbf{k}+\mathbf{q}+\mathbf{q}';\,t)\,\varphi_{\mathbf{q}'s'}^{+}(t)\,b_{\mathbf{q}s}^{+}(t),\quad B^{+}(\mathbf{k},\,0)]\rangle\rangle, \\ &\Phi_2(\mathbf{k},\,s\mathbf{q};\,\mathbf{k}'\,|\,t) \equiv \\ &\equiv -i\Theta(t)\,\langle\langle[B(\mathbf{k}+\mathbf{q};\,t)\,B^{+}(\mathbf{k}'+\mathbf{q};\,t)\,B(\mathbf{k}',\,t),\quad B^{+}(\mathbf{k},\,0]\rangle\rangle) \end{aligned}\right\} \tag{6.9}$$

are new, more complicated Green functions.

If we differentiate Green functions (6.9) with respect to time, we obtain a new system of equations in which the Green functions Φ_i are expressed in terms of Green functions of a still higher order. Thus, we can obtain an infinite set of equations that define initial Green functions (6.4).

In (6.5), the Green functions $P_1(t)$ is under the sum over **q** with the factor $1/\sqrt{N}$. When $N \to \infty$, therefore, we can with asymptotic accuracy ignore the first term on the right of (6.8).

If we wish to terminate the infinite chain of equations at Green functions (6.7), we can in (6.9) replace the mean values of the operator products by the products of the mean values, i.e., let

$$\langle\langle[B(\mathbf{k}+\mathbf{q}+\mathbf{q}';\,t)\,\varphi_{\mathbf{q}'s'}^{+}(t)\,b_{\mathbf{q}s}^{+}(t),\;B(\mathbf{k},\,0)]\rangle\rangle \approx \approx \langle\langle\varphi_{s'\mathbf{q}'}^{+}(t)\,b_{\mathbf{q}s}^{+}(t)\rangle\rangle\,\langle\langle[B(\mathbf{k}+\mathbf{q}+\mathbf{q}';\,t),\;B^{+}(\mathbf{k},\,0)]\rangle\rangle, \tag{6.10a}$$

$$\langle\langle[B(\mathbf{k}+\mathbf{q};\;t)\,B^{+}(\mathbf{k}'+\mathbf{q};\;t)\,B(\mathbf{k}',\,t),\;B^{+}(\mathbf{k},\,0)]\rangle\rangle \approx \approx \langle\langle B(\mathbf{k}+\mathbf{q};\;t)\,B^{+}(\mathbf{k}'+\mathbf{q};\,t)\rangle\rangle\,\langle\langle[\bar{B}(\mathbf{k}',\,t),\;B^{+}(\mathbf{k},\,0)]\rangle\rangle. \tag{6.10b}$$

Further, we have

$$\left.\begin{aligned} &\langle\langle B(\mathbf{k}+\mathbf{q};\;t)\,B^{+}(\mathbf{k}'+\mathbf{q};\,t)\rangle\rangle \approx \delta_{\mathbf{k}\mathbf{k}'}(1+\bar{n}_{\mathbf{k}+\mathbf{q}}), \\ &\langle\langle\varphi_{\mathbf{q}'s'}^{+}(t)\,b_{\mathbf{q}s}^{+}(t)\rangle\rangle \approx \delta_{ss'}\delta_{\mathbf{q}',-\mathbf{q}}(1+\bar{\nu}_{\mathbf{q}s}), \end{aligned}\right\} \tag{6.11}$$

where $\bar{n}_{k+q}$ is the average number of excitons with the wave vector $k + \mathbf{q}$ in the system; and $\bar{\nu}_{qs}$ is the average number of phonons of branch s with the wave vector $\mathbf{q}$.

If we substitute (6.10) into (6.9) and take into account (6.11) and definitions (6.4), we can transform (6.8) to the simple form

$$i\frac{dP_1(t)}{dt} = \{E(\mathbf{k}+\mathbf{q}) - \Omega_s(\mathbf{q})\}P_1(t) +$$

$$+\frac{1}{\sqrt{N}}F_s(\mathbf{k},\mathbf{q})\{(1+\bar{\nu}_{\mathbf{q}s}) - (1+\bar{n}_{\mathbf{k}+\mathbf{q}})\}G_r(\mathbf{k},t). \tag{6.12}$$

Making similar simplifications with the equations for the Green functions P_2 and P_3, we find

$$i\frac{dP_2(t)}{dt} = \{E(\mathbf{k}+\mathbf{q}) + \Omega_s(\mathbf{q})\}P_2(t) +$$

$$+\frac{1}{\sqrt{N}}F_s(\mathbf{k},\mathbf{q})\{\bar{\nu}_s(\mathbf{q}) + (1+\bar{n}_{\mathbf{k}+\mathbf{q}})\}G_r(\mathbf{k},t), \tag{6.13}$$

$$i\frac{dP_3(t)}{dt} = \{E(\mathbf{k}) - E(\mathbf{k}-\mathbf{q})\}P_3(t) +$$

$$+\frac{1}{\sqrt{N}}F_s(\mathbf{k}-\mathbf{q},\mathbf{q})\{\bar{n}_{\mathbf{k}-\mathbf{q}} - \bar{n}_{\mathbf{k}}\}\mathcal{D}_r(s\mathbf{q},t). \tag{6.14}$$

If we desire greater accuracy, instead of the Green functions [type (6.9)] that contain the products of four operators, we should simplify the higher-order Green functions that are obtained by differentiating (6.9) and similar equations with respect to time. Here, we shall limit ourselves to approximation (6.10)-(6.11). It reflects the most important properties of a system of weakly interacting excitons and phonons.

Equations (6.5), (6.6), and (6.12)-(6.14) form a closed system of equations that determine Green functions (6.4) in the approximation used. Let us move to the energy representation in this system. For this, we let

$$2\pi G_r(\mathbf{k},t) = \int_{-\infty}^{\infty} G_r(\mathbf{k},k_0)e^{-ik_0t}\,dk_0,$$

$$2\pi P_1(t) = \int_{-\infty}^{\infty} P_1(k_0)\exp(-ik_0t)\,dk_0,$$

$$2\pi P_2(t) = \int_{-\infty}^{\infty} P_2(k_0)\exp(-ik_0t)\,dk_0,$$

$$2\pi P_3(t) = \int_{-\infty}^{\infty} P_3(q_0) \exp(-iq_0 t)\, dq_0,$$

$$2\pi \mathscr{D}_r(s\mathbf{q}, t) = \int_{-\infty}^{\infty} \mathscr{D}_r(s\mathbf{q}, q_0) \exp(-iq_0 t)\, dq_0,$$

$$2\pi \delta(t) = \int_{-\infty}^{\infty} \exp(-ik_0 t)\, dk_0.$$

If we substitute these values into system (6.5), (6.6), and (6.12)-(6.14) and eliminate the functions $P_1(k_0)$, $P_2(k_0)$, and $P_3(q_0)$, we obtain a system of equations for the exciton and phonon Green functions in the energy representation in the second order of perturbation theory:

$$\{k_0 - E(\mathbf{k}) - M^{(2)}(\mathbf{k}, k_0)\} G_r^{(2)}(\mathbf{k}, k_0) = 1, \tag{6.15}$$

$$M^{(2)}(\mathbf{k}, k_0) = \frac{1}{N} \sum_{s, \mathbf{q}} |F_s(\mathbf{k}, \mathbf{q})|^2 \times$$

$$\times \left\{ \frac{1 + \bar{\nu}_{\mathbf{q}s} + \bar{n}_{\mathbf{k}+\mathbf{q}}}{k_0 - E(\mathbf{k}+\mathbf{q}) - \Omega_s(\mathbf{q})} + \frac{\bar{\nu}_{s\mathbf{q}} - \bar{n}_{\mathbf{k}+\mathbf{q}}}{k_0 - E(\mathbf{k}+\mathbf{q}) + \Omega_s(\mathbf{q})} \right\}, \tag{6.16}$$

$$\{q_0 - \Omega_s(\mathbf{q}) - \Pi(s\mathbf{q}, q_0)\} \mathscr{D}_r(s\mathbf{q}, q_0) = 1, \tag{6.17}$$

$$\Pi(s, \mathbf{q}; q_0) = \frac{1}{N} \sum_{\mathbf{k}} |F_s(\mathbf{k}, -\mathbf{q})|^2 \frac{\bar{n}_{\mathbf{k}-\mathbf{q}} - \bar{n}_{\mathbf{k}}}{q_0 - E(\mathbf{k}) + E(\mathbf{k}-\mathbf{q})}. \tag{6.18}$$

Expressions (6.15) and (6.16) coincide with expressions (5.25) and (5.26), which were found earlier by means of the temperature Matsubara functions. This is easily seen if in the sum of (5.26) we substitute $-\mathbf{q}$ for $\mathbf{q}$ and use Eqs. (6.2).

The expressions $M(\mathbf{k}, k_0)$ and $\Pi(s, \mathbf{q}; q_0)$ in (6.15) and (6.17) are called the exciton mass operator and the phonon polarization operator, respectively. These operators take into account exciton–phonon interaction. In the absence of exciton–phonon interaction ($F_s = 0$), both operators $M(\mathbf{k})$ and $\Pi(\mathbf{q})$ are equal to zero and the Green functions $G_r(\mathbf{k})$ and $\mathscr{D}(s, \mathbf{q})$ become the Green functions for free excitons and phonons. With interaction, these functions differ from the free-particle Green functions chiefly in $\mathbf{k}$, k_0 and $\mathbf{q}$, q_0, which satisfy the inequalities

$$|k_0 - E(\mathbf{k})| \lesssim |M(\mathbf{k}, k_0)|, \quad |q_0 - \Omega_s(\mathbf{q})| \lesssim |\Pi(s, \mathbf{q}; q_0)|. \tag{6.19}$$

In (6.16) and (6.18) let

$$k_0 = q_0 = \omega + i\eta, \qquad \eta \to 0,$$

where ω is the real frequency. Then, using symbolic identity (2.43), we can transform expressions (6.16) and (6.18) to

$$M^{(2)}(\mathbf{k}, k_0) = \Delta^{(2)}(\mathbf{k}, \omega) - \frac{i}{2}\gamma^{(2)}(\mathbf{k}, \omega), \tag{6.20}$$

$$\Pi^{(2)}(s, \mathbf{q}, q_0) = \Delta^{(2)}(s, \mathbf{q}; \omega) - \frac{i}{2}\gamma^{(2)}(s, \mathbf{q}; \omega), \tag{6.21}$$

where the real functions Δ and γ are

$$\Delta(\mathbf{k}, \omega) = \frac{\mathcal{P}}{N}\sum_{s,\,\mathbf{q}} |F_s(\mathbf{k}, \mathbf{q})|^2 \left\{\frac{1 + \bar{\nu}_{s\mathbf{q}} + \bar{n}_{\mathbf{k}+\mathbf{q}}}{\omega - E(\mathbf{k}+\mathbf{q}) - \Omega_s(\mathbf{q})} + \right.$$
$$\left. + \frac{\bar{\nu}_{\mathbf{q}s} - \bar{n}_{\mathbf{k}+\mathbf{q}}}{\omega - E(\mathbf{k}+\mathbf{q}) + \Omega_s(\mathbf{q})}\right\}, \tag{6.22}$$

$$\gamma(\mathbf{k}, \omega) = \frac{2\pi}{N}\sum_{s,\,\mathbf{q}} |F_s(\mathbf{k}, \mathbf{q})|^2 \{(\bar{\nu}_{\mathbf{q}s} - \bar{n}_{\mathbf{k}+\mathbf{q}})\,\delta[\omega - E(\mathbf{k}+\mathbf{q}) + \Omega_s(\mathbf{q})] +$$
$$+ (1 + \bar{\nu}_{\mathbf{q}s} + \bar{n}_{\mathbf{k}+\mathbf{q}})\,\delta[\omega - E(\mathbf{k}+\mathbf{q}) - \Omega_s(\mathbf{q})]\}, \tag{6.23}$$

$$\Delta(s, \mathbf{q}; \omega) = \frac{\mathcal{P}}{N}\sum_{\mathbf{k}} |F_s(\mathbf{k}, -\mathbf{q})|^2 \frac{\bar{n}_{\mathbf{k}-\mathbf{q}} - \bar{n}_{\mathbf{k}}}{\omega - E(\mathbf{k}) + E(\mathbf{k}-\mathbf{q})}, \tag{6.24}$$

$$\gamma(s, \mathbf{q}; \omega) = \frac{2\pi}{N}\sum_{\mathbf{k}} |F_s(\mathbf{k}, -\mathbf{q})|^2 (\bar{n}_{\mathbf{k}-\mathbf{q}} - \bar{n}_{\mathbf{k}})\,\delta[\omega - E(\mathbf{k}) + E(\mathbf{k}-\mathbf{q})] \tag{6.25}$$

To obtain a closed system, we must add to Eqs. (6.22)-(6.25) equations for the average numbers of excitons and phonons in the system. By definition,

$$\bar{n}_{\mathbf{k}} = \{G_<(\mathbf{k}, t)\}_{t=0}, \quad \bar{\nu}_{\mathbf{q}s} = \{\mathcal{D}_<(s, \mathbf{q}; t)\}_{t=0}, \tag{6.26}$$

where $G_<(\mathbf{k}, t)$ and $\mathcal{D}_<(s, \mathbf{q}; t)$ are correlation functions defined by

$$G_<(\mathbf{k}, t) = \langle\langle B^+(\mathbf{k}, t)\, B(\mathbf{k}, 0)\rangle\rangle; \quad \mathcal{D}_<(s, \mathbf{q}; t) = \langle\langle b^+_{\mathbf{q}s}(t)\, b_{\mathbf{q}s}(0)\rangle\rangle. \tag{6.27}$$

The Fourier transforms of correlation functions (6.27) are related to the imaginary parts of the retarded Green functions by the simple relations

$$\left.\begin{aligned}\{\exp[(\omega - \mu)\beta] - 1\}\, G_<(\mathbf{k}, \omega) &= -2\,\mathrm{Im}\, G_r(\mathbf{k}, \omega),\\ \{\exp(\omega\beta) - 1\}\,\mathcal{D}_<(s, \mathbf{q}; \omega) &= -2\,\mathrm{Im}\,\mathcal{D}_r(s, \mathbf{q}; \omega).\end{aligned}\right\} \tag{6.28}$$

Taking (6.16) and (6.18) into account, from Eqs. (6.15) and (6.17) follow the expressions for the retarded Green functions

$$G_r(\mathbf{k}, \omega) = \left\{\omega - E(\mathbf{k}) - \Delta(\mathbf{k}, \omega) + \frac{i}{2}\gamma(\mathbf{k}, \omega)\right\}^{-1}, \tag{6.29}$$

$$\mathscr{D}_r(s, \mathbf{q}; \omega) = \left\{\omega - \Omega_s(\mathbf{q}) - \Delta(s, \mathbf{q}; \omega) + \frac{i}{2}\gamma(s, \mathbf{q}; \omega)\right\}^{-1}. \qquad (6.30)$$

From (6.28)-(6.30) we find explicit expressions for the Fourier transforms of the correlation functions

$$G_<(\mathbf{k}, \omega) = \frac{\gamma(\mathbf{k}, \omega)\{\exp[(\omega - \mu)\beta] - 1\}^{-1}}{[\omega - E(\mathbf{k}) - \Delta(\mathbf{k}, \omega)]^2 + \frac{1}{4}\gamma^2(\mathbf{k}, \omega)},$$

$$\mathscr{D}_<(s, \mathbf{q}; \omega) = \frac{\gamma(s, \mathbf{q}; \omega)\{\exp(\omega\beta) - 1\}^{-1}}{[\omega - \Omega_s(\mathbf{q}) - \Delta(s, \mathbf{q}; \omega)]^2 + \frac{1}{4}\gamma^2(s, \mathbf{q}; \omega)}.$$

Substituting these values into the equations

$$2\pi G_<(\mathbf{k}, t) = \int_{-\infty}^{\infty} G_<(\mathbf{k}, \omega)\exp(-i\omega t)\,d\omega,$$

$$2\pi\mathscr{D}_<(s, \mathbf{q}; t) = \int_{-\infty}^{\infty} \mathscr{D}_<(s, \mathbf{q}; \omega)\exp(-i\omega t)\,d\omega,$$

we can, using (6.26), write final expressions for the distribution functions, i.e., the average numbers of excitons and phonons:

$$\bar{n}_{\mathbf{k}} = \frac{1}{2\pi}\int_{-\infty}^{\infty} \frac{\gamma(\mathbf{k}, \omega)\{\exp[(\omega - \mu)\beta] - 1\}^{-1}\,d\omega}{[\omega - E(\mathbf{k}) - \Delta(\mathbf{k}, \omega)]^2 + \frac{1}{4}\gamma^2(\mathbf{k}, \omega)}, \qquad (6.31)$$

$$\bar{\nu}_{\mathbf{q}s} = \frac{1}{2\pi}\int_{-\infty}^{\infty} \frac{\gamma(s, \mathbf{q}; \omega)\{\exp(\omega\beta) - 1\}^{-1}\,d\omega}{[\omega - \Omega_s(\mathbf{q}) - \Delta(s, \mathbf{q}; \omega)]^2 + \frac{1}{4}\gamma^2(s, \mathbf{q}; \omega)}. \qquad (6.32)$$

If to the six equations (6.22)-(6.25), (6.31), and (6.32) we add

$$n_{\text{ex}} = \sum_{\mathbf{k}} \bar{n}_{\mathbf{k}}, \qquad (6.33)$$

which determines the number of excitons in the system, we obtain a system of equations in the unknowns

$$\Delta(\mathbf{k}, \omega),\ \gamma(\mathbf{k}, \omega),\ \Delta(s, \mathbf{q}; \omega),\ \gamma(s, \mathbf{q}; \omega),\ \bar{n}_{\mathbf{k}},\ \bar{\nu}_{\mathbf{q}s},\ \mu.$$

For a given temperature and a given number of excitons, the values

$$\Delta(\mathbf{k}, \omega),\ \gamma(\mathbf{k}, \omega),\ \Delta(s, \mathbf{q}; \omega),\ \text{and}\ \gamma(s, \mathbf{q}; \omega)$$

determine exciton and phonon Green functions (6.29) and (6.30). It is assumed here that the dispersion laws $E(\mathbf{k})$ and $\Omega_s(\mathbf{q})$ for free excitons and phonons are known.

The study of exciton–phonon interaction is considerably simplified with states containing a small number of excitons. At a low exciton density

$$\bar{n}_{\mathbf{k}} \ll 1; \tag{6.34}$$

therefore, according to (6.24) and (6.25),

$$\Delta(s, \mathbf{q}; \omega) \approx \gamma(s, \mathbf{q}; \omega) \approx 0, \tag{6.35}$$

and phonon Green function (6.30) coincides with the free-phonon Green function. Under condition (6.34), from (6.32) follows the distribution function for phonons that are governed by Bose statistics:

$$\bar{\nu}_{\mathbf{q}s} = \{\exp[\beta\Omega_s(\mathbf{q})] - 1\}^{-1}. \tag{6.36}$$

Under conditions (6.34) expressions (6.22) and (6.23) are simplified and transformed to

$$\Delta(\mathbf{k}, \omega) = \frac{\mathcal{P}}{N} \sum_{s, \mathbf{q}} |F_s(\mathbf{k}, \mathbf{q})|^2 \left\{ \frac{1 + \bar{\nu}_{\mathbf{q}s}}{\omega - E(\mathbf{k} + \mathbf{q}) - \Omega_s(\mathbf{q})} + \right.$$
$$\left. + \frac{\bar{\nu}_{s\mathbf{q}}}{\omega - E(\mathbf{k} + \mathbf{q}) + \Omega_s(\mathbf{q})} \right\}, \tag{6.37}$$

$$\gamma(\mathbf{k}, \omega) = \frac{2\pi}{N} \sum_{s, \mathbf{q}} |F_s(\mathbf{k}, \mathbf{q})|^2 \{(1 + \bar{\nu}_{\mathbf{q}s}) \delta[\omega - E(\mathbf{k} + \mathbf{q}) - \Omega_s(\mathbf{q})] +$$
$$+ \bar{\nu}_{\mathbf{q}s} \delta[\omega - E(\mathbf{k} + \mathbf{q}) + \Omega_s(\mathbf{q})]\}. \tag{6.38}$$

If the infinite system of equations for the Green function is terminated with calculation of the time derivatives of function (6.9), then we obtain the energy representation of the Green function and the mass operator in the fourth approximation of perturbation theory. For example, at small exciton densities (when $\bar{n}_{\mathbf{k}} \ll 1$), the fourth-order Green function is

$$G^{(4)}(\mathbf{k}, k_0) = \{k_0 - E(\mathbf{k}) - M^{(4)}(\mathbf{k}, k_0)\}^{-1}, \tag{6.39}$$

where $k_0 = \omega + i\eta$, $\eta \to +0$; and

$$M^{(4)}(\mathbf{k}, \omega) = \sum_{s, \mathbf{q}} |F_s(\mathbf{k}, \mathbf{q})|^2 \times$$
$$\times \left\{ \frac{1 + \bar{\nu}_{\mathbf{q}s}}{\omega - E(\mathbf{k} + \mathbf{q}) - \Omega_s(\mathbf{q}) - M^{(2)}[\mathbf{k} + \mathbf{q}, \omega - \Omega_s(\mathbf{q})]} + \right.$$
$$\left. + \frac{\bar{\nu}_{\mathbf{q}s}}{\omega - E(\mathbf{k} + \mathbf{q}) + \Omega_s(\mathbf{q}) - M^{(2)}[\mathbf{k} + \mathbf{q}, \omega + \Omega_s(\mathbf{q})]} \right\} \tag{6.40}$$

is the exciton mass operator in the fourth approximation. Here, $M^{(2)}(\mathbf{k}, \omega)$ is the mass operator in the second approximation. If we substitute $M^{(4)}(\mathbf{k}, \omega)$ for $M^{(2)}(\mathbf{k}, \omega)$ in (6.40), we can determine the mass operator in the sixth approximation, etc. Thus, we can obtain an approximate integral equation for the exciton mass operator ($\bar{n}_k \ll 1$) without allowing for the variation of the functions $F_s(\mathbf{k}, \mathbf{q})$ in higher approximations:

$$M(\mathbf{k}, \omega) = \sum_{s, \mathbf{q}} |F_s(\mathbf{k}, \mathbf{q})|^2 \times$$
$$\times \left\{ \frac{1 + \bar{\nu}_{\mathbf{q}s}}{\omega - E(\mathbf{k} + \mathbf{q}) - \Omega_s(\mathbf{q}) - M[\mathbf{k} + \mathbf{q}, \omega - \Omega_s(\mathbf{q})]} + \right.$$
$$\left. + \frac{\bar{\nu}_{\mathbf{q}s}}{\omega - E(\mathbf{k} + \mathbf{q}) + \Omega_s(\mathbf{q}) - M(\mathbf{k} + \mathbf{q}, \omega + \Omega_s(\mathbf{q}))} \right\}. \tag{6.41}$$

7. The Dielectric Constant of Simple Molecular Crystals with Allowance for Interaction Retardation

In the calculations in Chapters II and III of the electronic excitations of molecular crystals whose molecules were fixed at the lattice points, we took into account only the Coulomb interaction between electric charges. This same approximation is used in calculating the electronic excitations of atoms and molecules. As we have seen, electronic excitations (excitons) can embrace large regions in molecular crystals. In this case, the role of the retardation of the interactions between the electric charges of the crystal must be explained.

This problem has been investigated most completely by Onishchuk [21] by the classical example of the interaction of oscillators (corresponding to molecular dipole transitions to excited states) with an electromagnetic field and by Onishchuk and Davydov [22] using quantum equations and the Green function method. To determine the retardation of molecular interaction, in [21, 22] a crystal model was used in which the interaction between oscillators (or molecules in quantum theory) was accomplished through a field containing both shortwave and longwave components. Earlier, Fano [23], Hopfield [24], Agranovich [25, 26], and Agranovich and Konobeev [27] studied retardation effects in the theory of excitons taking into account only the longwave (macroscopic) field (see also Section 5 of Chapter III).

Since a crystal is a heterogeneous medium, a question arises about the justification for introducing the dielectric-constant tensor, which characterizes in the Maxwell equations the properties of a homogeneous medium. This question has already been discussed by a number of authors [28-31]. We shall also consider this problem on the basis of the method used in [21, 22].

For simplicity, we shall use a model of a molecular crystal that contains one molecule in each unit cell. The molecules are rigidly fixed at the lattice points **n**. We shall take into account the molecular excited states the transitions to which from the ground state are allowed in the dipole approximation. These molecular excitations will be characterized by the transition frequency[5] Ω_f and the dipole moment $\mathbf{d}_f = \mathbf{e}_f \mathbf{d}_f$, where $\mathbf{e}_f$ is a unit vector. The square of the dipole moment is related to the oscillator strength F_f and frequency Ω_f of the transition and the electron charge e and mass m by the relation

$$\mathbf{d}_f^2 = \frac{e^2 \hbar F_f}{2m\Omega_f}.$$

If $B_f^+(\mathbf{n}, t)$ and $B_f(\mathbf{n}, t)$ are the Bose creation and annihilation operators (we shall use the Heisenberg representation) for an electronic excitation f in a molecule located at a lattice point **n**, then the electronic excitations of the molecules in the crystal are determined by the Hamiltonian operator

$$H_M = \sum_{\mathbf{n}, f} \hbar\Omega_f B_f^+(\mathbf{n}, t) B_f(\mathbf{n}, t). \quad (7.1)$$

In this same representation, the operator of the electric dipole moment for transition of the molecule to the f-th excited state has the form

$$\mathbf{p}_f(\mathbf{n}, t) = \mathbf{e}_f d_f [B_f^+(\mathbf{n}, t) + B_f(\mathbf{n}, t)]. \quad (7.2)$$

The Hamiltonian operator for the free electromagnetic field that carries the interactions between molecules can be written as

$$H_\gamma = \frac{1}{8\pi} \int \{16\pi^2 c^2 \mathbf{\Pi}^2(\mathbf{r}, t) + [\operatorname{rot} \mathbf{A}(\mathbf{r}, t)]^2\} d^3 r, \quad (7.3)$$

where $\mathbf{A}(\mathbf{r}, t)$ and $\mathbf{\Pi}(\mathbf{r}, t) = \frac{1}{4\pi c^2} \frac{\partial \mathbf{A}}{\partial t}$ are, respectively, the vec-

[5]The frequency Ω_f characterizes the electronic excitation of a molecule in the crystal. If $\Delta\varepsilon_f$ is the free-molecule excitation energy, then $\hbar\Omega_f \approx \Delta\varepsilon_f + D_f$, where D_f is the change in the interaction energy of all of the molecules with the given one in its transition to the f-th excited state.

tor-potential operator and the operator of its conjugate momentum, which satisfy the commutation relation

$$[\mathbf{A}_j(\mathbf{r}, t), \ \mathbf{\Pi}_l(\mathbf{r}', t)] = i\hbar\delta_{jl}\delta(\mathbf{r}-\mathbf{r}').$$

Here, we shall use a potential calibration in which the scalar potential is zero. In this case, the electric- and magnetic-field strengths **E** and **H** are expressed in terms of the vector potential by the relations

$$\mathbf{E} = -\frac{1}{c}\frac{\partial \mathbf{A}}{\partial t}, \qquad \mathbf{H} = \operatorname{rot}\mathbf{A}.$$

Finally, the interaction operator for the electromagnetic field with the molecules of the crystal (see Section 5, Chapter III) has the form

$$H_{\text{M-Y}} = -\frac{i}{c}\sum_{\mathbf{n}, f}\mathbf{e}_f\mathbf{A}(\mathbf{n}, t)\, d_f\Omega_f[B_f^+(\mathbf{n}, t) - B_f(\mathbf{n}, t)] + \\ + \sum_{\mathbf{n}}\frac{e^2S}{2mc^2}|\mathbf{A}(\mathbf{n}, t)|^2. \tag{7.4}$$

The total Hamiltonian of the crystal is equal to the sum of operators (7.1)-(7.4). The time dependence of the field and the dipole-moment operators is expressed in terms of the total Hamiltonian by the relation $i\hbar(\partial a/\partial t) = [H, a]$.

Applying this relation twice, we obtain the operator equations of motion

$$\left[\frac{1}{c^2}\frac{\partial^2}{\partial t^2} + \operatorname{rot}\operatorname{rot}\right]\mathbf{A}(\mathbf{r}, t) = \frac{4\pi}{c}\frac{\partial \mathbf{P}(\mathbf{r}, t)}{\partial t}, \tag{7.5}$$

$$\left[\frac{\partial^2}{\partial t^2} + \Omega_f^2\right]p_f(\mathbf{n}, t) = -\frac{e^2}{mc}F_f\left(\mathbf{e}_f\cdot\frac{\partial \mathbf{A}(\mathbf{n}, t)}{\partial t}\right), \tag{7.6}$$

where

$$\mathbf{P}(\mathbf{r}, t) = \sum_{\mathbf{n}, f}\mathbf{e}_f p_f(\mathbf{n}, t)\,\delta(\mathbf{r}-\mathbf{n}) \tag{7.7}$$

is the specific-polarization operator of the crystal. From Eq. (7.5) follows the additional operator equation

$$\operatorname{div}\left\{-\frac{1}{c}\frac{\partial \mathbf{A}(\mathbf{r}, t)}{\partial t} + 4\pi\mathbf{P}(\mathbf{r}, t)\right\} = 0, \tag{7.8}$$

which means that now we need consider only those states of the system in which the mean values of the operator on the left of (7.8) are equal to zero.

The electromagnetic field and polarization are produced in the crystal by external influences. Let us assume that these influences are from extrinsic macroscopic currents,

$$j_{\text{at}}(\mathbf{r}, t) = j(\mathbf{Q}, \omega)\exp\{i(\mathbf{Q}\mathbf{r} - \omega t)\}, \qquad Qa \ll 1, \tag{7.9}$$

which were switched on adiabatically in the infinite past ($t = -\infty$). This is accomplished by addition of the small positive imaginary value $i\eta$ to the frequency, i.e., by substituting $\omega + i\eta$ for ω. In the final formulas, we shall pass to the limit $\eta \to +0$.

Extrinsic currents (7.9) produce electric fields and polarization in the crystal whose mean values by time t are determined with the aid of retarded Green functions by the expressions (see [14], Section 28)

$$\langle A_j(\mathbf{r}, t)\rangle = -\frac{1}{\hbar c}\int \langle\langle A_j(\mathbf{r}, t) \mid A_l(\mathbf{r}', t')\rangle\rangle\, j_l(\mathbf{r}', t')\, d^3r' dt', \tag{7.10}$$

$$\langle p_f(\mathbf{n}, t)\rangle = -\frac{1}{\hbar c}\int \langle\langle p_f(\mathbf{n}, t) \mid A_l(\mathbf{r}', t')\rangle\rangle\, j_l(\mathbf{r}', t')\, d^3r' dt', \tag{7.11}$$

where

$$\langle\langle a(\mathbf{r}, t) \mid b(\mathbf{r}', t')\rangle\rangle \equiv -i\Theta(t - t')\langle 0 \mid [a(\mathbf{r}, t),\, b^+(\mathbf{r}', t')] \mid 0\rangle$$

for the retarded Green function of the Bose operators $a(\mathbf{r}, t)$ and $b(\mathbf{r}, t)$, which are given in the Heisenberg representation. Averaging is over the vacuum state of the system, which satisfies condition (7.8). In (7.10) and (7.11) and everywhere below, summation is over the twice-encountered tensor indices.

For crystals with infinite dimensions, the Green functions in (7.10) and (7.11) satisfy the conditions of translational invariance. For example,

$$\langle\langle A_j(\mathbf{r} + \mathbf{n}, t) \mid A_l(\mathbf{r}' + \mathbf{n}, t')\rangle\rangle = \langle\langle A_j(\mathbf{r}, t) \mid A_l(\mathbf{r}', t')\rangle\rangle. \tag{7.12}$$

Using motion equations (7.5) and (7.6), we find equations for the retarded Green functions:

$$\frac{1}{c^2}\frac{\partial^2}{\partial t^2}\langle\langle A_j(\mathbf{r}, t) \mid A_l(\mathbf{r}', t')\rangle\rangle + \operatorname{rot}_j \operatorname{rot}\langle\langle \mathbf{A}(\mathbf{r}, t) \mid \mathbf{A}_l(\mathbf{r}', t')\rangle\rangle =$$

$$= 4\pi\hbar\delta_{jl}\delta(t - t')\delta(\mathbf{r} - \mathbf{r}') + \frac{4\pi}{c}\frac{\partial}{\partial t}\langle\langle P_j(\mathbf{r}, t) \mid A_l(\mathbf{r}', t')\rangle\rangle,$$

$$\left[\frac{\partial^2}{\partial t^2} + \Omega_f^2\right]\langle\langle p_f(\mathbf{n}, t) \mid A_l(\mathbf{r}', t')\rangle\rangle = -\frac{e^2 F_f}{mc}\mathbf{e}_f\frac{\partial}{\partial t}\langle\langle \mathbf{A}(\mathbf{n}, t) \mid \mathbf{A}_l(\mathbf{r}', t')\rangle\rangle.$$

If we multiply these equations by $\mathbf{j}^{ex}(\mathbf{r}', t')$ and integrate with respect to the variables $\mathbf{r}'$ and t', we obtain, taking into account (7.10) and (7.11), equations for the mean fields and polarizations:

$$\left.\begin{aligned} \left\{\frac{1}{c^2}\frac{\partial^2}{\partial t^2} + \operatorname{rot}\operatorname{rot}\right\}\langle \mathbf{A}(\mathbf{r}, t)\rangle &= \\ = -4\pi \mathbf{j}^{ex}(\mathbf{r}, t) &+ \frac{4\pi}{c}\frac{\partial}{\partial t}\langle \mathbf{P}(\mathbf{r}, t)\rangle, \\ \left[\frac{\partial^2}{\partial t^2} + \Omega_f^2\right]\langle p_f(\mathbf{n}, t)\rangle &= -\frac{e^2 F_f}{mc}\mathbf{e}_f\frac{\partial}{\partial t}\langle \mathbf{A}(\mathbf{n}, t)\rangle, \end{aligned}\right\} \tag{7.13}$$

where

$$\langle \mathbf{P}(\mathbf{r}, t)\rangle = \sum_{\mathbf{n}, f}\mathbf{e}_f\langle p_f(\mathbf{n}, t)\rangle\,\delta(\mathbf{r}-\mathbf{n}) \tag{7.14}$$

is the mean specific polarization of the crystal.

Further, we shall assume that the crystal, which has N molecules, and the field occupy a large volume V in the form of a parallelepiped with edges $N_i\mathbf{a}_i$ (i = 1, 2, 3), where $\mathbf{a}_i$ are the three lattice basis vectors. The states of the crystal and the field satisfy cyclicity conditions with large periods $N_i\mathbf{a}_i$. In this case,

$$N = N_1N_2N_3; \quad V = vN,$$

where v is the volume of a unit cell.

Extrinsic currents (7.9) have a fixed frequency ω, so the fields and polarizations induced by them have the same frequency. To separate clearly the longwave (macroscopic) and shortwave (microscopic) components of the field and polarization, it is convenient in (7.13) to convert to Fourier transforms by means of the relations

$$\left.\begin{aligned} \langle \mathbf{A}(\mathbf{r}, t)\rangle &= \frac{1}{V}\sum_{\mathbf{k}}^{\infty} e^{i(\mathbf{kr}-\omega t)}\,\mathbf{A}(\mathbf{k}, \omega), \\ \langle \mathbf{P}(\mathbf{r}, t)\rangle &= \frac{1}{V}\sum_{\mathbf{k}}^{\infty} e^{i(\mathbf{kr}-\omega t)}\,\mathbf{P}(\mathbf{k}, \omega), \end{aligned}\right\} \tag{7.15}$$

where $\mathbf{k}$ is the wave vector, which takes in the infinite $\mathbf{k}$-space the discrete values

$$\mathbf{k} = \sum_{i=1}^{3}\frac{2\pi}{N_i}\nu_i\mathbf{b}_i, \qquad \nu_i = 0, \pm 1, \ldots, \infty,$$

where $\mathbf{b}_i$ are the basis vectors of the reciprocal lattice; $\mathbf{b}_j\mathbf{a}_i = \delta_{ij}$.

Taking the second equation of (7.15) and considering (7.14), we obtain

$$\mathbf{P}(\mathbf{k}, \omega) = \sum_f \mathbf{e}_f p_f(\mathbf{k}, \omega), \tag{7.16}$$

where

$$p_f(\mathbf{k}, \omega) = \sum_{\mathbf{n}} \int e^{-i(\mathbf{kn}-\omega t)} \langle p_f(\mathbf{n}, t)\rangle \, dt. \tag{7.17}$$

If

$$\mathbf{g} = 2\pi \sum_{i=1}^{3} \mathbf{b}_i m_i \quad (m_i = 0, \pm 1, \dots)$$

are the reciprocal-lattice vectors (multiplied by 2π), then, considering the equation $\exp(i\mathbf{ng}) = 1$, we obtain from (7.17) the important property

$$p_f(\mathbf{k} + \mathbf{g}, \omega) = p_f(\mathbf{k}, \omega), \tag{7.18}$$

which allows all $p_f(\mathbf{k}, \omega)$ to be reduced to their values in the first Brillouin zone in the k space. Transforming (7.17), taking (7.18) into account, we find

$$\langle p_f(\mathbf{n}, t)\rangle = \frac{1}{N} \sum_{\mathbf{k}} e^{i(\mathbf{kn}-\omega t)} p_f(\mathbf{k}, \omega). \tag{7.19}$$

In (7.19), the sum $\sum_{\mathbf{k}}$ indicates, unlike the sum $\sum_{\mathbf{k}}^{\infty}$ in (7.15), summation over the N values of k in the first Brillouin zone.

If we substitute expressions (7.15) and (7.19) into (7.13), we find, considering (7.16),

$$\left[\mathbf{k}^2 - \frac{\omega^2}{c^2}\right] \mathbf{A}(\mathbf{k}, \omega) - \mathbf{k}(\mathbf{k}\mathbf{A}(\mathbf{k}, \omega)) = \\ = \frac{4\pi}{c} \mathbf{j}(\mathbf{Q}, \omega)\, \delta_{\mathbf{kQ}} - \frac{4\pi i\omega}{c} \sum_f \mathbf{e}_f p_f(\mathbf{k}, \omega), \tag{7.20}$$

$$[\Omega_f^2 - \omega^2]\; p_f(\mathbf{k}, \omega) = \frac{ie^2 F_f \omega}{mcv} \sum_g \mathbf{e}_f \mathbf{A}(\mathbf{k} + \mathbf{g}, \omega). \tag{7.21}$$

Isolating the longwave field $\mathbf{A}(\mathbf{Q}, \omega)$ and the fields and polarizations related to it, we can rewrite system (7.20)-(7.21):

$$\left(Q^2 - \frac{\omega^2}{c^2}\right) \mathbf{A}(\mathbf{Q}, \omega) - \mathbf{Q}(\mathbf{Q} \cdot \mathbf{A}(\mathbf{Q}, \omega)) + i\frac{4\pi\omega}{c} \sum_f \mathbf{e}_f p_f(\mathbf{Q}, \omega) =$$
$$= \frac{4\pi}{c} \mathbf{j}(\mathbf{Q}, \omega), \tag{7.22}$$

$$(\Omega_f^2 - \omega^2) p_f(\mathbf{Q}, \omega) = \frac{ie^2 F_f \omega}{mcv} \sum_{\mathbf{g}=0}^{\infty} \mathbf{e}_f \mathbf{A}(\mathbf{Q} + \mathbf{g}, \omega), \tag{7.23}$$

$$\left[(\mathbf{Q} + \mathbf{g})^2 - \frac{\omega^2}{c^2}\right] \mathbf{A}(\mathbf{Q} + \mathbf{g}, \omega) - [\mathbf{g} + \mathbf{Q}]([\mathbf{Q} + \mathbf{g}] \cdot \mathbf{A}(\mathbf{Q} + \mathbf{g}, \omega)) =$$
$$= -i\frac{4\pi\omega}{c} \sum_f \mathbf{e}_f p_f(\mathbf{Q}, \omega). \tag{7.24}$$

In our formulation of the problem, the longwave field $\mathbf{A}(\mathbf{Q}, \omega)$, the polarization $p_f(\mathbf{Q}, \omega)$, and the shortwave fields $\mathbf{A}(\mathbf{Q} + \mathbf{g}, \omega)$ are absent without an external current. Further, it follows from (7.24) that shortwave fields can be created only by the polarization $p_f(\mathbf{Q}, \omega)$. Therefore, they can be eliminated from (7.23) and (7.24), since they are expressed through $p_f(\mathbf{Q}, \omega)$.

For brevity, let us temporarily introduce the vector $\mathbf{q} \equiv \mathbf{Q} + \mathbf{g}$ and represent each microfield as the sum of two components

$$\mathbf{A}(\mathbf{q}, \omega) = \mathbf{A}^{\parallel}(\mathbf{q}, \omega) + A^{\perp}(\mathbf{q}, \omega), \tag{7.25}$$

where $\mathbf{A}^{\parallel}(\mathbf{q}, \omega)$ and $\mathbf{A}^{\perp}(\mathbf{q}, \omega)$ are the parallel and perpendicular (with respect to the vector $\mathbf{q}$) components of the microfield. Then, from (7.24) we find

$$\left.\begin{aligned} \frac{\omega}{c} \mathbf{A}^{\parallel}(\mathbf{q}, \omega) &= i4\pi \sum_f \mathbf{e}_f^{\parallel}(\mathbf{q}) p_f(\mathbf{q}, \omega), \\ \left(\mathbf{q}^2 - \frac{\omega^2}{c^2}\right) \mathbf{A}^{\perp}(\mathbf{q}, \omega) &= -i\frac{4\pi\omega}{c} \sum_f \mathbf{e}_f^{\perp}(\mathbf{q}) p_f(\mathbf{q}, \omega), \end{aligned}\right\} \tag{7.26}$$

where

$$\mathbf{e}_f^{\parallel}(\mathbf{q}) \equiv \frac{\mathbf{q}(\mathbf{e}_f \mathbf{q})}{\mathbf{q}^2} \tag{7.27}$$

is the component of the vector $\mathbf{e}_f$ that is parallel to the vector $\mathbf{q}$ and

$$\mathbf{e}_f^{\perp}(\mathbf{q}) \equiv -\frac{[\mathbf{q} \times [\mathbf{q} \times \mathbf{e}_f]]}{\mathbf{q}^2} = \mathbf{e}_f - \mathbf{e}^{\parallel}(\mathbf{q}) \tag{7.28}$$

is the component of the vector $\mathbf{e}_f$ that is perpendicular to $\mathbf{q}$.

If we substitute (7.26) into (7.25), we find an explicit expression for the microfields in terms of the amplitudes of the vectors of the molecular dipole transitions:

$$\frac{\omega}{c}\,\mathbf{A}(\mathbf{q},\omega) = i4\pi \sum_f p_f(\mathbf{Q},\omega)\,\frac{\mathbf{q}(\mathbf{q}\mathbf{e}_f) - \mathbf{e}_f\frac{\omega^2}{c^2}}{\mathbf{q}^2 - \left(\frac{\omega}{c}\right)^2}, \tag{7.29}$$

where $\mathbf{q} \equiv \mathbf{Q} + \mathbf{g},\ \mathbf{g} \neq 0$.

Since the extrinsic currents were switched on adiabatically in the infinite past, we must substitute $\omega + i\eta$ for ω in the denominator of this expression (and in all subsequent expressions of this type). This substitution formally ensures the proper circumvention of the pole in integration. The physical meaning of this substitution is that microfields must be absent when molecular transitions (polarization of the medium) are absent.

After substitution of (7.29) on the right of (7.23), we find

$$\{[\Omega_f^2 - \omega^2]\,\delta_{ff'} + \Lambda_{ff'}(\mathbf{Q},\omega)\}\,p_f(\mathbf{Q},\omega) = \frac{ie^2F_f\omega}{mvc}(\mathbf{e}_f\cdot\mathbf{A}(\mathbf{Q},\ \omega)), \tag{7.30}$$

where

$$\Lambda_{ff'}(\mathbf{Q},\omega) = \omega_p^2 F_f \sum_{\mathbf{g}\neq 0} \frac{(\mathbf{e}_f\cdot\mathbf{Q}+\mathbf{g})(\mathbf{e}_{f'}\cdot\mathbf{Q}+\mathbf{g}) - \mathbf{e}_f\mathbf{e}_{f'}\frac{\omega^2}{c^2}}{(\mathbf{Q}+\mathbf{g})^2 - \frac{\omega^2}{c^2}} \tag{7.31}$$

is the "force" matrix, which characterizes the molecular interactions in terms of the microfields; and $\omega_p^2 = 4\pi e^2/mv$ is the square of the plasma frequency.

Equations (7.22) and (7.30) form a complete system of macroscopic equations that relate the macroscopic field $\mathbf{A}(\mathbf{Q},\omega)$ and the amplitudes of the molecular-transition vectors $p_f(\mathbf{Q},\omega)$ to the macroscopic currents that are created. If we eliminate $p_f(\mathbf{Q},\omega)$ from these equations, we obtain for the crystal a complete system of macroscopic Maxwell equations that relate macroscopic extrinsic current (7.9) to the longwave (macroscopic) field that it excites. This elimination is especially simple when the off-diagonal elements of matrix (7.31) can be ignored, i.e., when the mutual effect of the various molecular excited states is negligible. In this approximation,

$$\Lambda_{ff'} = \delta_{ff'} \Lambda_f,$$

and, according to (7.30), we find

$$p_f(\mathbf{Q}, \omega) = i \frac{\omega\omega_p^2 F_f}{4\pi c} \frac{\mathbf{e}_f \mathbf{A}(\mathbf{Q}, \omega)}{\Omega_f^2 - \omega^2 + \Lambda_f(\mathbf{Q}, \omega)}.$$

If we substitute this value into (7.22), we obtain

$$\left[\delta_{lj}\mathbf{Q}^2 - \frac{\omega^2}{c^2}\varepsilon_{lj}(\mathbf{Q}, \omega) - Q_l Q_j\right] A_j(\mathbf{Q}, \omega) = \frac{4\pi}{c} j_l(\mathbf{Q}, \omega), \quad (7.32)$$

where

$$\varepsilon_{lj}(\mathbf{Q}, \omega) = \delta_{lj} + \sum_f \frac{\omega_p^2 F_f e_{f;l} e_{f;j}}{\Omega_f^2 - \omega^2 + \Lambda_f(\mathbf{Q}, \omega)}. \quad (7.33)$$

If we compare (7.32) with (I.4.11), we can see that the tensor ε_{lj} is the dielectric-constant tensor, which determines the macroscopic fields produced in the crystal by extrinsic macroscopic currents. The effect of all of the macroscopic fields that participate in the interaction between molecules is included in "force" matrix (7.33). The interaction-retardation effect is characterized in (7.33) by the terms containing the factor ω^2/c^2. Naturally, ignoring the retardation amounts to the formal passage to the limit $\omega/c \to 0$.

Now we shall assume that only transverse extrinsic currents act in the crystal, i.e.,

$$\mathbf{j}(\mathbf{Q}, \omega) = \mathbf{j}^{\perp}(\mathbf{Q}, \omega), \quad \text{where } (\mathbf{j}^{\perp}(\mathbf{Q}, \omega) \cdot \mathbf{Q}) = 0.$$

In this case, according to (7.22), the longitudinal macroscopic (longwave) field $\mathbf{A}^{\parallel}(\mathbf{Q}, \omega)$ is created only by polarization of the medium, i.e.,

$$\mathbf{A}^{\parallel}(\mathbf{Q}, \omega) = i \frac{4\pi c}{\omega} \sum_f \mathbf{e}_f^{\parallel}(\mathbf{Q}) p_f(\mathbf{Q}, \omega), \quad (7.34)$$

$$\left(\mathbf{Q}^2 - \frac{\omega^2}{c^2}\right) \mathbf{A}^{\perp}(\mathbf{Q}, \omega) + i \frac{4\pi\omega}{c} \sum_f \mathbf{e}_f^{\perp}(\mathbf{Q}) p_f(\mathbf{Q}, \omega) = \frac{4\pi}{c} \mathbf{j}^{\perp}(\mathbf{Q}, \omega). \quad (7.35)$$

The vectors $\mathbf{e}_f^{\parallel}(\mathbf{Q})$ and $\mathbf{e}_f^{\perp}(\mathbf{Q})$ are defined by (7.27) and (7.28).

Let us isolate on the right of (7.30) the longitudinal part of the macroscopic field. Then, using (7.34), we can transform (7.30) to

$$\sum_{f} [(\Omega_j^2 - \omega^2)\,\delta_{ff'} + \Gamma_{ff'}(\mathbf{Q}, \omega)]\, p_{j'}(\mathbf{Q}, \omega) = \frac{ie^2 F_f \omega}{mvc}\, \mathbf{e}_f \mathbf{A}^{\perp}(\mathbf{Q}, \omega), \quad (7.36)$$

where the new force matrix is defined by the expression

$$\Gamma_{ff'}(\mathbf{Q}, \omega) = \Lambda_{ff'} + \omega_p^2 F_f \frac{(\mathbf{e}_f \mathbf{Q})(\mathbf{e}_{f'} \mathbf{Q})}{\mathbf{Q}^2}. \quad (7.37)$$

In approximation of the independent molecular excited states, when

$$\Gamma_{ff'} = \Gamma_f \delta_{ff'}, \quad (7.37\text{a})$$

from Eq. (7.36), we find

$$p_f(\mathbf{Q}, \omega) = i\,\frac{\omega_p^2 \omega}{4\pi c}\, \frac{\mathbf{e}_f \mathbf{A}^{\perp}(\mathbf{Q}, \omega)}{\Omega_f^2 - \omega^2 + \Gamma_f(\mathbf{Q}, \omega)}.$$

Substituting this value into Eq. (7.35), we transform it to

$$\left[-\frac{\omega^2}{c^2}\,\varepsilon_{jl}^{\perp}(\mathbf{Q}, \omega) + \mathbf{Q}^2 \delta_{jl}\right] A_l^{\perp}(\mathbf{Q}, \omega) = \frac{4\pi}{c}\, j_j^{\perp}, \quad (7.38)$$

where

$$\varepsilon_{jl}^{\perp}(\mathbf{Q}, \omega) = \eta_{jl} + \sum_f \frac{\omega_p^2 F_f e_{f;\,j}^{\perp}(\mathbf{Q})\, e_{f;\,l}^{\perp}}{\Omega_f^2 - \omega^2 + \Gamma_f(\mathbf{Q}, \omega)}, \quad (7.39)$$

when $\eta_{lm} = \delta_{lm} - s_l s_m$, $s = \mathbf{Q}/Q$.

If we compare (7.38) with (I.4.8), we see that the values of (7.39) should be considered as components of the transverse dielectric-constant tensor. With the aid of (7.38), the transverse dielectric-constant tensor determines the macroscopic field produced in the crystal by an extrinsic transverse macroscopic current with given frequency and wave vector. The force matrix Γ_f in (7.39) takes into account all interactions between the molecular dipole transitions of the crystal that are carried by the longitudinal macroscopic field and all shortwave fields (with allowance for retardation).

It is easy to see that the components (7.39) of the tensor $\varepsilon^{\perp}$ are determined by Eq. (7.38) with accuracy to the terms $\gamma s_l s_m$, where γ is an arbitrary constant. We selected the constant γ such that

$$s_l \varepsilon_{lm}^{\perp}(\mathbf{Q}, \omega) = 0.$$

The concept of the "transverse" dielectric-constant tensor was introduced by Pekar [32] and Agranovich and Ginzburg [29].

The transverse vectors $\mathbf{A}^{\perp}(\mathbf{Q}, \omega)$ and $\mathbf{j}^{\perp}(\mathbf{Q}, \omega)$ in (7.38) lie in a plane perpendicular to the wave vector $\mathbf{Q}$. Therefore, transverse dielectric constant (7.39) is a tensor that acts on the vectors lying in this plane (see Section 3, Chapter I).

In the above method for calculating the dielectric constant, the entire interaction between molecular transitions was taken into account through the field. This method results in the appearance in the matrices $\Gamma_f(\mathbf{Q}, \omega)$ and $\Lambda_f(\mathbf{Q}, \omega)$ of nonphysical terms for the self-energies of the dipoles. To eliminate these terms, it is convenient to transform the matrices Γ_f and Λ_f to sums over the points of the space lattice. For this transformation, we introduce the auxiliary matrix

$$L_f(\mathbf{Q}, \omega) = \frac{\hbar}{2\Omega_f}\left\{\Lambda_f(\mathbf{Q}, \omega) + \omega_p^2 F_f \frac{(\mathbf{Q}\mathbf{e}_f)^2 - (\omega^2/c^2)}{Q^2 - (\omega^2/c^2)}\right\} =$$

$$= \frac{\hbar}{2\Omega_f}\left\{\Gamma_f(\mathbf{Q}, \omega) + \omega_p^2 F_f \frac{\frac{\omega^2}{c^2}\left[\frac{(\mathbf{e}_f\mathbf{Q})^2}{Q^2} - 1\right]}{Q^2 - (\omega^2/c^2)}\right\}. \quad (7.40)$$

Considering the explicit form of matrix (7.31), we can see that the matrix $L_f(\mathbf{Q}, \omega)$ can be written as

$$L_f(\mathbf{Q}, \omega) = -\{\partial_f [e^{i\mathbf{Q}\mathbf{r}} S(\mathbf{r})]\}_{r=0},$$

where

$$\left.\begin{aligned} &\partial_f \equiv d_f^2 \left[(\mathbf{e}_f\nabla)^2 + \frac{\omega^2}{c^2}\right], \qquad S(\mathbf{r}) = \sum_{\mathbf{g}} S_{\mathbf{g}} \exp(i\mathbf{g}\mathbf{r}), \\ &S_{\mathbf{g}} = \frac{4\pi}{v}\left[(\mathbf{g} + \mathbf{Q})^2 - \frac{\omega^2}{c^2}\right]^{-1}. \end{aligned}\right\} \quad (7.41)$$

In expression (7.41), the sum over $\mathbf{g}$ is taken over all $\mathbf{g}$ including $\mathbf{g} = 0$. If we take into account that $\exp(i\mathbf{g}\mathbf{n}) = 1$, then it follows from (7.41) that the function $S(\mathbf{r})$ is periodic in $\mathbf{r}$, i.e.,

$$S(\mathbf{r} + \mathbf{n}) = S(\mathbf{r}).$$

Therefore, this sum can be represented as a sum over all lattice points:

$$S(\mathbf{r}) \equiv \sum_{\mathbf{g}} S_{\mathbf{g}} \exp(i\mathbf{gr}) = \sum_{\mathbf{n}} \varphi(\mathbf{r}-\mathbf{n}). \tag{7.42}$$

The functions $v^{-1/2}\exp(i\mathbf{gr})$ (v is the volume of a unit cell) form a complete orthonormal set of functions:

$$\frac{1}{v}\int_{v} \exp[i(\mathbf{g}-\mathbf{g}')\mathbf{r}]\,d^3r = \delta_{\mathbf{gg}'},$$

$$\frac{1}{v}\sum_{\mathbf{g}} \exp[i\mathbf{g}(\mathbf{r}-\mathbf{r}')] = \delta(\mathbf{r}-\mathbf{r}');$$

therefore, considering $\exp(i\mathbf{gn}) = 1$, we have

$$S_{\mathbf{g}} = \frac{1}{v}\int_{V} \exp(-i\mathbf{gr})\,\varphi(\mathbf{r})\,d^3r,$$

or

$$\varphi(\mathbf{r}) = \frac{v}{(2\pi)^3}\int S_{\mathbf{g}} \exp(i\mathbf{gr})\,d^3g = \frac{\exp\left\{-i\mathbf{Qr} + i\frac{\omega}{c}|\mathbf{r}|\right\}}{|\mathbf{r}|}. \tag{7.43}$$

When calculating integral (7.43), we substitute $\omega - i\eta$ for ω in the expression $S_{\mathbf{g}}$, which is defined by (7.41). This substitution ensures the proper circumvention of the pole.

Using (7.42) and (7.43), we can write matrix elements (7.40) as a sum of two terms:

$$L_f(\mathbf{Q},\omega) = L_f^{(1)} + iL_f^{(2)}, \tag{7.44}$$

where

$$\left.\begin{aligned} L_f^{(1)} &= -\left\{\partial_f \sum_{\mathbf{n}}{}' \frac{e^{i\mathbf{Qn}}}{|\mathbf{r}-\mathbf{n}|}\cos\left[\frac{\omega}{c}|\mathbf{r}-\mathbf{n}|\right]\right\}_{\mathbf{r}=0}, \\ L_f^{(2)} &= -\left\{\partial_f \sum_{\mathbf{n}}{}' \frac{e^{i\mathbf{Qn}}}{|\mathbf{r}-\mathbf{n}|}\sin\left[\frac{\omega}{c}|\mathbf{r}-\mathbf{n}|\right]\right\}_{\mathbf{r}=0}. \end{aligned}\right\} \tag{7.45}$$

The prime indicates that terms with $\mathbf{n} = 0$ are absent in the sums. We eliminate these terms, because they correspond to an infinite field self-energy.

Without allowance for retardation ($\omega/c = 0$), after elimination of the self-energy we obtain

$$\lim_{\frac{\omega}{c}\to 0} L_f = \lim \frac{\hbar\Gamma_f(\mathbf{Q}, \omega)}{2\Omega_f} = -\left\{d_f(\mathbf{e}_f\nabla)^2 \sum{}' \frac{e^{i\mathbf{Q}\mathbf{n}}}{|\mathbf{r}-\mathbf{n}|}\right\}_{\mathbf{r}=0}. \quad (7.46)$$

This expression coincides with expression (II.5.14) for the resonance-interaction matrix when the latter is applied to crystals with one molecule in a unit cell ($\alpha = \beta = 1$).

In the presence of retardation ($\omega/c \neq 0$), the imaginary part of (7.44) in the frequency range $c\omega \neq |\mathbf{Q}|$ is determined by the second auxiliary terms inside the braces in (7.40). According to (7.31) and (7.37), the matrices $\Lambda_{ff'}$ and $\Gamma_{ff'}$ are real at frequencies not equal to the x-ray quantum frequencies, i.e., when $c\omega \neq |\mathbf{Q}+\mathbf{g}|$. When calculating the matrices $\Lambda_{ff'}$ and $\Gamma_{ff'}$ therefore, only the real part need be retained in (7.44).

It follows from expression (7.45) that the retardation effect for the microfields that carry the interactions between molecules is proportional to

$$\alpha = \Delta L \cdot \left(\frac{\omega a}{c}\right)^2,$$

where ΔL is the width of the exciton band. When

$$\omega / c = 3 \cdot 10^4 \text{ cm}^{-1} \quad \text{and} \quad a = 10^{-7} \text{ cm}$$

α is 10^{-5} of the width of the exciton band. Thus, the discrete structure of the crystal and the retardation have little effect on the resonance-interaction matrix. At the longwave limit $\omega a/c \ll 1$, they are entirely negligible.

Equations (7.32) and (7.38) express the total and transverse dielectric constants in terms of the macroscopic values of the fields produced by the total and transverse extrinsic currents, respectively. In some cases, it is convenient to express the dielectric constants directly in terms of the photon Green functions. To derive the corresponding relations, let us transform the photon Green function in expression (7.10) by means of the equation

$$\langle\langle A_j(\mathbf{r}, t) | A_l(\mathbf{r}', t')\rangle\rangle = \langle\langle A_j(\mathbf{r}, t-t') | A_l(\mathbf{r}', 0)\rangle\rangle =$$

$$= \frac{1}{2\pi V} \sum_{\mathbf{k}} \int^{\infty} d\omega\, e^{i(\mathbf{k}\mathbf{r}-\omega[t-t'])} \langle\langle A_j(\mathbf{k}, \omega) | A_l(\mathbf{r}', 0)\rangle\rangle. \quad (7.47)$$

Further, let us introduce a new photon retarded Green function by means of the expression

$$\mathbf{T}_{jl}(\mathbf{k}, \omega \mid \mathbf{r}) = e^{i\mathbf{kr}} \langle\langle A_j(\mathbf{k}, \omega) \mid A_l(\mathbf{r}, 0) \rangle\rangle. \tag{7.48}$$

According to (7.11) and (7.47), the new photon Green function satisfies the condition of translational invariance under the variable $\mathbf{r}$:

$$\mathbf{T}_{jl}(\mathbf{k}, \omega \mid \mathbf{r}) = \mathbf{T}_{jl}(\mathbf{k}, \omega \mid \mathbf{r} + \mathbf{n}).$$

From the translational invariance of function (7.48) it follows that it can be represented as a sum over the reciprocal-lattice vectors:

$$\mathbf{T}_{jl}(\mathbf{k}, \omega \mid \mathbf{r}) = \sum_{\mathbf{g}} e^{i\mathbf{gr}} \mathbf{T}_{jl}^{(\mathbf{g})}(\mathbf{k}, \omega). \tag{7.49}$$

If we substitute expressions (7.47), (7.49), and (7.9) into mean-field definition (7.10) and take into account Fourier transform (7.15), we obtain

$$A_j(\mathbf{Q} + \mathbf{g}, \omega) = -\frac{1}{\hbar c} \mathbf{T}_{jl}^{(\mathbf{g})}(\mathbf{Q} + \mathbf{g}, \omega)\, j_l(\mathbf{Q}, \omega). \tag{7.50}$$

It follows from Eq. (7.50) that the macroscopic field $A_j(\mathbf{Q}, \omega)$ is created by the macroscopic current $j_l(\mathbf{Q}, \omega)$ only by means of the longwave component $\mathbf{T}_{lj}^{(0)}(\mathbf{Q}, \omega)$ of photon Green function (7.49). If we substitute $A_j(\mathbf{Q}, \omega)$ from Eq. (7.50) into Eq. (7.32), we obtain the desired equation for the longwave photon Green function

$$\left\{\mathbf{Q}^2\delta_{jl} - \frac{\omega^2}{c^2}\varepsilon_{jl}(\mathbf{Q}, \omega) - Q_j Q_l\right\} \mathbf{T}_{lm}^{(0)} = -4\pi\hbar\delta_{jm}. \tag{7.51}$$

Equation (7.51) describes the relationship of the total dielectric constant of the medium with the retarded Green function of the longwave electromagnetic field in a calibration in which the scalar potential is zero (Coulomb calibration). This relation was obtained in another way by Dzyaloshinskii and Pitaevskii [33.].

If the longwave phonon Green function for a transverse macroscopic field is defined by the relation

$$A_j^{\perp}(\mathbf{Q}, \omega) = -\frac{1}{\hbar c} \mathbf{T}_{jl}^{(0)\perp}(\mathbf{Q}, \omega)\, j_l^{\perp}(\mathbf{Q}, \omega),$$

then from (7.38) we obtain an equation for the transverse longwave phonon Green function:

$$\left[Q^2\delta_{jl} - \frac{\omega^2}{c^2}\,\varepsilon_{jl}^{\perp}(\mathbf{Q},\omega)\right] \mathbf{T}_{lm}^{(0)\perp}(\mathbf{Q},\omega) = -4\pi\hbar\delta_{jm}. \tag{7.52}$$

In this case, it follows from (7.48) and (7.49) that

$$T_{jl}^{(0)\perp} = \frac{1}{V}\int e^{i\mathbf{Qr}} \langle\langle A_j^{\perp}(\mathbf{Q},\omega) \mid A^{\perp}{}_l(\mathbf{r},0)\rangle\rangle\, d^3r. \tag{7.53}$$

According to (III.5.4), the transverse vector-potential operator is expressed in terms of the creation $a_l^{\dagger}(\mathbf{Q}, t)$ and annihilation $a_l(\mathbf{Q}, t)$ operators for photons with the wave vector $\mathbf{Q}$ and polarization l by the relation

$$A_l^{\perp}(\mathbf{r},t) = \sqrt{\frac{2\pi c}{V\,|\mathbf{Q}|}}\,\gamma_l(\mathbf{Q},t)\,e^{i\mathbf{Qr}}, \tag{7.54}$$

where

$$\gamma_l(\mathbf{Q},t) = a_l(\mathbf{Q},t) + a_l^{+}(-\mathbf{Q},t).$$

If we substitute into expression (7.53) the value

$$\langle\langle A_j^{\perp}(\mathbf{Q},\omega) \mid A_l^{\perp}(\mathbf{r},0)\rangle\rangle = \int e^{-i(\mathbf{Qr}'-\omega t)} \langle\langle A_j^{\perp}(\mathbf{r}',t) \mid A_l^{\perp}(\mathbf{r},0)\rangle\rangle\, d^3r'\,dt$$

and take into account expression (7.54), we find

$$T_{jl}^{(0)\perp}(\mathbf{Q},\omega) = \frac{2\pi\hbar c}{|\mathbf{Q}|}\,\Gamma_{jl}(\mathbf{Q},\omega), \tag{7.55}$$

where

$$\Gamma_{jl}(\mathbf{Q},\omega) = \int e^{i\omega t}\Gamma_{jl}(\mathbf{Q},t)\,dt$$

is the Fourier transform of the transverse photon retarded Green function, which was considered in Section III of this chapter and is defined by the equation

$$\Gamma_{jl}(\mathbf{Q},t) = \langle\langle\gamma_j(\mathbf{Q},t) \mid \gamma_l(\mathbf{Q},0)\rangle\rangle \equiv -i\Theta(t)\langle 0 \mid [\gamma_j(\mathbf{Q},t), \gamma_l^{+}(\mathbf{Q},0)] \mid 0\rangle.$$

If we substitute (7.55) into (7.52), we find

$$\left\{Q^2\delta_{jl} - \frac{\omega^2}{c^2}\,\varepsilon_{jl}^{\perp}(\mathbf{Q},\omega)\right\} \Gamma_{lm}^{(0)}(\mathbf{Q},\omega) = -\frac{2\,|\mathbf{Q}|}{c}\,\delta_{jm}, \tag{7.56}$$

which relates the transverse dielectric constant to the photon retarded Green function. In accordance with relation (3.13), it follows from (7.56) that the components of the transverse dielectric-constant tensor

$$\varepsilon_{jl}^{\perp}(\mathbf{Q}, \omega) = \frac{c^2 Q^2}{\omega^2}\delta_{jl} + \frac{2|\mathbf{Q}|c}{\omega^2}\{\Gamma^0(\mathbf{Q}, \omega)\}_{jl}^{-1} \tag{7.57}$$

are expressed in terms of the components of the inverse transverse photon Green function. As was noted in Section 3, the need to invert the Fourier transform of the transverse photon retarded Green function complicates the practical use of (7.57) in calculating the transverse dielectric-constant tensor.

8. The Dielectric Constant of Complicated Molecular Crystals with Allowance for Retardation

The method developed in the preceding section for determining the total retarded interaction between the dipole moments of quantum transitions can be extended to crystals that contain several molecules in a unit cell. Let a unit cell contain σ molecules whose positions are determined by the vectors

$$\mathbf{r}_{\mathbf{n}\alpha} = \mathbf{n} + \boldsymbol{\rho}_\alpha,$$
$$\alpha = 1, 2, \ldots, \sigma,$$

where $\mathbf{n}$ is the lattice vector. The dipole moments of the quantum transitions of the molecules occupying the position $\boldsymbol{\rho}_\alpha$ in a unit cell will be denoted by the vectors

$$\mathbf{d}_{f\alpha} = d_f \mathbf{e}_{f\alpha},$$

where $\mathbf{e}_{f\alpha}$ are unit vectors that determine the direction of the dipole moment in transition of a molecule to the f-th excited state;

$$d_f^2 = \frac{\hbar e^2 F_f}{2m\Omega_f},$$

and F_f is the oscillator strength of an intramolecular transition of the frequency

$$\Omega_f = \hbar^{-1}(\Delta\varepsilon_f + D_f).$$

By analogy with the calculations made in the preceding section, we find equations that relate the Fourier components of the amplitudes of the dipole quantum transitions in the molecules $p_{f\alpha}(\mathbf{Q}, \omega)$, the vector potentials $\mathbf{A}(\mathbf{Q}+\mathbf{g}, \omega)$, and the extrinsic macroscopic current $\mathbf{j}(\mathbf{Q}, \omega)$ for fixed wave vector and frequency. According to (7.20) and (7.21), we now have

$$\left(k^2-\frac{\omega^2}{c^2}\right)\mathbf{A}(\mathbf{k}, \omega)-\mathbf{k}(\mathbf{k}\cdot\mathbf{A}(\mathbf{k}, \omega))=$$
$$=4\pi j(\mathbf{Q}, \omega)\,\delta_{\mathbf{kQ}}-i\frac{4\pi\omega}{c}\sum_f e^{-i\mathbf{k}\rho_\alpha}\,\mathbf{e}_{f\alpha}p_{f\alpha}(\mathbf{k},\omega), \tag{8.1}$$

$$(\Omega_f^2-\omega^2)\,p_{f\alpha}(\mathbf{k}, \omega)=\frac{ie^2\omega F_f}{mcv}\sum_{\mathbf{g}} e^{i(\mathbf{k}+\mathbf{g})\rho_\alpha}\,(\mathbf{e}_{f\alpha}\cdot\mathbf{A}(\mathbf{k}+\mathbf{g}, \omega)). \tag{8.2}$$

In these equations and everywhere below, summation is over the twice-encountered tensor indices.

From Eq. (8.1) [see the similar transformations in the derivation of (7.29)] we find

$$\frac{\omega}{e}\,\mathbf{A}(\mathbf{q}, \omega)=4\pi i\sum_f e^{-i\mathbf{q}\rho_\beta}\,p_{f\beta}(\mathbf{Q}, \omega)\,\frac{\mathbf{q}(\mathbf{q}\mathbf{e}_{f\beta})-\mathbf{e}_{f\beta}\frac{\omega^2}{c^2}}{\mathbf{q}^2-\frac{\omega^2}{c^2}}, \tag{8.3}$$

where $\mathbf{q}=\mathbf{Q}+\mathbf{g}$, where $\mathbf{g}\neq 0$. After substitution of microfields (8.3) into (8.2), we obtain

$$[(\Omega_f^2-\omega^2)\,\delta_{ff'}\delta_{\alpha\beta}+\Lambda_{f\alpha;\,f'\beta}]\,p_{f'\beta}(\mathbf{Q}, \omega)=\frac{ie^2\omega F_f}{mvc}\,e^{i\mathbf{Q}\rho_\alpha}(\mathbf{e}_{f\alpha}\cdot\mathbf{A}(\mathbf{Q}, \omega)), \tag{8.4}$$

where the force matrix is determined by the relation

$$\Lambda_{f\alpha;\,f'\beta}(\mathbf{Q}, \omega)=\omega_p^2F_f\sum_{\mathbf{g}\neq 0}\frac{(\mathbf{e}_{f\alpha}\cdot\mathbf{Q}+\mathbf{g})\,(\mathbf{e}_{f'\beta}\cdot\mathbf{Q}+\mathbf{g})-\mathbf{e}_{f\alpha}\mathbf{e}_{f'\beta}\frac{\omega^2}{c^2}}{(\mathbf{Q}+\mathbf{g})^2-\frac{\omega^2}{c^2}}\times$$
$$\times\exp\{i(\mathbf{Q}+\mathbf{g})(\rho_\alpha-\rho_\beta)\}. \tag{8.5}$$

This matrix characterizes the interactions (with allowance for retardation) between the dipole moments of the molecular transitions carried by all of the shortwave fields. Just as in the case of crystals with one molecule in a unit cell, expression (8.5) contains an infinite self-energy for the dipoles corresponding to electron transitions in the molecules. This infinite energy is easily eliminated by the method discussed in Section 7. We shall assume that this elimination has been accomplished.

Let us ignore the off-diagonal elements of force matrix (8.5) that pertain to different molecular transitions, i.e., let

$$\Lambda_{f\alpha;\, f'\beta} = \delta_{ff'}\Lambda^{f}_{\alpha\beta}. \tag{8.6}$$

The remaining matrix $\Lambda^{f}_{\alpha\beta}(\mathbf{Q}, \omega)$ at fixed f, ω, and $\mathbf{Q}$ contains σ^2 elements. We can always find a unitary matrix $u(f, \mathbf{Q})$ of order σ that diagonalizes the matrix $\Lambda^{f}_{\alpha\beta}$. In order to calculate the matrix elements $u_{\alpha\beta}$, we must solve the system of equations

$$\left.\begin{aligned} u^{*}_{\alpha\mu}(f, \mathbf{Q})\, u_{\beta\mu}(f, \mathbf{Q}) &= \delta_{\alpha\beta}, \\ u_{\alpha\mu}(f, \mathbf{Q})\, \Lambda^{f}_{\alpha\beta} u^{*}_{\beta\mu}(f, \mathbf{Q}) &= \Lambda_{\mu}(f, \omega, \mathbf{Q}). \end{aligned}\right\} \tag{8.7}$$

If we assume that the solutions of this system of equations are found, from (8.4), we obtain

$$u_{\beta\mu}(f, \mathbf{Q})\, P_{f\beta}(\mathbf{Q}, \omega) = \frac{ie^2 F_f}{mvc} \frac{\omega u_{\alpha\mu}(\mathbf{e}_{f\alpha}\cdot\mathbf{A}(\mathbf{Q}, \omega))\, e^{i\mathbf{k}\mathbf{Q}_\alpha}}{\Omega_f^2 - \omega^2 + \Lambda_{\mu}(f, \omega, \mathbf{Q})}. \tag{8.8}$$

With the aid of (8.8), Eq. (8.1) at $\mathbf{k} = \mathbf{Q}$ is transformed to

$$\left[Q^2\delta_{lj} - \frac{\omega^2}{c^2}\varepsilon_{lj}(\mathbf{Q}, \omega) - Q_l Q_j\right] A_j(\mathbf{Q}, \omega) = \frac{4\pi}{c} j_l(\mathbf{Q}, \omega), \tag{8.9}$$

where

$$\varepsilon_{lj}(\mathbf{Q}, \omega) = \delta_{lj} + \sum_{f,\,\mu} \frac{8\pi\Omega_f d_l(\mu, f, \mathbf{Q})\, d^{*}_j(\mu, f, \mathbf{Q})}{\hbar v\,[\Omega_f^2 - \omega^2 + \Lambda_{\mu}(f, \omega, \mathbf{Q})]} \tag{8.10}$$

are the components of the total dielectric-constant tensor; and

$$\mathbf{d}(\mu, f, \mathbf{Q}) = u_{\alpha\mu}(f, \mathbf{Q})\, \mathbf{e}_{f\alpha} d_{f\alpha} e^{i\mathbf{Q}\mathbf{Q}_\alpha}$$

is the vector of the dipole moment of all molecules in a unit cell, which corresponds to the μ-th solution of system (8.7).

If free charges are absent and the extrinsic current is transverse, i.e.,

$$\mathbf{j}(\omega, \mathbf{Q}) = \mathbf{j}^{\perp}(\omega, \mathbf{Q}),$$

then at $\mathbf{k} = \mathbf{Q}$ Eq. (8.1) decomposes to two equations:

$$\mathbf{A}^{\parallel}(\mathbf{Q}, \omega) = i\frac{4\pi c}{\omega}\sum_f \mathbf{e}^{\parallel}_{f\alpha}(\mathbf{Q})\, p_{f\alpha}(\mathbf{Q}, \omega)\, e^{-i\mathbf{Q}\rho_\alpha}, \tag{8.11}$$

$$\left[Q^2 - \frac{\omega^2}{c^2}\right]\mathbf{A}^{\perp}(\mathbf{Q}, \omega) + i\frac{4\pi}{c}\sum_f \mathbf{e}^{\perp}_{f\alpha}(\mathbf{Q})\, p_{f\alpha}(\mathbf{Q}, \omega)\, e^{-i\mathbf{Q}\rho_\alpha} = \frac{4\pi}{c}\mathbf{j}^{\perp}(\mathbf{Q}, \omega), \tag{8.12}$$

where $\mathbf{e}^{\parallel}_{f\alpha}(\mathbf{Q})$ and $\mathbf{e}^{\perp}_{f\alpha}(\mathbf{Q})$ are the longitudinal and transverse components of the unit vector $\mathbf{e}_{f\alpha}$, which determines the dipole moment of the transition in molecule α to the f-th excited state.

Let us isolate the longitudinal component of the vector potential on the right of (8.4). Then, using (8.11), we obtain

$$\begin{aligned}[(\Omega_f^2 - \omega^2)\,\delta_{ff'}\delta_{\alpha\beta} + \Gamma_{f\alpha;\, f'\beta}(\mathbf{Q}, \omega)]\, p_{f'\beta}(\mathbf{Q}, \omega) &= \\ = i\frac{e^2\omega F_f}{mvc}(\mathbf{e}^{\perp}_{f\alpha}\cdot\mathbf{A}^{\perp}(\mathbf{Q}, \omega))\, e^{i\mathbf{Q}\rho_\alpha}, \end{aligned} \tag{8.13}$$

where the new force matrix

$$\Gamma_{f\alpha;\, f'\beta}(\mathbf{Q}, \omega) = \Lambda_{f\alpha;\, f'\beta} + \omega_p^2 F_f \frac{(\mathbf{e}_{f\alpha}\cdot\mathbf{Q})(\mathbf{e}_{f'\beta}\cdot\mathbf{Q})}{Q^2}\exp(i\mathbf{Q}\rho_{\alpha\beta}),$$
$$\rho_{\alpha\beta} = \rho_\alpha - \rho_\beta, \tag{8.14}$$

characterizes the molecular interactions carried by all shortwave fields and the longitudinal part of the macroscopic field. If we ignore interaction retardation, force matrix (8.14) becomes

$$\Gamma^{\mathrm{cou}}_{f\alpha;\, f'\beta}(\mathbf{Q}) = \omega_p^2 F_f \sum_{\mathbf{g}\geqslant 0}\frac{(\mathbf{e}_{f\alpha}\cdot\mathbf{Q}+\mathbf{g})\,(\mathbf{e}_{f'\beta}\cdot\mathbf{Q}+\mathbf{g})}{(\mathbf{Q}+\mathbf{g})^2}\exp\{i(\mathbf{Q}+\mathbf{g})\rho_{\alpha\beta}\}, \tag{8.15}$$

which takes into account the total Coulomb interaction between the electric charges of the molecules of the crystal.

Solution of Eq. (8.13) comes down to diagonalization of force matrix (8.14). Here, we shall consider the case when approximation of the independent molecular excited states is justified, i.e., when

$$\Gamma_{f\alpha;\, f'\beta} = \delta_{ff'}\Gamma^{f}_{\alpha\beta}(\mathbf{Q}, \omega). \tag{8.16}$$

In this approximation, for fixed f, ω, and $\mathbf{Q}$ the matrix $\Gamma^{f}_{\alpha\beta}(\mathbf{Q}, \omega)$ is diagonalized by means of the unitary matrix $v_{\alpha\mu}$, whose components satisfy the system of equations

$$\left.\begin{array}{l} v^*_{\alpha\mu}(f, \mathbf{Q})\, v_{\beta\mu}(f, \mathbf{Q}) = \delta_{\alpha\beta}, \\ v_{\alpha\mu}(f, \mathbf{Q})\, \Gamma^f_{\alpha\beta} v^*_{\alpha\mu}(f, \mathbf{Q}) = \Gamma_\mu(f, \mu, \mathbf{Q}). \end{array}\right\} \quad (8.17)$$

If the solutions of system (8.17) are known, then the solutions of Eq. (8.13) under condition (8.16) have the form

$$v_{\beta\mu}(f, \mathbf{Q})\, p_{f\beta}(\mathbf{Q}, \omega) = i\, \frac{2\omega\Omega_f\, d_f\, (\mathbf{d}^\perp(f, \mu, \mathbf{Q}) \cdot \mathbf{A}^\perp(\mathbf{Q}, \omega))}{\hbar c v\, [\Omega_f^2 - \omega^2 + \Gamma_\mu(f, \omega, \mathbf{Q})]}, \quad (8.18)$$

where

$$\mathbf{d}^\perp(f, \mu, \mathbf{Q}) = v_{\alpha\mu}(f, \mathbf{Q})\, \mathbf{e}^\perp_{f\alpha}(\mathbf{Q}) d_f\, e^{i\mathbf{Q}\rho_\alpha}$$

is the transverse dipole-moment vector for a quantum transition in a unit cell, which corresponds to the μ-th solution of system (8.17). The value

$$P^f_\mu(\mathbf{Q}, \omega) = v_{\beta\mu}(f, \mathbf{Q})\, p_{f\beta}(\mathbf{Q}, \omega)$$

according to (8.18), characterizes the amplitude of the dipole moment of a unit volume of the crystal produced by a transverse extrinsic current.

To obtain the transverse dielectric constant of a crystal containing σ molecules in a unit cell, it is sufficient to substitute expression (8.18) into Eq. (8.12). Thus, we obtain an equation that directly relates the transverse electromagnetic field to the macroscopic current that produces it:

$$\left[Q^2\delta_{lj} - \frac{\omega^2}{c^2}\, \varepsilon^\perp_{lj}(\mathbf{Q}, \omega)\right] A^\perp_j(\mathbf{Q}, \omega) = \frac{4\pi}{c}\, j^\perp_l(\mathbf{Q}, \omega), \quad (8.19)$$

where

$$\varepsilon^\perp_{lj}(\mathbf{Q}, \omega) = \eta_{lj} + \sum_{f, \mu} \frac{8\pi\Omega_f\, d^\perp_l(\mu, f, \mathbf{Q})\, d^\perp_j(\mu, f, \mathbf{Q})}{\hbar v\, [\Omega_f^2 - \omega^2 + \Gamma_\mu(f, \omega, \mathbf{Q})]} \quad (8.20)$$

is the transverse dielectric constant.

As has been noted above, force matrices (8.14) and (8.15) contain an infinite dipole-transition self-energy. For simplicity, let us consider how the self-energy is eliminated from Coulomb matrix (8.15). Let

$$L^{f}_{\alpha\beta}(\mathbf{Q}) = \frac{\hbar}{2\Omega_f} \Gamma^{\text{cou}}_{f\alpha;\, f\beta}(\mathbf{Q}). \tag{8.21}$$

Then, according to (8.15), we can write

$$L^{f}_{\alpha\beta}(\mathbf{Q}) = -d_f^2\{(\mathbf{e}_{f\alpha}\cdot\nabla)(\mathbf{e}_{f\beta}\cdot\nabla)\, e^{-i\mathbf{Qr}} S_{\alpha\beta}(\mathbf{r})\}_{r=0}, \tag{8.22}$$

where

$$S_{\alpha\beta}(\mathbf{r}) = \sum_{\mathbf{g}} S^{(\mathbf{g})}_{\alpha\beta}\cdot \exp(i\mathbf{gr}), \tag{8.23}$$

$$S^{(\mathbf{g})}_{\alpha\beta}(\mathbf{r}) = \frac{4\pi}{v}\,\frac{\exp\{i(\mathbf{Q}+\mathbf{g})\,\rho_{\alpha\beta}\}}{(\mathbf{Q}+\mathbf{g})^2 + i\eta}, \qquad \eta \to 0.$$

Considering the periodic (with the lattice constant) properties of function (8.23), we can (see the similar transformations in Section 7) transform it to a sum over the lattice vectors:

$$S_{\alpha\beta}(\mathbf{r}) = \sum_{\mathbf{n}} \frac{\exp(i\mathbf{Qn})}{|\mathbf{r}-\mathbf{n}-\rho_{\alpha\beta}|}\,. \tag{8.24}$$

Elimination of the self-energy terms in expression (8.24) boils down to rejecting the term with $\mathbf{n} = 0$ in the matrix elements $S_{\alpha\alpha}$. The thus-defined matrix precisely coincides with resonance-interaction matrix (II.5.15), which was used to calculate the exciton bands in crystals containing σ molecules in a unit cell.

9. Elementary Excitations in a Crystal with Complete Allowance for Retardation

As was shown in Section 5 of Chapter I, when retardation is ignored, the poles of the transverse dielectric-constant tensor determine the dependence of the exciton energy upon the wave vector. In the absence of retardation [in approximation (7.37a)], according to Eqs. (7.44) and (7.40), the force matrix contained in the definition of the transverse dielectric constant is directly expressed in terms of matrix (7.44) at $\omega = 0$,

$$\Gamma^{\text{cou}}_{f}(\mathbf{Q}, \omega) = \frac{2\Omega_f}{\hbar} L_f(\mathbf{Q}, 0).$$

If we substitute this expression into Eq. (7.39), we find that the exciton frequencies, i.e., the ω values at which $\varepsilon^{\perp}(\mathbf{Q}, \omega)$ has poles, are

$$\omega_{ex} = \sqrt{\Omega_f^2 + \frac{2\Omega_f}{\hbar} L_f(\mathbf{Q}, 0)}. \tag{9.1}$$

In complete agreement with Section 2 of Chapter II, therefore, the dependence of the exciton energy upon the wave vector is expressed by

$$E_f(\mathbf{Q}) \approx \hbar\Omega_f + L_f(\mathbf{Q}, 0) = \Delta\varepsilon + D_f + L_f(\mathbf{Q}, 0).$$

The elementary excitations of a crystal that contain an electronic excitation and an electromagnetic field are the solutions of Eqs. (7.38) without their right sides. For simplicity, let us study these equations for optically isotropic crystals. In an optically isotropic crystal, for any sense of the wave vector the two-dimensional tensor $\varepsilon^{\perp}(\mathbf{Q}, \omega)$ is diagonal, i.e.,

$$\varepsilon_{\xi\eta}^{\perp} = \varepsilon^{\perp}\delta_{\xi\eta},$$

where ξ is any direction perpendicular to the wave vector. In this case, when $\mathbf{j}^{\perp} = 0$, Eqs. (7.38) reduce to the scalar equation

$$\left[\frac{\omega^2}{c^2}\varepsilon^{\perp}(\mathbf{Q}, \omega) - \mathbf{Q}^2\right] A_{\xi}^{\perp}(\mathbf{Q}, \omega) = 0, \tag{9.2}$$

where, according to (7.39),

$$\varepsilon^{\perp}(\mathbf{Q}, \omega) = 1 + \sum_f \frac{\omega_p^2 F_f}{\Omega_f^2 - \omega^2 + \Gamma_f(\mathbf{Q}, \omega)}. \tag{9.3}$$

From the condition of nontrivial solvability of Eq. (9.2), with allowance for (9.3), we find the elementary-excitation dispersion equation

$$\frac{c^2\mathbf{Q}^2}{\omega^2} = 1 + \sum_f \frac{\omega_p^2 F_f}{\Omega_f^2 - \omega^2 + \Gamma_f(\mathbf{Q}, \omega)}. \tag{9.4}$$

Let us ignore longitudinal excitons, for which retardation does not play a part. For transverse excitons $(\mathbf{e}_f\mathbf{Q}) = 0$ and the force matrix [in approximation (7.37a)] is given by

$$\Gamma_f(\mathbf{Q}, \omega) = \frac{2\Omega_f}{\hbar} L_f(\mathbf{Q}, \omega) + \frac{\omega^2\omega_p^2 F_f}{\mathbf{Q}^2c^2 - \omega^2}. \tag{9.5}$$

If we ignore retardation of the shortwave fields in the force matrix, i.e., let

$$\Gamma_f(\mathbf{Q}, \omega) = \frac{2\Omega_f}{\hbar} L_f(\mathbf{Q}, 0),$$

then Eq. (9.4) exactly coincides with (III.5.22), with which we studied elementary excitations with allowance for retardation only of the macroscopic field – photoexcitons.

Let us see how the elementary excitations change when the retardation of the shortwave fields is taken into account. We shall assume that one of the molecular frequencies Ω_f differs considerably from the others. Then, in the range of frequencies ω close to this frequency, we can isolate the resonance term in the sum in Eq. (9.4) and transform the equation to

$$\frac{c^2\mathbf{Q}^2}{\omega^2} = \varepsilon_0 + \frac{\omega_p^2 F_f}{\Omega_f^2 - \omega^2 + \Gamma_f(\mathbf{Q}, \omega)}. \tag{9.6}$$

This equation determines the frequencies of the elementary excitations of the crystal, which are a superposition of an electronic excitation and the shortwave and macroscopic fields that interact with it. Equation (9.6) is very complicated, because the desired frequency ω also appears in it [see expressions (9.5) and (7.45)] under a large number of cosines. It was shown in Section 7 that with good approximation we can let

$$L_f(\mathbf{Q}, \omega) = L_f(\mathbf{Q}, 0).$$

In this approximation, when (9.1) is taken into account, Eq. (9.6) is brought to the form

$$\left(\frac{c^2\mathbf{Q}^2}{\omega^2} - \varepsilon_0\right)\{(\mathbf{Q}^2c^2 - \omega^2)\,[\omega_{\text{ex}}^2(\mathbf{Q}) - \omega^2] + \omega^2 B\} = (\mathbf{Q}^2c^2 - \omega^2)\,B, \tag{9.7}$$

where $B = \omega_p^2 F_f$.

When B = 0 (interaction between excitons and electromagnetic fields is absent), Eq. (9.7) determines three types of independent elementary excitations: excitons $\omega^2 = \omega_{\text{ex}}^2(\mathbf{Q})$, macroscopic electromagnetic waves $\omega^2 = c^2\mathbf{Q}^2/\varepsilon_0$, and microfields $\omega^2 = c^2\mathbf{Q}^2$. When $B \neq 0$, all of these excitations are mixed and form three new types of elementary excitations. A detailed analysis of the properties of these elementary excitations for various crystals has yet to be made.

Chapter V

The Dielectric Constant of Molecular Crystals with Allowance for Lattice Vibrations

1. Theory of the Width of Exciton Absorption Bands in One-Dimensional Molecular Crystals

The theory of light absorption by molecular crystals with allowance for lattice vibrations was first developed by Davydov [1] for a one-dimensional crystal model and by Davydov, Rashba, and Lubchenko [2, 3] for three-dimensional crystals. These studies were qualitative and did not take into account the frequency shift caused by exciton-phonon interaction. A theory of light absorption by Wannier excitons with allowance for lattice vibrations was developed by Toyozawa [4]. As was shown in Section 1 of Chapter IV, there is a considerable difference between the interaction of Wannier excitons and the interaction of Frenkel excitons with optical lattice vibrations. Therefore, Toyozawa's results cannot be carried over directly to molecular crystals.

Lately, because of the wide use of methods of quantum electrodynamics in the theory of solids, the number of theoretical studies of the structure of light-absorption bands in crystals has increased considerably. Most often, these studies are made with a one-dimensional crystal model [5, 6]. Suna [6] used the Green function method in a single-phonon approximation to study the absorption and luminescence spectra of one-dimensional crystals with allowance for exciton–phonon interaction. Owing to a considerable simplification (see below) of the exciton–phonon interaction operator, Suna was able in the single-phonon approximation (for an Einstein lattice-vibration model) to obtain a mass operator for the exciton Green function that became infinite at certain frequencies. This, on the one hand, does not allow higher approximations to be considered and, on the other, leads to some results that are physically meaningless.

Below, using the results of [7], we shall present a method of calculating the dielectric constant and the refractive index and attenuation factor for electromagnetic waves in a one-dimensional crystal and their dependence upon temperature. For simplicity, we shall consider a one-dimensional crystal with the lattice constant a that contains N ($\gg 1$) identical anisotropic molecules with one in each unit cell. Let us consider the exciton states for which local deformation of the crystal does not occur and interaction amounts only to phonon creation and annihilation. We shall assume that the interaction of excitons with the acoustic branch of lattice vibrations is considerably weaker than with the optical branch. The case of "strong coupling" (see Section 1 of Chapter IV and [8]) requires special consideration, since local excitations rather than excitons are produced with strong coupling.

To simplify the calculations, we shall take into account only the excitons of one band and the phonons of one optical branch – rotational molecular vibrations. Such a model was considered in Section 5 of Chapter II and Section 1 of Chapter IV. In a one-dimensional crystal, the excitons and phonons have, respectively, the wave vectors **k** and **q**, which are directed along the axis of the crystal. They take N equidistant values in the range $-\pi/a$, π/a. If $B^+(\mathbf{k})$, $B(\mathbf{k})$, $b_\mathbf{q}^+$, and $b_\mathbf{q}$ are the exciton and phonon creation and annihilation operators, then in the second-quantization representation the energy operators for free excitons and phonons are determined by ($\hbar = 1$)

$$\left.\begin{aligned} H_{\mathrm{ex}} &= \sum_{\mathbf{k}} E(\mathbf{k}) B^+(\mathbf{k}) B(\mathbf{k}), \\ H_{\mathrm{ph}} &= \sum_{\mathbf{q}} \Omega(\mathbf{q}) b_\mathbf{q}^+ b_\mathbf{q}, \end{aligned}\right\} \tag{1.1}$$

where $\mathbf{k}\|\mathbf{q}\|\mathbf{a}$. Further, according to (II.5.13),

$$E(\mathbf{k}) = \Delta\varepsilon + D - \frac{1}{2} L \cos \mathbf{ka}, \qquad L = \frac{e^2 F}{m\omega a^3}(3\cos^2\delta - 1) \tag{1.2}$$

is the exciton energy. Below, we shall consider the case of excitons with a positive effective mass when $\mathbf{ka} \ll 1$, i.e., when $L > 0$;

$$\Omega(\mathbf{q}) = \Omega(0)\left(1 + \frac{1}{2}\xi \cos \mathbf{qa}\right) \tag{1.3}$$

is the phonon energy. The interaction operator for excitons and high-frequency (optical) phonons is, in accordance with (IV.1.38) taken as

$$H_{\text{ex-ph}} = \frac{1}{\sqrt{N}} \sum_{\mathbf{k},\,\mathbf{q}} F(\mathbf{k},\ \mathbf{q})\, B^{+}(\mathbf{k}+\mathbf{q})\, B(\mathbf{k})\,(b_{\mathbf{q}} + b^{+}_{-\mathbf{q}}), \tag{1.4}$$

where

$$F(\mathbf{k},\ \mathbf{q}) = 3\gamma L \frac{\sin 2\vartheta}{1 - 3\cos^2\vartheta} \frac{\cos\left[\left(\mathbf{k} + \frac{1}{2}\mathbf{q}\right)\mathbf{a}\right] \cos\left(\frac{1}{2}\mathbf{qa}\right)}{\sqrt{1 + \frac{1}{2}\xi \cos \mathbf{qa}}}. \tag{1.5}$$

Here, γ^2 is the square of the amplitude of the zero-point rotational vibrations of the molecules in the crystal. For that which follows, it will be convenient to convert to the dimensionless variables

$$x = \mathbf{ak} \text{ and } y = \mathbf{aq}.$$

Further, all energies are expressed in units equal to the width of the exciton band. Then, Eqs. (1.2), (1.3), and (1.5) take the form

$$\left.\begin{aligned} E(x) &= E_0 + \frac{1}{2} - \frac{1}{2}\cos x, \\ \Omega(y) &= \Omega_0 + \frac{1}{2}\xi \cos y, \end{aligned}\right\} \tag{1.6}$$

$$F(x,\ y) = \rho \frac{\cos\left(x + \frac{1}{2}y\right) \cos\left(\frac{1}{2}y\right)}{\sqrt{1 + \frac{1}{2}\xi \cos y}}; \qquad \rho^2 = \frac{9\gamma^2 \sin(2\vartheta)}{(1 - 3\cos^2\vartheta)^2}. \tag{1.7}$$

In the new units, the square of the exciton-phonon coupling function $F^2(x, y)$ is proportional to the square of the amplitude of the zero-point rotational vibrations. This function vanishes at the boundaries of the Brillouin zone ($y = \pm\pi$). In [6], Suna makes the function $F(x, y)$ a constant, which considerably simplifies the calculations. As we shall see below, this choice of the coupling function results in nonanalytic expressions for the exciton mass operator even in the single-phonon approximation.

This difficulty is eliminated for all $x + y \neq 0$ by taking into account the dependence of $F(x, y)$ upon the variables x and y, which is given by expression (1.7). When $x + y = 0$, however, the mass operator of the exciton Green function also has infinite values. To eliminate this difficulty, in [7], instead of coupling function (1.7), we considered the function

$$F_\alpha(x+y) = \rho\Phi_\alpha^{1/2}(x+y)\frac{\cos(x+y)+\cos y}{2\sqrt{1+\frac{1}{2}\xi\cos y}}, \tag{1.8}$$

where

$$\Phi_\alpha(z) \equiv 1 - \exp\{-\alpha\,|\sin^2 z|\}.$$

When $\alpha \to \infty$, the function $\Phi_\alpha(z) \to 1$ and expression (1.8) coincides with expression (1.7). Calculations were made for $\alpha = 50$. In this case, $1 - \Phi_{50}(z) \leq 0.002$ for all $z \geq 0.35$. When $z = 0$, however, $\Phi_{50}(z) = 0$. The function $\Phi_{50}(z)$ forbids those exciton–phonon interactions as a result of which a quiescent exciton[1] is formed, i.e., an exciton with a zero wave vector, and does not change for all practical purposes the interactions in which are formed excitons with the wave vectors $\mathbf{k}' = \mathbf{k} + \mathbf{q} \geq 0.1(\pi/a)\mathbf{d}$.

As was shown in Section 2 of Chapter IV, the conditions of the passage of light through a crystal are determined by the form of

---

[1]The need for an artificial change in the coupling function for finite states of excitons with a zero vector ($\mathbf{k}' = \mathbf{k} + \mathbf{q} = 0$) is partially due to the fact that calculation of the mass operator requires that we move from summation to integration with respect to $z = a\mathbf{k}'$ [see (1.12)]. The number of finite states $N(z)$ per unit energy interval is always finite in real systems. With transition to the continuous limit, when the exciton energy is represented by the function $\varepsilon(z) = A = bz^2$, the number of finite states

$$N(z) = \frac{dz}{d\varepsilon(z)} = \frac{1}{2bz}$$

increases infinitely as $z = a(k + q) \to 0$.

the exciton retarded Green function. Let us calculate the retarded Green function for excitons that interact with phonons at temperature T. According to definition (IV.2.19), the exciton retarded Green function

$$G(\mathbf{k};t) = -i\Theta(t)\langle\langle[B(\mathbf{k},t),\ B^{+}(\mathbf{k},0)\rangle\rangle,$$

where the brackets $\langle\langle\ldots\rangle\rangle$ indicate statistical averaging with the density matrix at temperature T with the total Hamiltonian

$$H = H_{\mathrm{ex}} + H_{\mathrm{ph}} + H_{\mathrm{ex\text{-}ph}}.$$

The operators B(**k**, t) are given in the Heisenberg representation, so that (we use units in which $\hbar = 1$)

$$B(\mathbf{k},t) = e^{iHt}B(\mathbf{k})\,e^{-iHt}.$$

Using the method for calculating the retarded Green functions discussed in Section 6 of Chapter IV, we find that at low exciton densities the Green function in the l-th approximation in the energy representation is

$$G_l(\mathbf{k},\omega) = [\omega - E(\mathbf{k}) - M_l(\mathbf{k},\omega)]^{-1}, \tag{1.9}$$

where $M_l(\mathbf{k}, \omega)$ is the exciton mass operator, which is defined (for $l = 1, 2$) by means of the sequential system of equations

$$M_l(\mathbf{k},\omega) =$$
$$= \frac{1}{N}\sum_{\mathbf{q}}|F(\mathbf{k},\mathbf{q})|^2\left\{\frac{1+\nu_{\mathbf{q}}}{\omega-\Omega(\mathbf{q})-E(\mathbf{k}+\mathbf{q})-M_{l-1}(\mathbf{k}+\mathbf{q},\,\omega-\Omega(\mathbf{q}))}+\right.$$
$$\left.+\frac{\nu_{\mathbf{q}}}{\omega+\Omega(\mathbf{q})-E(\mathbf{k}+\mathbf{q})-M_{l-1}(\mathbf{k}+\mathbf{q},\,\omega+\Omega(\mathbf{q}))}\right\}, \tag{1.10}$$

where

$$\nu_{\mathbf{q}} = \left[\exp\left(\frac{\Omega(\mathbf{q})}{kT}\right)-1\right]^{-1}.$$

Equations (1.10) can also be used in higher approximations ($l \geq 3$) if it is assumed that the coupling function F(**k**, **q**), i.e., the vertex part of the zeroth approximation, approximates the vertex part $F^{\mathrm{acc}}(\mathbf{k},\mathbf{q})$ in the higher approximations. The validity of such an approximation has been demonstrated by Migdal [9] (with accuracy to a small value on the order of $(m/M)^{1/2}$, where m is the electron mass and M is the nuclear mass) in a study of electron–

phonon interaction (at T = 0). Agranovich and Konobeev [10] have also shown that for exciton–phonon interaction in three-dimensional crystals, the vertex part can with great accuracy be replaced by the function F(**k**, **q**). This result remains valid for one-dimensional crystals when the coupling function is in the form of (1.8) for sufficiently large but finite α.

When solving Eqs. (1.10) by the method of successive approximations, in the zeroth approximation (exciton–phonon interaction is absent) we must let $M_0 = -i\eta$, where η is a small positive value determined by the rule for circumvention of the pole of the free-exciton Green function [see (IV.4.9)]. At zero temperature, $\nu_q = 0$ and the l-th approximation in solution of system (1.10) corresponds to allowing for processes in which l phonons participate. Below, we shall use the phrase "l-phonon approximation" even for calculations at nonzero temperature.

The mass operator $M_l(k, \omega)$ is a complex function of the real variables k and ω. The real part of the mass operator determines the change in exciton energy, so that

$$E_l(\mathbf{k}, \omega) = E(\mathbf{k}, \omega) + \operatorname{Re} M_l(\mathbf{k}, \omega),$$

and the imaginary part of the mass operator

$$\gamma_l(\mathbf{k}, \omega) = -\operatorname{Im} M_l(\mathbf{k}, \omega)$$

characterizes the attenuation.

Usually,

$$\xi \ll \Omega_0 \ll E_0,$$

Therefore, the dependence of $\Omega(\mathbf{q})$ upon **q** can be ignored in calculation of the mass operator in (1.10). Now, we move from the sum over **q** to the integral, using the relation

$$\frac{1}{N}\sum_{\mathbf{q}} \ldots = \frac{1}{2\pi}\int_0^{\pi} dy \ldots, \qquad \text{at } \mathbf{q} \parallel \mathbf{a}. \tag{1.11}$$

Then, after substitution of the explicit form of function (1.8), Eq. (1.10) takes the form

$$M_l^{\alpha}(x, \omega) = \frac{\rho^2}{2\pi}\left\{(1+\nu_0)\int_0^{\pi} \frac{[\cos x + \cos z]^2 \, \Phi_{\alpha}(z)\, dz}{\cos z + A_l(\Omega_0, z)} + \right.$$

$$+ \nu_0 \int_0^\pi \frac{[\cos x + \cos z]^2 \Phi_\alpha(z)\,dz}{\cos z + A_l(-\Omega_0, z)} \Bigg\}, \qquad (1.12)$$

where

$$z = (x + y),$$

$$A_l(\Omega, z) = 2\left\{\omega - \Omega_0 - E_0 - \frac{1}{2} - M_{l-1}^{(\alpha)}(z, \omega - \Omega_0)\right\}. \qquad (1.13)$$

To calculate (1.12) in the single-phonon approximation ($l = 1$), we must let $M_0^\alpha = -i\eta$ in (1.13). Then,

$$A_1(\Omega_0) = 2\left(\omega - \Omega_0 - E_0 - \frac{1}{2} + i\eta\right), \qquad \eta \to +0,$$

and Eq. (1.12) takes the form

$$M_1^{(\alpha)}(x, \omega) = \frac{1}{2}\rho^2\{(1 + \nu_0)\,[[2\cos x - A_1(\Omega_0) + [\cos x - A_1(\Omega_0)]^2 \times$$
$$\times I_\alpha^{(1)}[A(\Omega_0)]]] - \nu_0\,[[2\cos x - A_1(-\Omega_0) + [\cos x - A_1(-\Omega_0)]^2 \times$$
$$\times I_\alpha^{(1)}[A_1(-\Omega_0)]]\}, \qquad (1.14)$$

where

$$I_\alpha^{(1)}[A_1(\Omega_0)] \equiv$$

$$\equiv \frac{1}{\pi}\int_0^\pi \frac{\Phi_\alpha(z)\,dz}{\cos z + A_1(\Omega_0)} = \begin{cases} \dfrac{A_1(\Omega_0)\,\Phi_\alpha[\arccos A_1(\Omega_0)]}{|A_1(\Omega_0)|\sqrt{A_1^2(\Omega_0) - 1}}, & \text{if } |A_1(\Omega_0)| \geqslant 1; \\ -i\,\dfrac{\Phi_\alpha[\arccos A_1(\Omega_0)]}{\sqrt{1 - A_1^2(\Omega_0)}}, & \text{if } |A_1(\Omega_0)| < 1. \end{cases} \qquad (1.15)$$

If we consider zero temperatures, then in the approximation $F(x, y) = \rho$, which was used by Suna [6], in the single-phonon approximation the mass operator is not a function of the exciton wave vector and reduces to integral (1.15) when $\alpha = \infty$:

$$M_1^s(\omega) = 2\rho^2 I_\infty^{(1)}[A_1(\Omega_0)]. \qquad (1.16)$$

The real and imaginary parts of the mass operator $M_1^s = \Delta_1^s - i\gamma_1^s$ as functions of ω are shown in Fig. 27 with the index 1. The mass operator M_1^s is a nonanalytic function of ω, since it undergoes discontinuities at the points $|A_1(\Omega_0)| = 1$, i.e., at $\omega = E_0 + \Omega_0$ and $\omega = E_0 + \Omega_0 + 1$, where E_0 is the bottom of the exciton band. This operator differs from zero and is real outside the range

$$F_0 + \Omega_0 \leqslant \omega \leqslant F_0 + \Omega_0 + 1. \tag{1.17}$$

The mass operator M_1^s is imaginary within range (1.17). Green function (1.9) in the single-phonon approximation has at $\mathbf{k} = \mathbf{Q}$ poles on the real axis ω at ω values defined by

$$\omega = E(\mathbf{Q}) + \operatorname{Re} M_1^s(\omega).$$

If these poles correspond to ω values that lie outside range (1.17), they are physically meaningless.

In [7], we calculated the mass operator in the single-phonon approximation using (1.14) for $\alpha = 50$ and $\alpha = \infty$. The real $\Delta^{(\infty)}$ and imaginary $\gamma^{(\infty)}$ parts of the mass operator for $\alpha = \infty$ are shown by curves 2 in Fig. 27. Curves 3 show the real Δ^{50} and imaginary γ^{50} parts of the mass operator at $\alpha = 50$. We see that curves 2 and 3 practically coincide for all ω except $\omega \approx E_0 + \Omega_0$, for which curves 2 take infinite values and curves 3 take finite values, which are indicated by the short horizontal lines.

The mass operators were calculated in higher approximations using the found values of $M_1^{50}(k, \omega)$ for the temperatures $kT = 0.02$ and 0.035 (in units of the exciton-band width). Formula (1.14), with $A_l(\Omega_0, z)$ replaced by $A_l(\Omega_0, 0)$ (for $l \geq 2$), was used to calculate $M_2^{50}(k, \omega)$ and $M_3^{50}(k, \omega)$. The obtained values of the mass

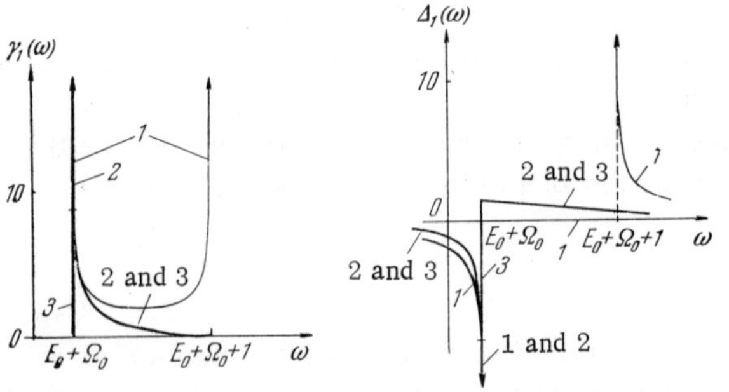

Fig. 27. Imaginary and real parts of mass operator in single-phonon approximation. Curves 1) calculations of Suna [6]; curves 2) calculations of Davydov and Nitsovich [7] at $\alpha = \infty$; curves 3) the same, for $\alpha = 50$.

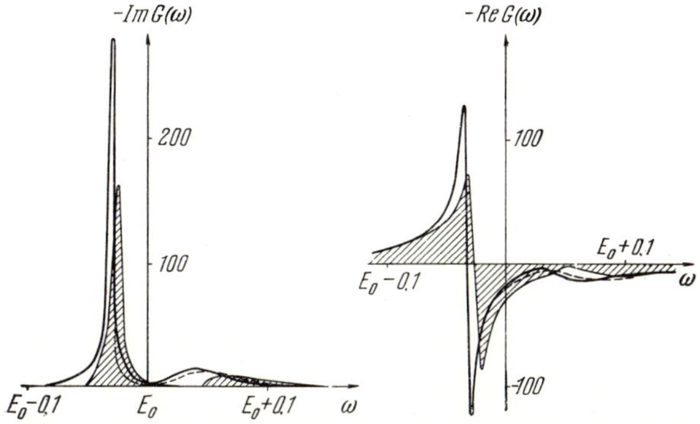

Fig. 28. Imaginary and real parts of exciton Green function. The areas under the first-approximation curves are shaded. The second and third approximations are given by the broken and solid curves, respectively.

operators determine, with the aid of (1.14), the exciton Green functions in the corresponding approximation. To illustrate the convergence of the approximations, Fig. 28 shows the real and imaginary parts of the Green functions for the three successive approximations (l = 1, 2, 3) at kT = 0.02, Ω_0 = 0.05, and ρ = 0.1. It follows from Fig. 28 that the second and third approximations differ little from one another. In calculating the exciton Green functions, therefore, we can limit ourselves to the third approximation.

The concepts of the dielectric constant ε, the refractive index, and the attenuation factor pertain to macroscopic three-dimensional media. These concepts are not directly applicable to one-dimensional crystals or to individual atoms and molecules. We can, however, consider a three-dimensional medium composed of one-dimensional crystals laid parallel to one another at distances that are smaller than the wavelength of light but great enough so that their mutual effect can be ignored. This medium will correspond to the oriented-gas model. In a certain approximation, this model can describe a medium composed of oriented polymers.

The dielectric constant of such a medium determines its macroscopic response to an external influence exerted by a longwave electromagnetic field. If the field is polarized along the vector of the dipole transition in a molecule, then in the frequency range corresponding to the molecular-transition frequency, according to Section 2 of Chapter IV, the dielectric constant is expressed in terms of the exciton Green function by the formula

$$\varepsilon(\mathbf{Q}, \omega) = \varepsilon_0 - \frac{4\pi \mathbf{d}^2 N_0}{L} G(\mathbf{Q}, \omega),$$

where ω is the photon energy (in units of the exciton-band width L); $\mathbf{Q}$ is the photon wave vector; N_0 is the number of molecules per unit volume of the medium; and ε_0 is the dielectric constant, which is determined by all of the electronic excited states of the molecules except the exciton state in question. The refractive index n and the attenuation factor $\varkappa$ are given by

$$(n + i\varkappa)^2 = \varepsilon.$$

Figure 29 shows the n and $\varkappa$ values for kT = 0.02 and 0.035 that were found in [7] as functions of the photon energy. The values $\varepsilon_0 = 2$, $4\pi \mathbf{d}^2 N_0/L = 1$, $\Omega_0 = 0.05$, and $\rho = 0.1$ were used in the calculations. The $\varkappa(\omega)$ curves determine the structure of the exciton absorption band in the medium. The distinctive feature of the absorption curves is their asymmetry. As temperature decreases, the principal maximum of the absorption curve increases and its shape approaches the shape of a Lorentz curve.

2. Dispersion and Absorption of Light by Three-Dimensional Molecular Crystals

Let us consider an orthorhombic molecular crystal with the basis vectors $\mathbf{a}_x$, $\mathbf{a}_y$, $\mathbf{a}_z$ ($|\mathbf{a}_x| = |\mathbf{a}_z| = a$). We shall assume that each unit cell contains one molecule and that one exciton band is separated from all of the other electronic excited states. The excitons are characterized by the energy E(**k**), the quasi-momentum **k** (in units in which $\hbar = 1$), and the molecular-transition dipole moment **d**. We shall assume that the dipole moment lies along the basis vector $\mathbf{a}_y$.

To simplify the calculations, we shall use a cylindrical coordinate system whose axis is along the dipole-moment vector. Let us assume that in the **k** space the first Brillouin zone can be ap-

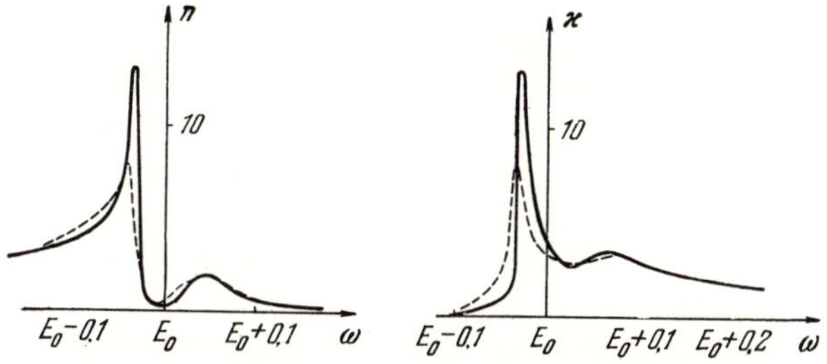

Fig. 29. Refractive index and attenuation factor. Solid lines are for kT = 0.02 and broken lines are for kT = 0.035.

proximated by a cylinder such that the wave vectors **k** in this zone take the values

$$-1 \leqslant y \leqslant 1, \quad 0 \leqslant \rho \leqslant 1, \qquad 0 \leqslant \varphi \leqslant 2\pi, \tag{2.1}$$

where

$$\pi y = k_y a_y, \quad \pi\rho = a\sqrt{k_x^2 + k_z^2}, \quad \varphi = \tan^{-1}\frac{k_x}{k_z}. \tag{2.2}$$

Further, we shall assume that the structure of the exciton band is defined by the function

$$E(y, \rho) = L\{e + \Xi(y, \rho)\}, \tag{2.3}$$

where

$$\Xi(y, \rho) = \frac{1}{2} + \frac{(\alpha - \rho^2 - y^2)(y^2 - \rho^2)}{2\alpha(y^2 + \rho^2)}, \qquad L = \frac{4\pi d^2}{a^2 a_y \varepsilon_0}. \tag{2.4}$$

According to (2.2), transverse excitons are characterized by $y = 0$. As $\rho \to 0$, we move toward the bottom of the exciton band, which has the energy $E_{\min} = eL$. The effective mass of transverse excitons is positive and equals $\alpha\pi^2/a^2L$. The width of the exciton band for transverse excitons is $L(2\alpha)^{-1}$. Longitudinal excitons correspond to $\rho = 0$. The peak of the band is reached as $y \to 0$ and corresponds to the energy

$$E_{\max} = (e + 1)L.$$

Therefore, L characterizes the total width of the exciton band.

The nonanalyticity of function (2.3) when $\rho = y = 0$, as was shown in Section 5 of Chapter II, is determined by the long-range resonance dipole–dipole interaction between the molecules in an infinite crystal. The structure of exciton band (2.3) resembles the structure of the first exciton band in an anthracene crystal that was calculated by Davydov and Sheka [11]. According to Section 1 of Chapter IV, in our model the excitons interact with high-frequency phonons that correspond to rotational molecular vibrations (deviations of $\mathbf{d}$ from the axis $\mathbf{a}_y$) with wave vectors in a plane perpendicular to the axis $\mathbf{a}_y$. We shall ignore the phonon dispersion, i.e., let

$$\Omega_{op}(\rho) = L\Omega_{op} \tag{2.5}$$

The dispersion law for low-frequency (acoustic) phonons that interact with excitons is taken as

$$\Omega_{ac}(\rho, y) = \sqrt{\rho^2 + y^2}\, L\Omega_{ac}. \tag{2.6}$$

According to Section 6 of Chapter IV, the retarded Green function for excitons that interact with phonons is defined by the equation

$$G(\mathbf{k}, \omega) = [\omega - E(\mathbf{k}) - M(\mathbf{k}, \omega)]^{-1}, \tag{2.7}$$

where $M(\mathbf{k}, \omega)$ is the exciton mass operator. For a low exciton density, the mass operator is calculated by the integral equation

$$M(\mathbf{k}, \omega) = \frac{v}{(2\pi)^3} \sum_s \int | F_s(\mathbf{k}, \mathbf{q}) |^2 \times$$

$$\times \left\{ \frac{\nu_s(\mathbf{q}) + 1}{\omega - E(\mathbf{k}+\mathbf{q}) - \Omega_s(\mathbf{q}) - M[\mathbf{k}+\mathbf{q}, \omega - \Omega_s(\mathbf{q})] + i\eta} + \right.$$
$$\left. + \frac{\nu_s(\mathbf{q})}{\omega - E(\mathbf{k}+\mathbf{q}) + \Omega_s(\mathbf{q}) - M[\mathbf{k}+\mathbf{q}, \omega + \Omega_s(\mathbf{q})] + i\eta} \right\} d^3q, \tag{2.8}$$

where the summation $\sum_s$ is over the two phonon branches (2.5) and (2.6); and

$$\nu_s(\mathbf{q}) = \left[\exp\left(\frac{\Omega_s(\mathbf{q})}{kT}\right) - 1\right]^{-1} \tag{2.8a}$$

is the average number of phonons of branch s with the wave vector $\mathbf{q}$. Equation (2.8), just as (1.10), is approximate, since under the integral the accurate vertex function $F_s^{acc}(\mathbf{k}, \mathbf{q})$ is replaced by the zeroth-approximation function $F_s(\mathbf{k}, \mathbf{q})$. As we already noted in Section 1 of this chapter, the validity of this approximation (at T = 0) was demonstrated by Migdal [9] in a study of electron–phonon interaction and by Agranovich and Konobeev [10] in a study of exciton–phonon interaction.

We shall be concerned with Green function (2.7) at $\mathbf{k}$ equal to the wave vector $\mathbf{Q}$ of light waves. For visible and ultraviolet light, $Qa \ll 1$. Further, considering the slight dependence of exciton energy (2.3) upon ρ and y at $\mathbf{k} = \mathbf{Q}$, we can calculate mass operator (2.8) at $\mathbf{k} = 0$. In this case, as was shown in Section 1 of Chapter IV, the coupling function $F_s(0, \mathbf{q})$ for excitons with high-frequency phonons can be approximated by the expression

$$|F_{op}(0, \mathbf{q})|^2 = g_{op} L^2 (1 - \rho^2)^4. \tag{2.9}$$

Accordingly, for the coupling function for excitons with low-frequency phonons we have

$$|F_{ac}(0, \mathbf{q})|^2 = g_{ac} L^2 (1 - \rho^2 - y^2) \sqrt{\rho^2 + y^2}. \tag{2.10}$$

Function (2.9) vanishes at the boundaries of the first Brillouin zone (2.1). Hence, the dispersion law for high-frequency phonons does not play an important role when $\rho \to 1$. This justifies our choice of expression (2.5). The value g_{ac} in (2.10) is proportional to the mean value of the ratio of the square of the zero-point translational molecular vibrations to the square of the lattice constant. The value g_{op} in (2.9) is proportional to the mean value of the square of the zero-point rotational vibrations.

First of all, let us calculate the exciton mass operator, bearing in mind interaction only with high-frequency phonons or only with low-frequency phonons. Then we shall consider the general case of interaction.

Interaction with High-Frequency Phonons. Using the identity

$$\lim_{\eta \to +0} (x + i\eta)^{-1} = \frac{\mathscr{P}}{x} - i\pi\delta(x),$$

applying the method of successive approximations to integral equation (2.8), and substituting (2.9) into (2.5), we obtain (in L units) an expression for the real part of the mass operator:

$$M(\mathbf{k}, \omega) = \Delta(\mathbf{k}, \omega) - i\gamma(\mathbf{k}, \omega), \tag{2.11}$$

$$\Delta_{op} = g_{op}\{(\nu_{op} + 1) B(\zeta - \Omega_{op}) + \nu_{op} B(\zeta + \Omega_{op})\}, \tag{2.12}$$

where

$$\zeta = \frac{\omega}{L} - e$$

is a variable that defines the frequency (for $\hbar = 1$) in L units, which is displaced by the value e such that the bottom of the exciton band corresponds to $\zeta = 0$;

$$B(\zeta) = \frac{\mathcal{P}}{8} \int \frac{(1-\rho^2)^4 \rho \, d\rho \, dy \, d\varphi}{\zeta - \Xi(\rho, y)}. \tag{2.13}$$

The imaginary part of mass operator (2.11) is given by the expression (also in L units)

$$\gamma_{op} = g_{op}\{(\nu_{op} + 1) A(\zeta - \Delta_{op}(\zeta) - \Omega_{op}) + \\ + \nu_{op} A(\zeta - \Delta_{op}(\zeta) + \Omega_{op})\}, \tag{2.14}$$

where

$$A(\zeta) = \frac{\pi}{8} \int (1-\rho^2)^4 \rho \delta[\zeta - \Xi(\rho, y)] \, d\rho \, dy \, d\varphi. \tag{2.15}$$

Integration in (2.13) and (2.15) is carried out for all values of the variables in range (2.1). The sign $\mathcal{P}$ indicates that the integral is taken in terms of the principal value. The results of numerical

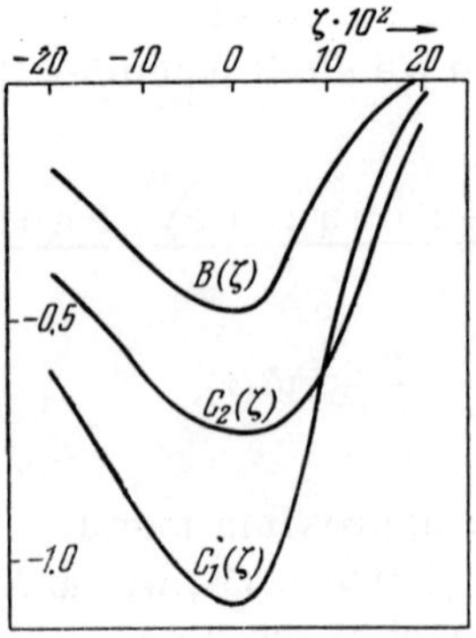

Fig. 30. Auxiliary functions defining the real part of the exciton mass operator.

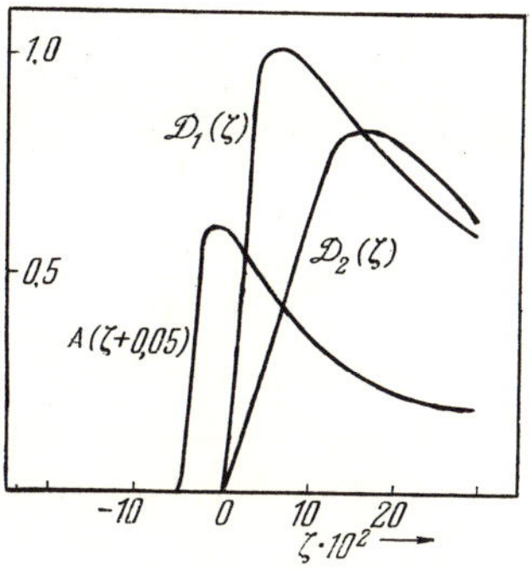

Fig. 31. Auxiliary functions defining the imaginary part of the exciton mass operator.

integration of the expressions $B(\zeta)$ and $A(\zeta)$ at $\Omega_{op} = 0.05$ made by Davydov and Myasnikov [12] are shown in Figs. 30 and 31.

Interaction with Low-Frequency Phonons. In [12], when calculating the mass operator for excitons that interact with low-frequency phonons, it was assumed that the limiting phonon frequency was $\Omega_{ac} = 0.03$. For temperatures that satisfy the inequality

$$kT > L\Omega_{ac}, \tag{2.16}$$

using (2.8) and (2.6), we can write

$$\nu_{ac}(\mathbf{q})\sqrt{\rho^2 + y^2} \approx \frac{kT}{L\Omega_{ac}}. \tag{2.17}$$

When $L = 10^3\ \text{cm}^{-1}$, this equation is satisfied at temperatures equal to and exceeding 20°K.

Considering (2.17) and ignoring the small acoustic frequencies in the denominators of the integrand of (2.8), we obtain the real part of mass operator (2.11) in L units:

$$\Delta_{ac}(\zeta) = g_{ac}\left\{\frac{2kT}{L\Omega_{ac}} C_1(\zeta) + C_2(\zeta)\right\}, \tag{2.18}$$

where

$$C_1(\zeta) = \frac{\mathcal{P}}{8}\int \frac{(1-\rho^2)^2\,\rho\, d\rho\, dy\, d\varphi}{\zeta - \Xi(\rho, y)}, \tag{2.19}$$

$$C_2(\zeta) = \frac{\mathcal{P}}{8}\int \frac{(1-\rho^2)^2\,\rho^2\,d\rho\,dy\,d\varphi}{\zeta - \Xi(\rho,\,y)}. \qquad (2.20)$$

The imaginary part of mass operator (2.11) in the same units is

$$\gamma_{ac}(\zeta) = g_{ac}\left\{\frac{2kT}{L\Omega_{ac}}\mathcal{D}_1[\zeta - \Delta_{ac}(\zeta)] + \mathcal{D}_2[\zeta - \Delta_{ac}(\zeta)]\right\}, \qquad (2.21)$$

where

$$\left.\begin{aligned} \mathcal{D}_1(\zeta) &= \frac{\pi}{8}\int (1-\rho^2)^2\,\rho\,\delta[\zeta - \Xi(\rho,\,y)]\,d\rho\,dy\,d\varphi, \\ \mathcal{D}_2(\zeta) &= \frac{\pi}{8}\int (1-\rho^2)^2\,\rho^2\delta[\zeta - \Xi(\rho,\,y)]\,d\rho\,dy\,d\varphi. \end{aligned}\right\} \qquad (2.22)$$

The results of numerical integration of the functions C_i and $\mathcal{D}_i$ performed in [12] are shown in Figs. 30 and 31.

If $L\Omega_{ac}^0$ corresponds to the frequency of the low-frequency phonons that interact most strongly with excitons, then at temperatures kT that satisfy the inequality

$$kT < L\Omega_{ac}^0, \qquad (2.23)$$

the real and imaginary parts of mass operator (2.11) are given by the expressions

$$\left.\begin{aligned} \Delta_{ac}^0(\zeta) &= g_{ac}\left[1 + 2e^{-\frac{L\Omega_{ac}^0}{kT}}\right] C_2(\zeta), \\ \gamma_{ac}^0(\zeta) &= g_{ac}\left[1 + 2e^{-\frac{L\Omega_{ac}^0}{kT}}\right] \mathcal{D}_2\,[\zeta - \Delta_{ac}^0(\zeta)]. \end{aligned}\right\} \qquad (2.24)$$

Simultaneous Interaction of Excitons with Low- and High-Frequency Phonons. If the interactions of excitons with low- and high-frequency phonons are comparable in strength, then in the approximation used in [12] the real part of mass operator (2.11) is

$$\Delta(\zeta) = \Delta_{op}(\zeta) + \Delta_{ac}(\zeta), \qquad (2.25)$$

where $\Delta_{op}(\zeta)$ and $\Delta_{ac}(\zeta)$ are calculated by expressions (2.12) and (2.18), or (2.24) for low temperatures. The imaginary part of mass operator (2.11) is

$$\gamma(\zeta) = \gamma_{op}^{(1)}(\zeta) + \gamma_{ac}^{(1)}(\zeta), \qquad (2.26)$$

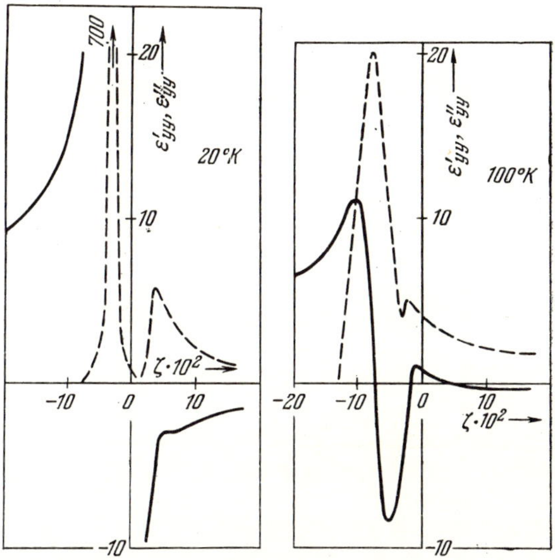

Fig. 32. Real (solid curves) and imaginary (broken curves) parts of dielectric constant in exciton absorption band for $g_{ac} \ll g_{op} = 0.1$.

where $\gamma_{op}^{(1)}(\zeta)$ and $\gamma_{ac}^{(1)}(\zeta)$ are calculated by expressions (2.14) and (2.21), or by (2.24) if in these expressions $\Delta_{op}(\zeta)$ and $\Delta_{ac}(\zeta)$ are replaced by (2.25).

If we express the dielectric constant

$$\varepsilon_{yy}(\mathbf{Q}, \omega) = \varepsilon'_{yy}(\mathbf{Q}, \omega) + i\varepsilon''_{yy}(\mathbf{Q}, \omega)$$

for an electromagnetic wave with frequency ω and wave vector **Q** perpendicular to the axis $\mathbf{a}_y$ in terms of Green function (2.11) with the aid of formula (IV.2.21), then at $\mathbf{Qa}_y \ll 1$ we obtain

$$\left.\begin{aligned} \varepsilon'_{yy}(\zeta) &= \varepsilon^0_{yy}(\zeta) + \frac{\Delta(\zeta) - \zeta}{[\zeta - \Delta(\zeta)]^2 + \gamma^2(\zeta)}, \\ \varepsilon''_{yy}(\zeta) &= \frac{\gamma(\zeta)}{[\zeta - \Delta(\zeta)]^2 + \gamma^2(\zeta)}. \end{aligned}\right\} \qquad (2.27)$$

The real $\varepsilon'(\zeta)$ and imaginary $\varepsilon''(\zeta)$ parts of the dielectric constant were calculated in [12] by formula (2.27) for kT = L/14 and L/70, which at L = 10^3 cm^{-1} correspond to 100 and 20°K. The calculation results are shown in Figs. 32, 33, and 34, where the solid lines are for $\varepsilon'(\zeta)$ and the broken lines are for $\varepsilon''(\zeta)$.

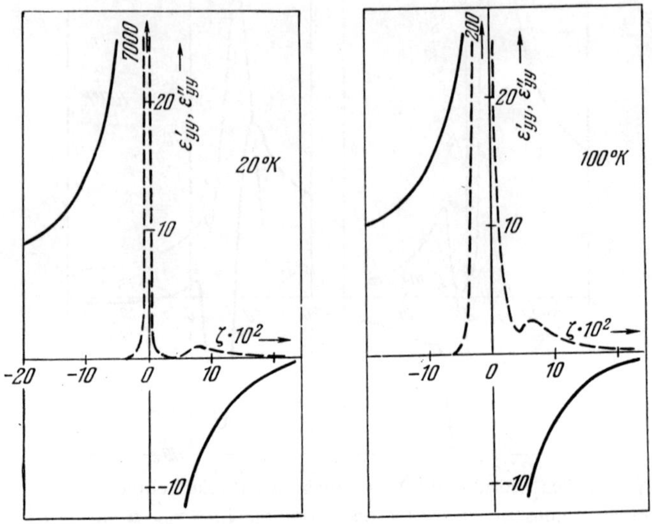

Fig. 33. The same as Fig. 32 but for $g_{ac} \ll g_{op} = 0.01$.

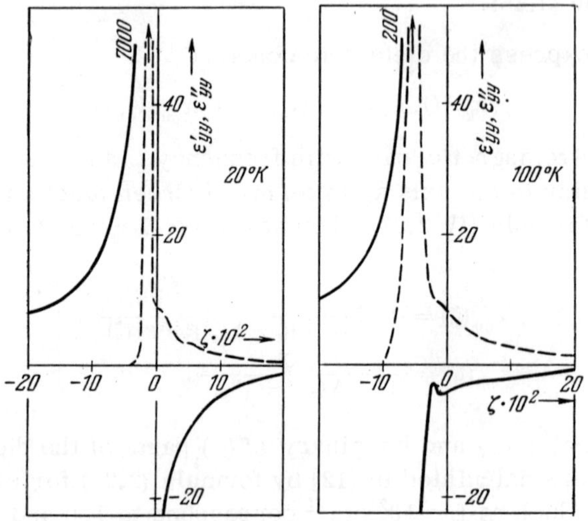

Fig. 34. The same as Fig. 32 but for $g_{ac} = g_{op} = 0.01$.

The refractive index n and the attenuation factor for the same temperature values were calculated using the relation

$$(n + i\varkappa)^2 = \varepsilon. \tag{2.28}$$

These results are shown in Figs. 35, 36, and 37. All of the calculations in [12] were made at $\varepsilon_{yy}^0 = 4$, $\Omega_{op} = 0.05$, and $\Omega_{ac} = 0.03$. In Figs. 32 and 35, $g_{ac} \ll g_{op} = 0.1$; in Figs. 33 and 36, $g_{ac} \ll g_{op} = 0.01$; and in Figs. 34 and 37, $g_{ac} = g_{op} = 0.01$.

It follows from the shape of the broken curves in Figs. 32, 33, and 34 that the imaginary part of the dielectric constant, which corresponds to the true light absorption in the crystal, has for low temperatures and small coupling constants g_{op} and g_{ac} the form of a sharp peak with a small structure to the right of it. If the interaction with low-frequency phonons is insignificant (Figs. 32 and 33), the half-width of the principal peak is a function only of the average number of high-frequency phonons that exist at the given temperature. As the temperature approaches absolute zero, the principal peak takes on a deltoid shape. Interaction with low-frequency phonons expands the shortwave part of the principal peak and makes it more symmetric (Fig. 34).

An increase in temperature and in the phonon-exciton coupling results in secondary peaks on the shortwave side of the principal

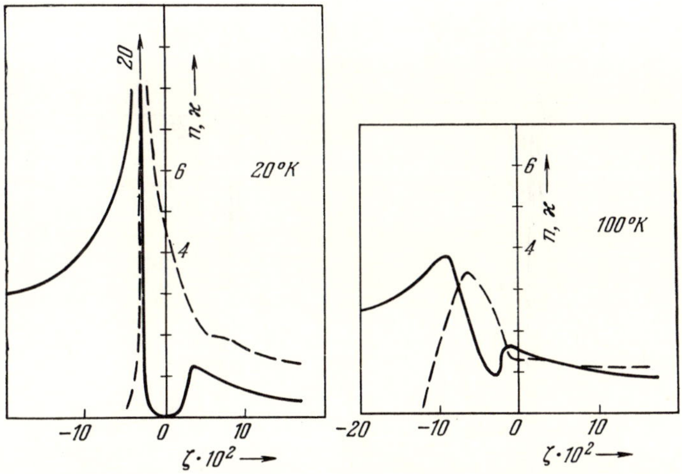

Fig. 35. Refractive index (solid curves) and attenuation factor of electromagnetic waves in crystal (broken curves) for the values in Fig. 32.

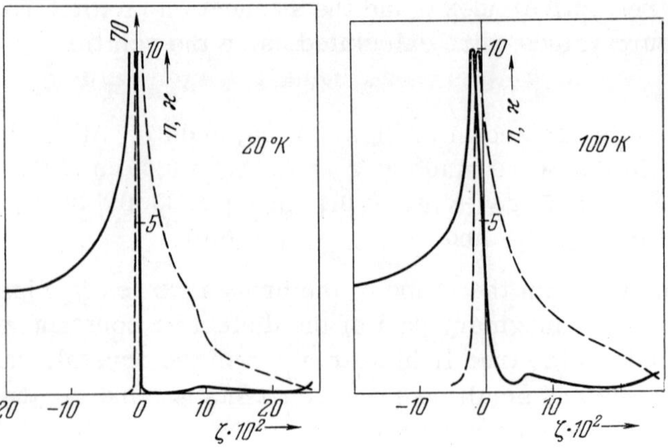

Fig. 36. The same as Fig. 35 but for the values in Fig. 33.

peak (Figs. 33 and 34). This increases the asymmetry of the absorption curve. It should be noted that although the secondary peaks are due to interaction with phonons, their position – even in the case of interaction of excitons only with high-frequency phonons of frequency Ω_{op} (Figs. 32 and 33) – is not equal to this frequency but is greatly dependent upon the structure of the exciton band. The distance between the principal and a secondary peak usually exceeds Ω_{op}.

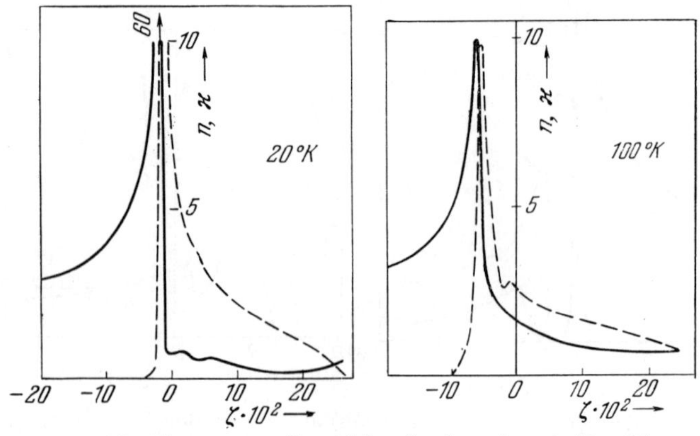

Fig. 37. The same as Fig. 35 but for the values in Fig. 34.

The asymmetry of the absorption curve $\varepsilon''(\xi)$, i.e., its deviation from a Lorentz curve, is due to the dependence of the real $\Delta(\xi)$ and imaginary $\gamma(\xi)$ parts of the mass operator upon the frequency ξ. Since in [4] Toyozawa did not take into account this dependence, he obtained Lorentz curves. To obtain asymmetric curves for light absorption in ionic crystals, Toyozawa examined interaction with low-frequency phonons in higher approximations [13].

The attenuation factor $\varkappa$ for electromagnetic waves (reduction of wave amplitude) is shown in Figs. 35-37. This factor cannot equal zero even in the absence of absorption. For example, at $\varepsilon'' = 0$ and $\varepsilon' < 0$ we have, according to (2.28), $n = 0$ and $\varkappa \neq 0$. In this case, the exponential decrease in the amplitude of an electromagnetic wave inside a crystal is due to reflection of the wave from its surface.

3. Dispersion and Absorption of Light in Strong Interaction of Electronic Excitations with Phonons

As was indicated in Section 1 of Chapter IV, the case of "strong coupling" of an electronic excitation with molecular vibrations are realized in crystals with a highly deformable lattice when the resonance interactions between the excited and unexcited molecules are weak. When the "strong coupling" conditions are satisfied, simultaneously with excitation of a molecule there is local lattice deformation, which destroys the translational symmetry of the crystal. These excited states are not exciton states, because they are not characterized by a particular quasi-momentum $\hbar\mathbf{k}$. It is convenient to call them *localized electronic excitations*.

In Chapter II, Section 7, it was noted that localized electronic excitations in crystals of aromatic compounds are usually associated with molecular levels consisting of combinations of an intramolecular electronic excitation and not totally symmetric vibrations of the atoms.

The interaction of localized electronic excitations with molecular vibrations in a crystal has been examined by Huang and Rhys [14], Davydov [15, 16], Davydov and Lubchenko [17], Lubchenko [18-20], Ratner and Zil'berman [21], Perlin [22], and others. Excited states due to excitations of "impurity" molecules introduced into the lattice were studied in these papers. But the results of these theoretical studies do not relate directly to localized electronic excitations in molecular crystals.

The distinctive feature of localized electronic excitations in crystals is that their formation in light absorption or their disappearance in luminescence is usually associated with many-phonon processes. In other words, in the formation or disappearance of such excitations, the lattice energy is changed by a value equal to the energy of tens and hundreds of phonons. The methods of conventional perturbation theory, in which the number of participating phonons is increased by one in each successive approximation, cannot be used to study such phenomena. A theory must be developed in which many-phonon processes are taken into account in the first stage of calculations. Frenkel [23] first called attention to the possibility of developing such a theory.

Below, we shall expound, on the basis of [15, 16], an elementary theory of this kind that allows us to determine the form of the absorption and luminescence bands of localized electronic excitations and their dependence upon temperature. For simplicity, we shall consider the case when a localized excitation corresponds to the transition of one molecule of the crystal to an excited state. We shall write the energy operator of the crystal in the form

$$H = H_{\text{vib}}(R) + H_{\text{mol}}(r) + V(r, R), \tag{3.1}$$

where R is the set of all degrees of freedom of the lattice; r is the set of internal degrees of freedom of the molecule that is excited; V(r, R) is the molecule-lattice interaction operator; and $H_{\text{vib}}(R) = T_R + U(R)$. Here, T_R is the kinetic-energy operator for molecular vibrations and U(R) is the potential-energy operator for molecular vibrations in a crystal without an electronic excitation.

If ε_l and ψ_l are the eigenvalues and eigenfunctions of the operator H_{mol}, then in an adiabatic approximation the operator

$$V_l(R) = \int \psi_l^* V(r, R)\, \psi_l\, dr - \int \psi_0^* V(r, R)\, \psi_0\, dr \tag{3.2}$$

is the additional potential energy, which, together with U(R), characterizes the motion of the molecules in the crystal when one of them is excited. Here, the vibration energy of the crystal is determined by the equation

$$[H_{\text{vib}}(R) + V_l(R) - (E - \varepsilon_l)]\,\Phi_l(R) = 0. \tag{3.3}$$

Let R_0 be the set of equilibrium positions of the molecules in a crystal without an electronic excitation, i.e.,

$$\min U(R) = U(R_0);$$

Then, with transition of one molecule to the l-th excited state, the potential energy is expressed by the formula

$$W_l(R) \equiv U(R) + V_l(R) = W_l(R_0) + (R - R_0)\,V_l^{(1)} + \frac{1}{2}(R - R_0)^2\,W_l^{(2)} + \ldots \tag{3.4}$$

In a crystal without an electronic excitation, $l = 0$, $V_l(R) = 0$, and $W_0(R) = U(R)$.

Let us introduce normal coordinates such that

$$T_R + \frac{1}{2}(R - R_0)^2 W_l^{(2)} = \frac{\hbar}{2}\sum_s \omega_{sl}\left(\xi_s^2 - \frac{\partial^2}{\partial \xi_s^2}\right), \tag{3.5}$$

where ω_{sl} is the vibration frequency of the molecules in a crystal in which one of the molecules is in the l-th excited state. The value $l = 0$ pertains to the ground state of a molecule. Simultaneously with transformation (3.5), let us transform the second term on the right of (3.4):

$$(R - R_0)\,V_l^{(1)} = \sum_s a_{sl}\xi_s, \qquad a_{s0} = 0. \tag{3.6}$$

Having substituted (3.4), (3.5), and (3.6) into (3.3), let us transform it to

$$\left\{\frac{\hbar}{2}\sum_s \omega_{sl}\left[-\frac{\partial^2}{\partial \xi^2} + (\xi_s - \xi_{sl})^2\right] - \mathscr{E}_l\right\}\Phi_l(\xi) = 0, \tag{3.7}$$

where

$$\xi_{sl} = -\frac{a_{sl}}{\hbar\omega_{sl}}, \qquad \xi_{sl} = 0$$

determines the displacements of the equilibrium positions of the molecules when one of the molecules goes to the l-th excited state.

The total energy of the crystal

$$E_l = \mathscr{E}_l + \left\{\varepsilon_l + W_l(R_0) - \frac{\hbar}{2}\sum_s \omega_{sl}\xi_{sl}^2\right\}. \tag{3.8}$$

Equation (3.7) decomposes into independent Schrödinger equations for harmonic oscillators. Its solution can be written as

$$\Phi_{l,\{n_s\}}(\xi) = \prod_s \varphi_{n_s}(\xi_s - \xi_{sl}), \tag{3.9}$$

where $\{n_s\} \equiv \ldots n_s \ldots n_{s'}' \ldots$ are the sets of quantum numbers of the oscillators, which determine the vibrational states of the molecules when one of them is in the l-th electronic state; and $\varphi_{n_s}(\xi_s - \xi_{sl})$ are the wave functions of the harmonic oscillators. The total wave function of the crystal can be written as

$$\Psi^0_{l\{n_s\}} = \psi_l(r)\,\Phi_{l\{n_s\}}(\xi). \tag{3.10}$$

We shall assume that the internal states of all of the molecules, except the one in the l-th excited state, are unchanged in this quantum transition. Therefore, their functions and energies are not written out in explicit form. The total energy of the molecule in question and of the lattice vibrations, which corresponds to function (3.10), is

$$E_{l\{n_s\}} = \hbar \sum_s \omega_{sl}\left(n_s + \frac{1}{2}\right) + \varepsilon_l + W_l(R_0) - \frac{\hbar}{2}\sum_s \omega_{sl}\xi^2_{sl}. \tag{3.11}$$

We shall be concerned with the energy change of the system with transition of a molecule from the ground (0) to the l-th excited state, which is characterized by the set of quantum numbers $\{n_s'\}$, i.e., $E_{l\{n_s'\}} - E_{0\{n_s\}}$. If this transition occurs without a change in the lattice-vibration quantum numbers ($n_s' = n_s$), then it is called a phononless electron transition. The energy change in a phononless electronic transition is, according to (3.11),

$$\hbar\Omega^{\{n_s\}}_{l0} \equiv E_{l\{n_s\}} - E_{0\{n_s\}} = \hbar\left[\Omega_l + \sum_s (\omega_{sl} - \omega_{s0})\, n_s\right], \tag{3.12}$$

where

$$\hbar\Omega_l \equiv \Delta\varepsilon_l + V_l(R_0) - \frac{\hbar}{2}\sum_s \omega_{sl}\xi^2_{sl} + \frac{\hbar}{2}\sum_s (\omega_{sl} - \omega_{s0}). \tag{3.13}$$

The difference $\hbar\Omega^{\{n_s\}}_{l0} - \Delta\varepsilon_l$ characterizes the change in lattice energy in a phononless electron transition, which is related to lattice deformation in the area of the excited molecule, and the change in the energy of the zero-point lattice vibrations [the last term in (3.13)]. It is very interesting that under the conditions $\omega_{sl} \neq \omega_{s0}$, the energy of a phononless transition is a function of the quantum numbers n_s, i.e., of the molecular-vibration states and therefore, of the temperature of the crystal, when n_s is replaced by the statistical average.

If the interaction operator for a molecule with an electromagnetic wave is written in the dipole approximation as

$$H' = -e\mathbf{r}\mathbf{E}_0 (e^{-i\omega t} + e^{i\omega t}), \tag{3.14}$$

where $\mathbf{E}_0$ is the electric-field strength of the wave at the place occupied by the molecule, then in the first approximation of perturbation theory the wave function of a crystal with an excited molecule can, with allowance for only one excited state, be written as

$$\Psi_{0\{n_s\}} = \Psi^0_{0\{n_s\}} \exp\left\{-iE_{0\{n_s\}}\frac{t}{\hbar}\right\} + \\ + \mathbf{d}_{l0}\mathbf{E}_0 \sum_{\{n'_s\}} \frac{\prod_s M^{0l}_{n_s n'_s} \psi_l(r) \prod_s \varphi_{n'_s}(\xi_s - \xi_{sl})}{\hbar\left[\Omega_{l0} + \sum_s (n'_s\omega_{sl} - n_s\omega_{s0}) - \omega - i\gamma\right]} \exp\left\{-i\left[\omega + \frac{E_{0\{n_s\}}}{\hbar}\right]t\right\}, \tag{3.15}$$

where

$$\mathbf{d}_{l0} = e\int \psi_l^* \mathbf{r} \psi_0 \, dr, \qquad M^{0l}_{n_s n'_s} = \int \varphi_{n_s}(\xi_s)\varphi_{n'_s}(\xi_s - \xi_{sl})\, d\xi_s \tag{3.16}$$

are the overlap integrals of the oscillator functions.

With electronic excitation of one molecule of the crystal, the molecules closest to it will be shifted by a finite value as a result of the change in the interaction forces. But in the normal lattice coordinates, the values ξ_{sl} that characterize these displacements have the order of magnitude $N^{-1/2}$, where N is the total number of degrees of freedom of all the molecules of the crystal. With accuracy to the terms ξ^2_{sl}, matrix elements (3.16) have the form

$$\left.\begin{aligned} M^{0l}_{n_s n'_s} &= 0, \quad \text{if } n'_s \neq n_s, \quad n_s \pm 1, \\ M^{0l}_{n_s n_s} &= 1 - \frac{1}{2}\left(n_s + \frac{1}{2}\right)\xi^2_{sl}, \\ M^{0l}_{n_s, n_s+1} &= -\sqrt{(n_s+1)/2}\,\xi_{sl}, \\ M^{0l}_{n_s, n_s-1} &= \sqrt{n_s/2}\,\xi_{sl}. \end{aligned}\right\} \tag{3.17}$$

Ignoring terms of a higher order of smallness in (3.17) is justified, because the final result contains a sum of the type $\sum_s |M_{n_s n'_s}|^2$, which contains N terms. When $N \to \infty$, therefore, only the quantities containing ξ_{sl}^2 are preserved.

Using (3.17), we can show that in the same approximation

$$\sum_{n''_s} M_{n_s n''_s} M_{n'_s n''_s} = \begin{cases} 0, & \text{if } n'_s \neq n_s; \\ 1, & \text{if } n'_s = n_s. \end{cases}$$

Let us determine the mean electric dipole moment of the molecule and crystal in state (3.15) and take its statistical average over all possible states of molecular vibration at a given temperature. Then, using the equation

$$\langle \mathbf{p} \rangle = \{\beta \mathbf{E}_0 e^{-i\omega t} + \beta^* \mathbf{E}_0 e^{i\omega t}\},$$

we find the polarizability tensor β of the molecule. If x, y, and z are the principal axes of this tensor, then

$$\beta_{xx} = \frac{|\mathbf{d}^x_{l0}|^2}{\hbar} \overline{\sum_{\{n'_s\}}} \frac{\prod_s |M_{n_s n'_s}|^2}{\Omega + \sum_s (n'_s - n_s)\,\omega_{sl} - \omega + i\gamma},$$

where $\Omega \equiv \Omega_{l_0}^{\{\bar{n}_s\}}$ is defined by expression (3.12), the line over the sum indicates averaging over all initial states $\{n_s\}$ of lattice vibrations, and γ indicates the finite width of the excitation level of a free molecule.

We shall introduce the refractive index n and the attenuation factor $\varkappa$ by means of the equation

$$(n + i\varkappa)^2 = \varepsilon_0 + 4\pi N_0 \beta_{xx},$$

where N_0 is the number of molecules per unit volume. Then, for an optically isotropic crystal we obtain [15, 16]

$$\varepsilon = \varepsilon' + i\varepsilon'',$$

where

$$\left.\begin{aligned} \varepsilon' &= n^2 - \varkappa^2 = \varepsilon_0 + \frac{4\pi N_0}{\hbar} |\mathbf{d}_{l0}|^2 F(\omega), \\ \varepsilon'' &= 2n\varkappa = \frac{4\pi N_0}{\hbar} |\mathbf{d}_{l0}|^2 S(\omega), \end{aligned}\right\} \quad (3.18)$$

where

$$F(\omega) = \mathrm{Im}(\Lambda), \quad S(\omega) = \mathrm{Re}\,\Lambda, \tag{3.19}$$

$$\Lambda = \int_0^\infty \exp\{i\mu[\Omega - \omega + i\gamma] + g(\mu)\}\, d\mu, \tag{3.20}$$

$$g(\mu) = \frac{1}{2}\sum_s \{(\bar{n}_s + 1)\, e^{i\mu\omega_{sl}} + \bar{n}_s e^{-i\mu\omega_{sl}} - (2\bar{n}_s + 1)\}\, \xi_{sl}^2, \tag{3.21}$$

$$\bar{n}_s = \left(\exp\left[\frac{\hbar\omega_{s0}}{kT}\right] - 1\right)^{-1}.$$

These formulas are also valid for the case when an impurity molecule introduced into the crystal is excited. In this case, N_0 in (3.18) must equal the number of impurity molecules per unit volume of the crystal. Sometimes, a case is realized approximately in which transition of an impurity molecule is not accompanied by appreciable lattice deformation ($\xi_{sl} = 0$). In this case, $g(\mu) = 0$ and

$$\Lambda(\omega) = \frac{\gamma + i[\Omega - \omega]}{[\Omega - \omega]^2 + \gamma^2}.$$

The curves of the dispersion and attenuation factor (3.18) in this limiting case coincide in shape with the corresponding curves for free molecules. Only a shift in the resonance frequency

$$\Delta \equiv \Omega - \frac{\Delta\varepsilon_l}{\hbar} = \frac{V_l}{\hbar} + \sum_s (\omega_{s0} - \omega_{sl})\, \bar{n}_s. \tag{3.22}$$

occurs. Above the Debye temperature, i.e., when

$$\hbar\Delta = V_l + kT \sum_s (\omega_{s0} - \omega_{sl}) / \omega_{s0}. \tag{3.22a}$$

When localized electronic excitations are produced in a crystal, the molecules are displaced from their equilibrium positions (local lattice deformation is very considerable) ($\xi_{sl}^2 \neq 0$). For low temperatures, it is convenient to transform function (3.21) to

$$g(\mu) = -g_0 + \sum_s [b_s \exp(i\mu\omega_{sl}) + c_s \exp(-i\mu\omega_s)], \tag{3.23}$$

where

$$b_s = \frac{1}{2}(\bar{n}_s + 1)\, \xi_{sl}^2, \quad c_s = \frac{1}{2}\bar{n}_s \xi_{sl}^2, \quad g_0 = \sum_s (b_s + c_s). \tag{3.24}$$

Substituting (3.23) into integral (3.20), let us transform it to

$$\Lambda = e^{-g_0} \int_0^{\infty} \exp\{i\mu[\Omega + \omega + i\gamma]\} \sum_{n,m} I_{n,m}(\mu)\, d\mu, \tag{3.25}$$

where

$$I_{n,m}(\mu) = \frac{1}{n!m!}\Big[\sum_s b_s \exp(i\mu\omega_{sl})\Big]^n \Big[\sum_s c_s \exp(-i\mu\omega_{sl})\Big]^m,$$

and n and m take positive integral values from zero to infinity.

Let us assume that interaction of a molecular excitation occurs only with lattice vibrations (one of the optical branches) for which $\omega_{sl} = \omega_0$. Then integral (3.25) is easily calculated, and we obtain

$$\Lambda \doteq e^{-g_0} \sum_{n,m} \frac{b^n c^m \{\gamma + i[\Omega + (n-m)\omega_0 - \omega]\}}{n!m!\{[\Omega - \omega + (n-m)\omega_0]^2 + \gamma^2\}}, \tag{3.26}$$

where

$$b = \sum_s b_s; \quad c = \sum_s c_s.$$

In the limiting case of low temperatures $c \approx 0$, so only the term with $m = 0$ is retained in the sum over m in (3.26). Therefore

$$\Lambda = e^{-g_0} \sum_{n=0}^{\infty} \frac{b^n \{\gamma + i[\Omega + n\omega_0 - \omega]\}}{n!\{[\Omega + n\omega_0 - \omega]^2 + \gamma^2\}}. \tag{3.27}$$

If we substitute (3.26) or (3.27) into (3.18), we obtain the real and imaginary parts of the dielectric constant. In this case, the true-absorption curve $\varepsilon''(\omega)$ is represented by a set of equidistant bands[2] at distances $(n - m)\omega_0$ from the frequency Ω. With a change in temperature, the position of frequency Ω is shifted slightly, since according to (3.22), the frequency shift is a function of $\bar{n}_s$. The intensity distribution of the bands corresponding to various n and m is changed at the same time. Each term in (3.26) corresponds

[2]This results was first obtained by Huang and Rhys [14].

to a local excitation in which n photons are created and m photons are annihilated simultaneously. At low temperatures, when $c \approx 0$, according to (3.27), the ratio of the maxima of two successive bands is

$$\frac{b^n (n+1)!}{b^{n+1} n!} = \frac{n+1}{b}. \tag{3.28}$$

if

$$b = \frac{1}{2} \sum_s (\bar{n}_s + 1) \xi_{sl}^2 < 1, \tag{3.29}$$

then each maximum is smaller than the preceding one. In this case, maximum absorption corresponds to a phononless transition ($n = 0$).

The phononless transition $n = m = 0$ is determined by

$$\Lambda_0 = \frac{\gamma + i(\Omega - \omega)}{[\Omega - \omega]^2 + \gamma^2} \exp(-g_0).$$

The imaginary part of the dielectric constant corresponding to the phononless transition is

$$\varepsilon'' = \frac{4\pi N_0}{\hbar} |\mathbf{d}_{l0}|^2 \frac{\gamma}{(\Omega - \omega)^2 + \gamma^2} e^{-g_0},$$

where

$$g_0 = \sum_s \left(\bar{n}_s + \frac{1}{2}\right) \xi_{sl}^2.$$

Therefore, the maximum of the phononless-transition curve decreases with an increase in temperature.

If at a certain low temperature inequality (3.29) is satisfied and the principal absorption maximum corresponds to a phononless transition (Fig. 38), b can become greater than unity when the temperature is increased. Then, according to (3.28), in the first peaks of the absorption band (when $n < b - 1$) the subsequent maxima will be greater than the preceding ones. With an increase in temperature, in addition to the increase in b, the c values become different from zero and, therefore, new maxima appear that corre-

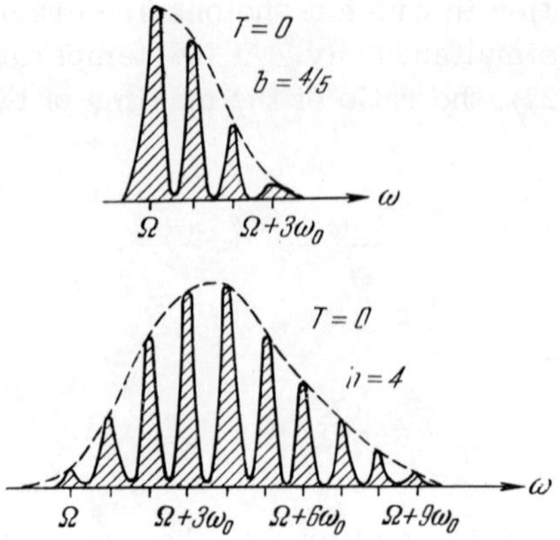

Fig. 38. Qualitative change in true absorption for various values of the parameter b, which determines the coupling of the electronic excitation with lattice vibrations.

spond to $m \neq 0$ in expression (3.36). Some of them (when $m > n$) will be situated on the longwave side of the phononless-transition frequency Ω. Because of the increase in b and c, the absorption region is expanded while the intensity of each maximum is increased, due to the factor $\exp(-g_0)$. The case in question, of the interaction of an electronic excitation with phonons of only one frequency, is an idealization. Besides phonons of frequency ω_0, phonons of other frequencies will be created and absorbed. The systems of bands that pertain to such transitions must be superimposed on the picture in Fig. 38 and will occupy the area bounded by the broken curve.

If phonons with various frequencies participate in the interaction of electronic excitations with lattice vibrations, then it is convenient to represent function (3.23) as follows:

$$g(\mu) = -g_0 + \sum_s (b_s + c_s)\cos \mu\omega_{sl} + i\sum_s (b_s - c_s)\sin \mu\omega_{sl}, \qquad (3.30)$$

where b_s, c_s, and g_0 are defined by expressions (3.24). In this case,

$$\Lambda = e^{-g_0}\int_0^\infty \exp\left\{i\mu\left[\Omega - \omega + \sum_s (b_s - c_s)\frac{\sin \mu\omega_{sl}}{\mu}\right] - \mu\gamma + \right.$$
$$\left. + \sum_s (b_s + c_s)\cos \mu\omega_{sl}\right\} d\mu. \qquad (3.31)$$

It follows from the integrand of (3.31) that Λ is chiefly a function of the small range of values $\mu \approx 0$. Therefore, let us replace integral (3.31) by an integral taken between the limits 0 and μ_0.

If $\omega_s \leq \omega_0$, we take $\mu_0 = 1/\omega_0$; then for $\mu < \mu_0$ we can transform (3.30) to

$$g(\mu) = i\mu A - \mu^2 B^2, \tag{3.32}$$

where

$$A = \frac{1}{2}\sum_s \omega_{sl}\xi_{sl}^2, \qquad B^2 = \frac{1}{2}\sum_s (2\bar{n}_s + 1)\,\omega_{sl}^2\xi_{sl}^2.$$

Under the condition

$$B^2 > \frac{1}{\mu_0^2} = \omega_0^2 \tag{3.33}$$

we can transform (3.31) to

$$\Lambda = \int_0^\infty \exp\{i\mu[\Omega + A - \omega] - \mu\gamma - \mu^2 B^2\}\, d\mu =$$

$$= \frac{\gamma\sqrt{\pi}}{2B}\,[1 - \Phi(z)]\exp(|z|^2), \tag{3.34}$$

where

$$z = (\gamma - i[\Omega_r - \omega])/2B;$$

$$\Omega_r = \Omega + A$$

is the resonance frequency; and $\Phi(z)$ is the Gaussian error function, which is defined by the integral

$$\Phi(z) = \frac{2}{\sqrt{\pi}}\int_0^z e^{-y^2}\,dy = \begin{cases} \dfrac{2}{\sqrt{\pi}}\left(z - \dfrac{z^3}{3} + \ldots\right), & \text{if } z < 1; \\ 1, & \text{if } z > 1. \end{cases} \tag{3.35}$$

Close to the resonance frequency $\left|\frac{\Omega_r - \omega}{2B}\right| < 1$, after separation of the real and imaginary parts in (3.34) and taking into account (3.18) and (3.19), we obtain, providing $B \gg \gamma$,

$$\varepsilon' = \varepsilon_0 + \mathcal{D}\,\frac{\Omega_r - \omega}{B^2}\exp\left[\frac{\gamma^2 - (\Omega_r - \omega)^2}{4B^2}\right], \tag{3.36}$$

$$\varepsilon'' = \frac{\gamma\sqrt{\pi}}{2}\,\frac{\mathcal{D}}{B}\exp\left[\frac{\gamma^2 - (\Omega_r - \omega)^2}{4B^2}\right], \tag{3.37}$$

where

$$\mathscr{D} \equiv \frac{4\pi N_0 \,|\, \mathbf{d}_l \,|^2}{\hbar}.$$

Far from resonance, we have

$$\left.\begin{aligned} \varepsilon' &= \varepsilon_0 + \frac{\mathscr{D}(\Omega_r - \omega)}{(\Omega_r - \omega)^2 + \gamma^2}, \\ \varepsilon'' &= \frac{\mathscr{D}\gamma}{(\Omega_r - \omega)^2 + \gamma^2}. \end{aligned}\right\} \tag{3.38}$$

Thus, when inequality (3.33) is satisfied, the curve $\varepsilon''(\omega)$ near the resonance frequency has the shape of a Gaussian curve. This result was also obtained by Pekar [24] for light absorption by impurity molecules.

The real dielectric constant $\varepsilon'(\omega)$ is represented by function (3.36), which varies more sharply than the usual dispersion curve for free molecules.

At not very low z, the cubic term in expansion (3.35) must also be taken into account. Then the curve $\varepsilon'(\omega)$ will contain a term that results in "asymmetry" of the curve of the real part of the dielectric constant versus frequency:

$$\varepsilon' = \varepsilon_0 + \frac{\mathscr{D}}{B^2}(\Omega_r - \omega)\left[1 - \frac{(\Omega_r - \omega)^2}{24B^2}\right]\exp\left\{-\frac{(\Omega_r - \omega)^2}{4B^2}\right\}.$$

The imaginary part of the dielectric constant is practically unchanged:

$$\varepsilon'' = \frac{\gamma\sqrt{\pi}\mathscr{D}}{2B}\left[1 + \frac{\gamma(\Omega_r - \omega)^2}{4\sqrt{\pi}B^3}\right]\exp\left\{-\frac{(\Omega_r - \omega)^2}{4B^2}\right\}.$$

4. Theory of Strong Coupling of Electronic Excitations with Phonons in the Second-Quantization Representation

In this section, we shall study in the second-quantization representation the case of strong coupling of electronic excitations with phonons. When the molecules are fixed in the positions R (adiabatic approximation), the energy operator of the crystal can, according to Section 1 of Chapter IV, be written as

$$H_a(R) = \sum_{n} [\Delta\varepsilon_n + D_n(R)] B_n^+ B_n + \sum_{n,m}' M_{nm}(R) B_m^+ B_n + W(R), \quad (4.1)$$

where

$$D_n(R) = \sum_{m(\neq n)} D_{nm}(R);$$

W(R) is the potential-energy operator of the unexcited molecules; and B_n^+ and B_n are the creation and annihilation operators for an electronic excitation in the molecule that occupies lattice point **n**. They satisfy the commutation relations

$$[B_n, B_m^+] = \delta_{n,m}.$$

The strong-coupling approximation is realized when

$$D_{nm}(R) \gg M_{nm}(R).$$

If these inequalities are satisfied, the electronic excitations of the individual molecules will be almost independent of one another and energy operator (4.1) can be replaced by the approximate expression

$$H_a(R) = \sum_{n} [\Delta\varepsilon_n + D_n(R)] B_n^+ B_n + W(R). \quad (4.2)$$

Let R = 0 correspond to the equilibrium positions of the molecules, for which W(R) has a minimum. Then, if we expand the functions $D_{nm}(R)$ and W(R) into series in powers of the deviations from the equilibrium positions and retain the first terms that are functions of R [linear in $D_{nm}(R)$ and quadratic in W(R)], we find,[3] after adding to operator (4.2) the molecular-vibration kinetic-energy operator, the Hamiltonian operator of the electronic vibrations in the crystal and the molecular vibrations.

In particular, if we are concerned with the electronic excited states in which the electronic excitation is concentrated on the molecule located at the point **n**, the Hamiltonian can be written as

$$H = H_{el} + H_{vib} + H_{int} \quad (4.3)$$

Here,

$$H_{el} = E_n B_n^+ B_n,$$

[3]By retaining the linear terms in the expansion of $D_{nm}(R)$ in R we take into account the change in the equilibrium positions relative to which molecular vibrations occur and ignore the change in the normal-vibration frequencies (see Section 3).

where

$$E_{\mathrm{n}} = \Delta\varepsilon_{\mathrm{n}} + D_{\mathrm{n}}(0)$$

is the electronic-excitation energy of the n-th molecule when all of the molecules of the crystal are rigidly fixed in their equilibrium positions R = 0. This energy differs from the excitation energy of a free molecule by the value $D_n(0)$, which takes into account the change in the interaction energy between this molecule and all other molecules of the crystal in its transition to an excited state. The molecular-vibration energy operator

$$H_{\mathrm{vib}} = \sum_{s,\,\mathbf{q}} \Omega_s(\mathbf{q}) \left(b_{s\mathbf{q}}^{+} b_{s\mathbf{q}} + \frac{1}{2} \right); \tag{4.4}$$

where $\Omega_s(\mathbf{q})$ are the normal-vibration frequencies (here and below we use units in which $\hbar = 1$); and $b_{s\mathbf{q}}^{+}$ and $b_{s\mathbf{q}}$ are the creation and annihilation operators for phonons of branch s with the wave vector **q**. They satisfy the commutation relations

$$[b_{s\mathbf{q}},\ b_{s\mathbf{q}}^{+}] = \delta_{ss'}\delta_{\mathbf{q}\mathbf{q}'}, \qquad [b_{s\mathbf{q}},\ b_{s'\mathbf{q}'}] = 0$$

and commute with the electronic-excitation operators B_{n}.

Finally,

$$H_{\mathrm{int}} = B_{\mathrm{n}}^{+} B_{\mathrm{n}} \sum_{s,\,\mathbf{q}} \chi(s,\mathbf{q}) (b_{s\mathbf{q}} + b_{s,\,-\mathbf{q}}^{+}) \tag{4.5}$$

is the interaction operator for electronic excitations with molecular vibrations. Here,

$$\chi(s,\mathbf{q}) = \chi^{*}(s,-\mathbf{q}) = \frac{1}{\sqrt{N}} \sum_{\mathrm{m},\,j}{}' \frac{e_s^{j}(\mathbf{q})}{\sqrt{2I_j\Omega_s(\mathbf{q})}} \left\{ \left[\frac{\partial}{\partial R_{\mathrm{n}}^{j}} + e^{i\mathbf{q}(\mathrm{m}-\mathrm{n})} \frac{\partial}{\partial R_{m}^{j}} \right] D_{\mathrm{nm}}(R) \right\}_{0}.$$

In the Heisenberg representation, the time dependence of the operators B_{n} and $b_{s\mathbf{q}}$ is determined by the system of equations

$$i\frac{dB_{\mathrm{n}}}{dt} = [B_{\mathrm{n}},\ H] = E_{\mathrm{n}}B_{\mathrm{n}} + B_{\mathrm{n}} \sum_{s,\,\mathbf{q}} \chi(s,\mathbf{q}) (b_{s\mathbf{q}} + b_{s,\,-\mathbf{q}}^{+}),$$

$$i\frac{db_{s\mathbf{q}}}{dt} = [b_{s\mathbf{q}},\ H] = \Omega_s(\mathbf{q})\, b_{s\mathbf{q}} + \chi^{*}(s,\mathbf{q})\, B_{\mathrm{n}}^{+}B_{\mathrm{n}}.$$

Hamiltonian (4.3) is unusual in that it diagonalizes for any value of the coupling function $\chi(s, \mathbf{q})$ for electronic excitations with molecular vibrations. This diagonalization is accomplished by converting from the operators B_n and b_{sq} to the new operators A_n and a_{sq}, which are given by the relations

$$B_n = SA_nS^+, \qquad b_{sq} = Sa_{sq}S^+, \tag{4.6}$$

where

$$S = \exp\{\hat{\sigma}A_n^+A_n\}, \tag{4.7}$$

$$\hat{\sigma} = \sum_{s,\,q} \Omega_s^{-1}(q)\,[\chi^*(s,\,q)\,a_{sq}^+ - \chi(s,\,q)\,a_{sq}]. \tag{4.8}$$

Using the rules for unitary transformations of the operators (see mathematical appendix 1), we find

$$\left.\begin{aligned} B_n &= A_n \exp(-\hat{\sigma}),\\ b_{sq} &= a_{sq} - \frac{\chi^*(s,\,q)}{\Omega_s(q)}\,A_n^+A_n. \end{aligned}\right\} \tag{4.9}$$

Since transformations (4.9) are unitary, the new operators satisfy the same commutation relations as the old ones, i.e.,

$$[A_n,\,A_m^+] = \delta_{n,\,m}, \qquad [a_{sq},\,a_{s'q'}^+] = \delta_{ss'}\,\delta_{qq'}, \ldots$$

After substitution of (4.9) into Hamiltonian (4.3), it is diagonalized with respect to the new operators:

$$H = \mathscr{E}_n A_n^+A_n + \sum_{s,\,q} \Omega_s(q)\left(a_{sq}^+a_{sq} + \frac{1}{2}\right), \tag{4.10}$$

where

$$\mathscr{E}_n = \Delta\varepsilon_n + D_n(0) - \sum_{s,\,q} \frac{|\chi(s,\,q)|^2}{\Omega_s(q)} \tag{4.11}$$

is the electronic-excitation energy of a molecule and the lattice-deformation energy around the excited molecule. In the deviation of operator (4.10), it was assumed that the molecule could be in only one excited state, i.e., the eigenvalues of the operators $A_n^+A_n$ are either 0 or 1; therefore, $(A_n^+A_n)^2 = A_n^+A_n$.

It follows from the form of energy operator (4.10) that the new operators A_n and a_{sq} characterize the steady elementary excitations of the crystal, which take into account the lattice deformation in the area of the excited molecule.

The time dependence of the new operators in the Heisenberg representation is determined by the simple expressions

$$\left.\begin{aligned} A_{\mathbf{n}}(t) &= e^{iHt} A_{\mathbf{n}} e^{-iHt} = A_{\mathbf{n}} \exp(-i\mathscr{E}_{\mathbf{n}} t), \\ a_{s\mathbf{q}}(t) &= e^{iHt} a_{s\mathbf{q}} e^{-iHt} = a_{s\mathbf{q}} \exp(-i\Omega_s(\mathbf{q})\, t). \end{aligned}\right\} \quad (4.12)$$

According to the general theory (Section 2, Chapter IV), the dielectric constant of an optically isotropic crystal is expressed by the formula

$$\varepsilon(\omega) = \varepsilon_0 - \frac{4\pi \mathbf{d}^2 \omega_f}{v\omega} G_r(\omega) \quad (4.13)$$

in terms of the Fourier transform of the retarded Green function corresponding to electronic excitations of the molecules. This Green function is defined in terms of the electronic-excitation operators $B_{\mathbf{n}}$ by the relation

$$G_r(t) = -i\Theta(t) \langle\langle [B_{\mathbf{n}}(t), B_{\mathbf{n}}^+(0)] \rangle\rangle, \quad (4.14)$$

where

$$B_{\mathbf{n}}(t) = e^{iHt} B_{\mathbf{n}} e^{-iHt} \quad (4.15)$$

is the Heisenberg representation of the annihilation operator for an excitation on the molecule **n**. Considering (4.9), we can express operator (4.15) in terms of the new operators

$$B_{\mathbf{n}}(t) = e^{-\hat{\sigma}(t)} A_{\mathbf{n}}(0) \exp(-i\mathscr{E}_{\mathbf{n}} t), \quad (4.16)$$

where

$$\hat{\sigma}(t) = \sum_{s,\,\mathbf{q}} \Omega_s^{-1}(\mathbf{q}) \left[\chi^*_{(s,\mathbf{q})} e^{i\Omega_s(\mathbf{q})\, t} a_{s\mathbf{q}}^+ - \chi(s, \mathbf{q}) e^{-i\Omega_s(\mathbf{q})\, t} a_{s\mathbf{q}}\right]. \quad (4.17)$$

If we substitute (4.16) into Green function (4.14), we find

$$G_r(t) = -i\Theta(t) \langle\langle e^{-\hat{\sigma}(t)} e^{\hat{\sigma}(0)} \rangle\rangle \exp(-i\mathscr{E}_{\mathbf{n}} t). \quad (4.18)$$

Considering the explicit form of operator (4.17), we can write

$$\langle\langle e^{\hat{\sigma}(t)} e^{-\hat{\sigma}(0)} \rangle\rangle = \prod_{s,\,\mathbf{q}} \langle\langle M_{s\mathbf{q}} \rangle\rangle, \quad (4.19)$$

where

$$\begin{aligned} M_{s\mathbf{q}} &= \exp\{\alpha^*_{s\mathbf{q}}(t)\, a^+_{s\mathbf{q}} - \alpha_{s\mathbf{q}}(t)\, a_{s\mathbf{q}}\} \exp\{\alpha_{s\mathbf{q}}(0)\, a_{s\mathbf{q}} - \alpha^*_{s\mathbf{q}}(0)\, a^+_{s\mathbf{q}}\}, \quad (4.20) \\ \alpha_{s\mathbf{q}}(t) &= -\Omega_s^{-1}(\mathbf{q})\, \chi(s, \mathbf{q}) \exp\{-i\Omega_s(\mathbf{q})\, t\}. \end{aligned}$$

Using formula (4.12) of the mathematical appendix, we obtain

$$\langle\langle M_{sq}\rangle\rangle = \exp\{\Omega_s^{-2}(\mathbf{q})\,|\,\chi(s,\mathbf{q})\,|^2\,[(\bar{n}_{sq}+1)\,e^{-i\Omega_s(\mathbf{q})\,t}+ \\ + \bar{n}_{sq}e^{i\Omega_s(\mathbf{q})\,t} - (2\bar{n}_{sq}+1)]\}.$$

Inserting the value found into Eq. (4.20) we obtain (4.21)

$$\langle\langle e^{-\hat{\sigma}(t)}e^{\hat{\sigma}(0)}\rangle\rangle = \exp g(t),$$

where

$$g(t) = \sum_{s,\,\mathbf{q}} \Omega_s^{-2}(\mathbf{q})\,|\,\chi(s,\mathbf{q})\,|^2\{(\bar{n}_{sq}+1)\,e^{-i\Omega_s(\mathbf{q})\,t}+ \\ + \bar{n}_{sq}\,e^{i\Omega_s(\mathbf{q})\,t} - (2\bar{n}_{sq}+1)\}. \tag{4.22}$$

The Fourier transform of retarded Green function (4.18), with allowance for (4.21), is given by the integral

$$G_r(\omega) = \int_{-\infty}^{\infty} e^{i\omega t-\eta_i t}G(t)\,dt = -i\int_0^{\infty} \exp\{i(\omega-\mathscr{E}_n+i\eta)\,t+g(t)\}\,dt. \tag{4.23}$$

If we introduce ξ_{sq} by means of the relations

$$\xi_{sq}^2 = 2\Omega_s^{-2}(\mathbf{q})\,|\,\chi(s,\mathbf{q})\,|^2, \tag{4.24}$$

it is easy to see that the Fourier transform of retarded Green function (4.23) is related to the function $\Lambda(\omega)$ of (3.19) by the simple relation

$$G_r(\omega) = -i\Lambda^*(\omega), \tag{4.25}$$

if the normal-vibration frequencies for electronic excitation are not taken into account in the function $\Lambda(\omega)$. If we substitute (4.25) into (4.13) and separate the real and imaginary parts, we obtain in the frequency range $\omega \sim \omega_f$ expressions that coincide with expressions (3.18).

5. Excitons in Thin Crystals

Very thin films (on the order of 0.1 μ) of single crystals are used for experimental study of the dispersion and absorption spectra of crystals in the strong-absorption region. In a study of light absorption in thin films of anthracene single crystals, Brodin [25, 26] found an oscillating dependence of the attenuation factor and refractive index of light waves upon film thickness. To interpret these results, the author [27] has proposed that the discreteness of the exciton energy and wave vector in thin crystal films be taken into account. As was shown in [27], this discreteness causes the intensity of interaction of light with the crystal to vary monotonically with the thickness of the crystal.

The theory of exciton states in thin single-crystal films is still in the initial stage of development. The distinctive features of the interaction of light with thin crystal films can evidently manifest themselves when the thickness of the film is comparable with the wavelength of the light. But with such thin films, the applicability of the conventional macroscopic concepts of the dielectric constant, attenuation factor, and refractive index requires special study.

To determine the interaction of thin crystals with light, let us consider a cubic crystal with the lattice constant a and one isotropic molecule in each unit cell. We shall assume that the crystal extends without bounds along the x and y axes and consists of N_3 molecular layers along the z axis. We shall study the possible energies of the exciton states of such a crystal, assuming that the molecules are rigidly fixed at the lattice points $\mathbf{n} = \sum_{i=1}^{3} n_i \mathbf{a}_i$, where n_i are integers. In films with a large area, the choice of the boundary conditions in the plane xy is not important. Therefore, we can use cyclicity conditions with the large periods $N_1\mathbf{a}_1$ and $N_2\mathbf{a}_2$, where N_1 and $N_2 \gg N_3$. The boundary conditions along the z axis come down to the requirement that molecules be absent for $n_3 \leq 0$ and $n_3 \geq N_3 + 1$. The total number of molecules in the region of the crystal in question is $N_1N_2N_3$.

The energy operator for the excited states of the crystal in the Heitler–London approximation in the second-quantization representation, according to (III.2.21), with allowance for only one molecular excited state, has the form

$$\Delta H = (\Delta\varepsilon + D)\sum_{\mathbf{n}} B_{\mathbf{n}}^{+}B_{\mathbf{n}} + \sum_{\mathbf{n},\mathbf{m}}{}' M_{\mathbf{nm}}B_{\mathbf{m}}^{+}B_{\mathbf{n}}, \tag{5.1}$$

where $\Delta\varepsilon$ is the molecular excitation energy and D is the difference between the interaction energies of an excited and unexcited molecule with all the remaining molecules of the crystal. For molecules that do not have a constant dipole moment, it is sufficient to take into account only interaction with the nearest molecules when calculating D. We shall assume, therefore, that D is not a function of the location of the excited molecule and take it out of the summation. The value M_{nm} is the matrix element of the excitation transfer between the molecules **n** and **m**. In infinite crystals, this matrix element is a function only of the difference **n** − **m**. Let us assume that this property is retained in a thin film. The operators B_n^+ and B_n are the creation and annihilation operators for an excitation in the molecule **n**. These operators satisfy the commutation relations

$$B_m B_n^+ - B_n^+ B_m = \delta_{nm}. \tag{5.2}$$

To find the energy of the electronic excitations described by Hamiltonian (5.1), this operator must be diagonalized. We can see that diagonalization of operator (5.1) is accomplished by canonical transformation of the operators B_n to the new operators $B(\mathbf{k}_\perp, k_3)$ by means of the unitary transformation

$$B_n = \sum_{k_3} \sum_{\mathbf{k}_\perp} B(\mathbf{k}_\perp, k_3)\, u_n(\mathbf{k}_\perp, k_3), \tag{5.3}$$

where

$$\left.\begin{aligned} u_n(\mathbf{k}_\perp, k_3) &= \sqrt{\frac{2}{N}}\, e^{i\mathbf{k}_\perp \mathbf{n}_\perp} \sin(k_3 n_3 a), \\ N &= N_1 N_2 (N_3 + 1), \quad n_3 = 1, 2, \ldots, N_3, \end{aligned}\right\} \tag{5.4}$$

$$\left.\begin{aligned} k_3 &= \frac{2\pi\nu_3}{a(N_3+1)}, \quad \nu_3 = 1, 2, \ldots, N_3, \\ \mathbf{k}_\perp &= \frac{2\pi}{a^2}[\nu_1 \mathbf{a}_1 + \nu_2 \mathbf{a}_2], \quad -\frac{N_i}{2} \leqslant \nu_i < \frac{N_i}{2}, \quad i = 1, 2 \end{aligned}\right\} \tag{5.5}$$

is the wave vector perpendicular to the axis $\mathbf{a}_3$; and

$$\mathbf{n}_\perp = \mathbf{a}_1 n_1 + \mathbf{a}_2 n_2, \quad n_i = 0, \pm 1, \ldots, +\frac{N_i}{2}; \quad i = 1, 2.$$

The functions of transformation (5.3) satisfy the orthonormality conditions

$$\left.\begin{aligned} &\sum_{n} u_n^*(\mathbf{k}'_\perp, k'_3)\, u_n(\mathbf{k}_\perp, k_3) = \delta_{\mathbf{k}_\perp \mathbf{k}'_\perp} \delta_{k_3 k'_3}, \\ &\sum_{k_3, \mathbf{k}_\perp} u_n^*(\mathbf{k}_\perp, k_3)\, u_m(\mathbf{k}_\perp, k_3) = \delta_{nm}. \end{aligned}\right\} \tag{5.6}$$

Substituting expression (5.3) into operator (5.1) and taking (5.6) into account, we find

$$\Delta H_{ex} = \sum_{k_3,\, \mathbf{k}_\perp} E(\mathbf{k}_\perp, k_3) B^+(\mathbf{k}_\perp, k_3) B(\mathbf{k}_\perp, k_3), \tag{5.7}$$

where

$$E(\mathbf{k}_\perp, k_3) = \Delta\varepsilon + D + L(\mathbf{k}_\perp, k_3), \tag{5.8}$$

$$L(\mathbf{k}_\perp, k_3) = \sum_{\mathbf{n},\,\mathbf{m}}{}' u^*_{\mathbf{m}}(\mathbf{k}_\perp, k_3) M_{\mathbf{nm}} u_{\mathbf{n}}(\mathbf{k}_\perp, k_3). \tag{5.9}$$

Using conditions (5.6), with the aid of (5.2) we can show that the new operators $B(\mathbf{k}_\perp, k_3)$ satisfy the commutation relations

$$B(\mathbf{k}_\perp, k_3) B^+(\mathbf{k}'_\perp, k'_3) - B^+(\mathbf{k}'_\perp, k'_3) B(\mathbf{k}_\perp, k_3) = \delta_{\mathbf{k}_\perp \mathbf{k}'_\perp} \delta_{k_3 k'_3}.$$

Therefore, the operators $B^+(\mathbf{k}_\perp, k_3)$ are the creation operators for collective electronic excitations, which are characterized by the vector $\mathbf{k}_\perp$, the number k_3, and energy (5.8). The operators $B(\mathbf{k}_\perp, k_3)$ annihilate the corresponding states. The operator $B^+(\mathbf{k}_\perp, k_3) B(\mathbf{k}_\perp, k_3)$ commutes with excitation-energy operator (5.7) and is the operator of the number of excited states $\mathbf{k}_\perp$, k_3. The eigenvalues of this operator are 0, 1, 2, ..., Λ.

The elementary-excitation energies $E(\mathbf{k}_\perp, k_3)$ form a quasi-continuous band with $N_1 N_2 N_3$ sublevels. Owing to the inequality N_1 and $N_2 \gg N_3$, these sublevels are divided into N_3 groups, which differ in k_3. Each group (fixed k_3) has $N_1 N_2$ closely situated sub-levels.

In the coordinate representation, the wave functions of the elementary excitations k_3, $\mathbf{k}_\perp$ have the form

$$\Phi(k_3, \mathbf{k}_\perp) = \sqrt{\frac{2}{N}} \sum_{\mathbf{n}} e^{i\mathbf{k}_\perp \mathbf{n}_\perp} \sin(k_3 n_3 a)\, \psi_{\mathbf{n}}, \tag{5.10}$$

where

$$\psi_{\mathbf{n}} = \varphi^f_{\mathbf{n}} \prod_{\mathbf{m}(\neq \mathbf{n})} \varphi^0_{\mathbf{m}}$$

are the wave functions that represent the crystal states in which the molecule at the point **n** is excited while the remaining molecules are in the ground state. When $\mathbf{k}_\perp = 0$ and $k_3 = \pi\nu_3/a(N_3+1)$, wave functions (5.10) are standing waves with $(\nu_3 - 1)$ nodes.

According to (5.10), the elementary excitations $\mathbf{k}_\perp$ and k_3 have a specific projection of the quasi-momentum $\hbar\mathbf{k}_\perp$ perpendicular to the axis $\mathbf{a}_3$. The quasi-momentum component does not have a definite value along the axis $\mathbf{a}_3$. Each state with a definite k_3 is a superposition of two states with the momentum projections $\pm\hbar k_3$. These elementary excitations embrace the entire region of the crystal, and we shall call them excitons, although $\hbar k_3$ does not have a definite value.

In the coordinate representation, the interaction operator for the crystal and a transverse electromagnetic field with the strength

$$\mathbf{E}(\mathbf{r}, t) = \mathbf{E}_0 \sin(\mathbf{Q}\mathbf{r})\, e^{-i\omega t} + \text{Hermitian conj.}, \qquad \mathbf{Q} \parallel \mathbf{a}_3, \quad \mathbf{E}_0\mathbf{Q} = 0, \tag{5.11}$$

can be written as

$$V(t) = we^{-i\omega t} + \text{Hermitian conj.}, \tag{5.12}$$

where

$$w = -\frac{ie}{m\omega c}(\mathbf{E}_0\hat{\mathbf{p}}_n)\sin(Qn_3a);$$

and $\hat{\mathbf{p}}_n$ is the momentum operator of all of the electrons of the molecule. In the occupation-number representation, this operator has the form

$$\hat{\mathbf{p}}_n = \{\langle\varphi_f|\,\hat{\mathbf{p}}_n\,|\varphi_0\rangle B_n^+ + \langle\varphi_0|\,\hat{\mathbf{p}}_n\,|\varphi_f\rangle B_n\}.$$

Since

$$\langle\varphi_f|\,\hat{\mathbf{p}}_n\,|\varphi_0\rangle = im\omega_f\langle\varphi_f|\,\mathbf{r}\,|\varphi_0\rangle,$$

expression (5.12) is transformed to

$$w = \frac{\omega_f}{\omega}(\mathbf{E}_0\mathbf{d})\sum_n (B_n^+ - B_n)\sin(Qn_3a), \tag{5.13}$$

where $\mathbf{d} = \langle\varphi^f|\,e\mathbf{r}\,|\varphi^0\rangle$ is the electric dipole moment of a quantum transition in the molecule. Owing to the assumed isotropy of the molecules, the vector $\mathbf{d}$ is parallel to the field vector $\mathbf{E}_0$. Using transformation (5.3), we find interaction operator (5.13) in the form

$$w = \frac{\omega_f}{\omega}\sqrt{\frac{N}{2}} \sum_{\mathbf{k}_\perp, k_3} \{B^+(\mathbf{k}_\perp, k_3) - B(-\mathbf{k}_\perp, k_3)\}\, \delta_{\mathbf{k}_\perp 0} \Delta(k_3 - Q), \quad (5.14)$$

where

$$\Delta(x-y) = \frac{\cos\{(N_3+1)(x-y)a/2\}\sin\{N_3(x-y)a/2\}}{(N_3+1)\sin\{(x-y)a/2\}} \quad (5.15)$$

is a function introduced in [27]. It is equal to unity when $x = y = 2\pi\nu_3/a(N_3+1)$ and vanishes when $|x-y| = 2\pi/aN_3$. With a further increase in the difference $|x-y|$, function (5.15) periodically reaches values not exceeding N_3^{-1}. When $N_3 \to \infty$, this function reduces to the Kronecker symbol $\delta_{k_3,Q}$.

Thus, according to (5.14), the operator is a nonmonotonic function of the film thickness. The interaction operator for a light wave with a thin crystal has its maximum value when the wave number Q of the wave

$$Q = k_3 \equiv \frac{2\pi\nu_3}{a(N_3+1)} \quad \text{where} \quad \nu_3 = 1, 2, \ldots, N_3. \quad (5.16)$$

In large crystals ($N_3 \gg 1$), the discreteness of N_3 in (5.16) is not of great importance, since adjacent k_3 differ little from one another. At small N_3, the discreteness of k_3 is very considerable. But at very small N_3, the sharp maximum of function (5.15), which determines the resonance with respect to the wave numbers, is smoothed.

Below we shall consider not very thin ($N_3 \geq 100$) single-crystal films for which the macroscopic concept of the dielectric constant can still be used. Let us calculate the dielectric constant of such a crystal using the retarded Green function method discussed in Section 2 of Chapter IV.

The operator of the specific electric moment of the crystal in the area of point n in the coordinate representation has the form

$$\mathbf{P}(\mathbf{n}) = \frac{e\mathbf{r}_\mathbf{n}}{v},$$

where v is the volume of a unit cell. In the Heisenberg representation and in the exciton occupation-number representation, taking (5.3) into account, it is transformed to [see similar transformations

(IV.2.17) for infinite crystals]

$$\widetilde{\mathbf{P}}(\mathbf{n}, t) = \frac{\mathbf{d}}{v}\sqrt{\frac{2}{N}} \sum_{k_3 \mathbf{k}_\perp} \{B^+(-\mathbf{k}_\perp, k_3; t) + B(\mathbf{k}_\perp, k_3; t)\} e^{i\mathbf{k}_\perp \mathbf{n}_\perp} \sin(k_3 n_3 a). \quad (5.17)$$

The mean value of specific-electric-moment operator (5.17) can be calculated by the Kubo formula [28, 29]

$$\langle \mathbf{P}(\mathbf{n}, t)\rangle = \mathrm{Sp}\,\{\widetilde{\rho}(t)\widetilde{\mathbf{P}}(\mathbf{n}, t)\}, \quad (5.18)$$

where

$$\widetilde{\rho}(t) = \rho_0 - i\int_{-\infty}^{t} [\widetilde{V}(\tau), \rho_0]\, d\tau;$$

$$\widetilde{V}(\tau) = \frac{(\mathbf{E}_0\mathbf{d})\,\omega_f}{\omega}\sqrt{\frac{N}{2}} \sum_{k_3 \mathbf{k}_\perp} \{B^+(-\mathbf{k}_\perp, k_3; \tau) - B(-\mathbf{k}_\perp, k_3; \tau)\}\, \delta_{\mathbf{k}_\perp 0}\, \Delta(k_3 - Q); \quad (5.19)$$

and ρ_0 is the density matrix, which determines the crystal states without a field at temperature T. In formulas (5.17) and (5.19) (at $\hbar = 1$)

$$B(\mathbf{k}_\perp, k_3, t) = e^{iHt} B(\mathbf{k}_\perp, k_3)\, e^{-iHt} \quad (5.20)$$

is the Heisenberg representation of the exciton operator; and H is the total Hamiltonian of the system, which includes the interaction of excitons with lattice-vibration phonons.

If we substitute (5.17) and (5.19) into (5.18), we find the mean value of the specific polarization produced in the crystal by wave (5.11) with allowance for only one excited state:

$$\langle \mathbf{P}(\mathbf{n}, t)\rangle = \frac{\mathbf{E}_0\mathbf{d}^2\omega_f}{\omega v} \sum_{k_3} \{G_r^+(k_3, -\omega) - G_r(k_3, \omega)\} \times \Delta(k_3 - Q)\sin(k_3 n_3 a)\, e^{-i\omega t} + \text{Hermitian conj}. \quad (5.21)$$

where $G_r(k_3, \omega)$ are the Fourier components of the retarded Green functions, which are defined by the relations

$$G_r(k_3, \omega) = \int_{-\infty}^{\infty} G_r(k_3, \tau)\, e^{i\omega\tau}\, d\tau, \quad (5.22)$$

$$G_r(k_3, \tau) = -i\Theta(\tau) \langle\langle [B(0k_3; \tau), B^+(0k_3; 0)] \rangle\rangle,$$

where

$$\langle\!\langle \ldots \rangle\!\rangle \equiv \mathrm{Sp}\,\{\rho_0 \ldots\}.$$

It follows from (5.21) that the specific polarizability of the film α (for normal light incidence), which is given by the relation

$$\langle \mathbf{P}(\mathbf{n}, t)\rangle = \alpha \mathbf{E}_0 \sin(Qn_3a)\, e^{-i\omega t} + \text{Hermitian conj.},$$

is equal to

$$\alpha = \frac{d^2\omega_f}{\omega v} \sum_{k_3=1}^{N_3} \{G_r^+(k_3, -\omega) - G_r(k_3, \omega)\}\, \Delta(k_3 - Q)\, \frac{\sin(k_3 n_3 a)}{\sin(Q n_3 a)}. \quad (5.23)$$

The corresponding component of the dielectric-constant tensor is $\varepsilon(Q, \omega) = \varepsilon_0 + 4\pi\alpha$, where ε_0 is the dielectric constant, as determined by all of the electronic states of the crystal except for the exciton states taken into account in (5.23).

At absolute zero, ignoring molecular vibrations, retarded Green function (5.22) is easily calculated (see the similar calculations in Section 4, Chapter IV). We obtain

$$G_r^0(k_3, \omega) = [\omega - E(0, k_3) + i\eta]^{-1}, \quad (5.24)$$

where $E(0, k_3)$ is exciton energy (5.8) at $\mathbf{k}_\perp = 0$. If we substitute (5.24) into (5.23) and take into consideration that $G_r^+(k, -\omega) \ll G_r(k, \omega)$ and $\omega_f \approx \omega$, we obtain

$$\alpha(Q, \omega) = \frac{d^2}{v} \sum_{k_3=1}^{N_3} \frac{\Delta(k_3 - Q)\,\dfrac{\sin(k_3 n_3 a)}{\sin(Q n_3 a)}}{E(0, k_3) - \omega + i\eta}. \quad (5.25)$$

If we introduce the attenuation factor $\gamma = 2\eta$, this expression coincides with accuracy to the factor $\sin(k_3 n_3 a)/\sin(Q n_3 a)$ with the expression found in [27] and later in [30]. According to (5.25), at fixed $\mathbf{Q}$ and variable N_3 (or, vice versa, at fixed N_3 and variable $\mathbf{Q}$), the polarizability α does not vary monotonically.

Taking lattice vibrations into account, at nonzero temperature exciton Green function (5.24) is replaced by the Green function

$$G_r(k_3, \omega) = [\omega - E(0, k_3) - M(0, k_3)]^{-1},$$

where $M(0, k_3)$ is the mass operator for excitons that interact with phonons. Under certain simplifying assumptions, the imaginary part of this operator

$$\frac{1}{2}\Upsilon(k_3) = -\operatorname{Im} M(0, k_3)$$

was calculated by Haug [30].

6. Elementary Theory of the Urbach Rule

In a study of the absorption coefficient of light in silver halide crystals, in 1953 Urbach [31] found the empirical formula

$$\varkappa(\omega) = \varkappa(\omega_0) \exp\left(-\frac{\sigma(\omega_0 - \omega)}{kT}\right), \tag{6.1}$$

which determines at $\omega < \omega_0$ the dependence of the absorption coefficient upon the photon energy[4] ω and the temperature of the crystal. When the parameter σ was close to unity, formula (6.1) gave a good description of the longwave edge of the absorption band over a wide range of frequencies and temperatures. Further, it was found that formula (6.1) gave a remarkably accurate description of the variation (by 4-6 orders of magnitude) of the absorption coefficient when there was a frequency change at the longwave edges of the first fundamental (excitons and local excitations) and impurity absorption bands for many crystals.

To explain the temperature dependence of the absorption coefficient, it is sometimes necessary to make the parameter σ slightly dependent upon temperature. For example, Matsui has shown [32] that formula (6.1) gives a good description in the range of 10^5–0.1 cm^{-1} of the absorption coefficient of light in anthracene if at temperatures of 79°, 200°, 253°, 293°, and 346°, σ is made equal to 0.727, 1.361, 1.500, 1.539, and 1.637, respectively, for the first absorption band in the b component of the spectrum and 0.630, 1.238, 1.400, 1.472, and 1.533, respectively, for the first band of the a component. The slight dependence of the parameter σ upon temperature was first noted by Mahr [33]. References to more recent experimental work can be found in Knox' book [34].

Numerous attempts have been made (see, for example, [35-39]) to give a theoretical explanation of formula (6.1), which is called

[4] In units in which $\hbar = 1$.

the Urbach rule. At the present time, however, there is no satisfactory theory for this rule. In [34], Knox remarks that the proof of the Urbach rule remains one of the important unsolved problems of the theory.

In this section, we shall present a simple theory [40] of the Urbach rule.[5] The main idea of the calculation is that the longwave edge of the absorption band and, therefore, the Urbach rule are dependent upon quantum transitions from the vibration sublevels of the crystal lattice to the level of the first electronic excitation. In our opinion, the lack of success of the previous theories in explaining the Urbach rule is due the fact that their authors attempted to relate this rule to the change in the first electronic-excitation band of the crystal. For example, in order to explain the Urbach rule for the longwave part of the exciton absorption band, they attempted to find exciton states whose levels were considerably below the energy of an exciton produced in direct transition. And in [39], an attempt was made to explain the longwave edge of the absorption band by introducing the concept of polaritons (mixed exciton-phonon states). They found very weak absorption that extended below the edge of the exciton absorption band. But the frequency dependence that they found differed considerably from that predicted by the Urbach rule.

Let us assume that the crystal contains a small quantity of identical impurity centers that do not have internal vibration sublevels. Let $\psi_0(\mathrm{r})$, $\psi_f(\mathrm{r})$, E_0, and E_f be the wave functions and energies of the ground and electronic excited states of an impurity molecule. If we ignore the change in the normal lattice-vibration frequencies when the impurity molecule is excited, the states of the crystal with the energies

$$E_0 + \sum_s (n_s + 1/2)\,\Omega_s \quad \text{and} \quad E_f + \sum_s (n_s' + 1/2)\,\Omega_s \tag{6.2}$$

will correspond in an adiabatic approximation to the wave functions

$$|0, \{n_s\}\rangle = \psi_0(r) \prod_s \varphi_{n_s}(\xi_s), \tag{6.3}$$

[5] A more rigorous theory is presented in [41].

$$|f, \{n_s'\}\rangle = \psi_f(r) \prod_s \varphi_{n_s'}(\xi_s - \xi_{s0}), \tag{6.4}$$

where $\{n_s\} = (\ldots n_s \ldots)$ is the set of quantum numbers n_s of normal lattice vibrations; $n_s\Omega_s$ and $\varphi_{n_s}(\xi_s)$ are the energy and wave function of the s-th normal lattice vibration with an unexcited impurity, which is characterized by the quantum number $n_s = 0, 1, \ldots$; $\varphi_{n_s}(\xi_s - \xi_{s0})$ is the wave function of the s-th normal lattice vibration with the impurity in the f-th electronic state; and ξ_{s0} characterizes the equilibrium-position displacements of the lattice vibrations when the impurity molecule is excited. They determine the coupling of an electronic excitation with lattice vibrations.

The longwave edge of the absorption band is determined by the quantum transition, under the influence of a light wave, from states (6.3) to the phononless[6] excited state $|f, \{0_s\}\rangle$. In the dipole approximation, these transitions are characterized by the matrix elements of the electric-dipole-moment operator of the impurity molecule, which, when (6.3) and (6.4) are taken into account, are

$$\langle f, \{0_s\} | e\mathbf{r} | 0, \{n_s\}\rangle = \mathbf{d}_{f0} \prod_s M_{n_s, 0_s}, \tag{6.5}$$

where $\mathbf{d}_{f0}$ is the vector of the electric dipole transition in the impurity molecule; and

$$M_{n_s, 0_s} = \int \varphi_{n_s}(\xi_s - \xi_{s0}) \varphi_{0s}(\xi_s) d\xi_s$$

are the overlap integrals of the wave functions of the s-th normal lattice vibration, which pertain to different electronic states of the impurity molecule. Here, with accuracy to the squares of the small displacements ξ_{s0}, we have

$$M^2_{n_s, 0_s} = \begin{cases} 1 - {}^1/_2\, \xi^2_{s0}, & \text{if } n_s = 0; \\ {}^1/_2\, \xi^2_{s0}, & \text{if } n_s = 1; \\ 0, & \text{if } n_s \neq 0, 1. \end{cases} \tag{6.6}$$

At temperature T, the probability that the state of the crystal is characterized by function (6.3) is proportional to

$$\exp\left(-\frac{\sum_s n_s \Omega_s}{kT}\right).$$

[6]Other excited states were taken into account in [41].

Therefore, the absorption coefficient for light of frequency ω, which is determined by the quantum transitions from all possible initial states (6.3) to one final state, can be written as

$$\varkappa(\omega) = A \sum_{\{n_s\}} |\langle f, \{0_s\} | er | 0, \{n_s\} \rangle|^2 \exp\left\{-\frac{\sum_s n_s\Omega_s}{kT}\right\} \delta\left(\omega_0 - \sum_s n_s(\Omega_s - \omega)\right),$$

where $\omega_0 = E_f - E_0$. If we substitute (6.5) into this expression and use the delta function

$$\delta(x) = \mathrm{Re}\, \frac{1}{\pi} \int_0^\infty e^{ix\mu}\, d\mu,$$

we find

$$\varkappa(\omega) = \frac{2A\,|\mathbf{d}_{f_0}|^2}{\pi} \mathrm{Re} \int_0^\infty D(\mu)\, e^{i\mu(\omega_0 - \omega)}\, d\mu, \tag{6.7}$$

where, taking (6.6) into account,

$$D(\mu) = \sum_{\{n_s\}} \prod_s M^2_{n_s, 0_s} \exp\left\{-\left(i\mu + \frac{1}{kT}\right) \sum_s n_s\Omega_s\right\} \approx$$

$$\approx e^{-a} \sum_{m=0}^{\infty} \frac{1}{m!} \left\{\sum_s \frac{1}{2} \xi^2_{s0} \exp\left[-\left(i\mu + \frac{1}{kT}\right)\Omega_s\right]\right\}^m; \quad a = \frac{1}{2} \sum_s \xi^2_{s0}. \tag{6.8}$$

Expression (6.8) is calculated most simply for an Einstein model of a crystal with one optical branch: $\omega_s = \omega_0$. In this case,

$$D(\mu) = e^{-a} \sum_m \frac{1}{m!} \left(\frac{1}{2} \sum_s \xi^2_{s0}\right)^m \exp\left[-\left(i\mu + \frac{1}{kT}\right) m\Omega_0\right]. \tag{6.9}$$

Substituting this into (6.7), we find

$$\varkappa(\omega) = \varkappa(\omega_0) \frac{a^n}{n!} \delta(\omega_0 - \omega - n\Omega_0)\, e^{-\frac{\omega_0 - \omega}{kT}}, \tag{6.10}$$

where n is an integer. Using the Stirling formula $\ln n! \approx n \ln n$ (at $n \gg 1$), we can transform (6.10) to

$$\varkappa(\omega) = B\delta(\omega_0 - \omega - n\Omega_0)\, e^{-\frac{(\omega_0 - \omega)}{kT}(1+\zeta)}, \tag{6.11}$$

where

$$\zeta = \frac{kT}{\Omega_0} \ln \frac{\omega_0 - \omega}{a\Omega_0}. \qquad (6.12)$$

The function ζ when $\omega_0 - \omega > a\Omega_0$ is slightly dependent upon the phonon energy ω, and over a wide range of ω it can be considered a parameter of the theory that is proportional to temperature, if we ignore the temperature variations of ω_0, Ω_0, and ξ_{s0}.

The absorption coefficient given by formula (6.11) differs from zero for the discrete values $\omega = \omega_0 - n\Omega_0$. If we allow for the width γ of the excited states, the delta function can be replaced by

$$\gamma\,[(\omega_0 - \omega - n\Omega_0)^2 + \gamma^2]^{-1}$$

which has maxima $1/\gamma$ at $\omega = \omega_0 - n\Omega_0$. In this case, at sufficiently small Ω_0 the absorption coefficient is a continuous function of ω with closely situated maxima, the dependence of which upon frequency coincides with the Urbach rule if we let $(1 + \zeta) = \sigma$.

Let us consider a crystal model with two optical branches $\Omega_1 = \Omega_{s1}$ and $\Omega_2 = \Omega_{s2}$. If we introduce

$$a_1 = \frac{1}{2} \sum_{s_1} \xi^2_{s_1 0} \quad \text{and} \quad a_2 = \frac{1}{2} \sum_{s_2} \xi^2_{s_2 0},$$

then expression (6.8) is transformed to

$$D(\mu) = e^{-a} \sum_{m,\,l} \frac{a_1^{m-l} a_2^{l}}{(m-l)!\,l!} \exp\left\{-\left(i\mu + \frac{1}{kT}\right)[(m-l)\,\Omega_1 + l\Omega_2]\right\}.$$

If we substitute this into (6.7), we obtain an expression for the absorption coefficient:

$$\varkappa(\omega) = B\delta\,(\omega_0 - \omega - (n-l)\,\Omega_1 - l\Omega_2)\, e^{-\frac{(\omega_0 - \omega)}{kT}(1+\zeta)}, \qquad (6.13)$$

which differs from zero for $\omega = \omega_0 - (n - l)\Omega_1 - l\,\Omega_2$ at any integral n and l. Here,

$$\zeta = \frac{kT}{\omega - \omega_0} \ln \sum_{l,\,n} \frac{a_1^{n-l} a_2^{l}}{(n-l)!\,l!},$$

where summation is over all l and n that satisfy the conditions

$$(n-l)\Omega_1 + l\Omega_2 = \omega_0 - \omega, \quad l \leqslant n.$$

Formula (6.13) extends to any number q of optical-vibration branches. If we let the frequencies Ω_i correspond to the parameters $a_i = \frac{1}{2}\sum_i \xi^2_{s_i 0}$, then

$$\zeta = \frac{kT}{\omega - \omega_0} \ln \sum_{l_1 l_2 \ldots l_q} \frac{a_1^{(l_1-l_2)} a_2^{(l_2-l_3)} \ldots a_q^{l_q}}{(l_1-l_2)!\,(l_2-l_3)!\ldots l_q!}.$$

In this case, the absorption coefficient is given by

$$\varkappa(\omega) = B\delta(\omega_0 - \omega - [l_1 - l_2]\,\Omega_1 - [l_2 - l_3]\,\Omega_2 - \ldots$$
$$\ldots - l_q\Omega_q)\, e^{-\frac{(\omega_i - \omega)}{kT}(1+\zeta)}. \qquad (6.14)$$

With allowance for the width of the excited states and a large number of small phonon frequencies, formula (6.13) becomes

$$\varkappa(\omega) = \varkappa(\omega_0)\, e^{-\frac{(\omega_0 - \omega)}{kT}(1+\zeta)},$$

which coincides with the Urbach rule (6.1).

Now let us consider local centers with internal vibrations. If the internal vibrations of an impurity molecule in the l-th electronic state are characterized by the energies ν_p (at $\nu_0 = 0$) and the wave functions $\Phi^l_p(R)$, $l = 0, f$; $p = 0, 1, \ldots$, then the absorption coefficient for $\omega < \omega_0$ can be written as

$$\varkappa(\omega) = \sum_p A\,|\langle f, 0|0, \nu_p\rangle|^2 \sum_{\{n_s\}} |\langle f, \{0_s\}|\mathbf{er}|0, \{n_s\}\rangle|^2 \times$$
$$\times\delta\Big(\omega_0 - \nu_p - \sum_s n_s\Omega_s - \omega\Big) \exp\left\{-\frac{\nu_p + \sum_s n_s\Omega_s}{kT}\right\}, \qquad (6.15)$$

where

$$\langle f, 0|0, \nu_p\rangle = \int \Phi_0^{*f}(R)\,\Phi_p^0(R)\,dR$$

is the overlap integral of the wave functions of the intramolecular vibrations.

If we convert in (6.15) to the integral representation of the delta function and substitute (6.5) and (6.6), we find

$$\varkappa(\omega) = \frac{2A}{\pi}\sum_p |\mathbf{d}_{f0}|^2\,|\langle f, 0|0, \nu_p\rangle|^2\, \mathrm{Re}\int_0^\infty D_p(\mu)\, e^{i\mu(\omega_f - \omega)}\, d\mu, \qquad (6.16)$$

where

$$D_p(\mu) = \exp\left(-a - \left[i\mu + \frac{1}{kT}\right]\nu_p\right)\sum_m \frac{1}{m!}\left\{\frac{1}{2}\sum_s \xi_{s0}^2 e^{-\left[i\mu + \frac{1}{kT}\right]\Omega_s}\right\}^m. \quad (6.17)$$

For an Einstein model ($\Omega_s = \Omega_0$) this expression takes the form

$$D_p(\mu) = \exp\left\{-\left(i\mu + \frac{1}{kT}\right)\nu_p - a\right\}\sum_m \frac{a^m}{m!} e^{-\left[i\mu + \frac{1}{kT}\right]\mu\Omega_0}. \quad (6.18)$$

After substituting (6.18) into (6.16) and integrating, we obtain

$$\varkappa(\omega) = B\sum_p |\langle f, 0|0, \nu_p\rangle|^2 e^{-\frac{\nu_p}{kT}}\delta(\omega_0 - \nu_p - \omega - n\Omega_0)e^{-\frac{\omega_0 - \nu_p - \omega}{kT}(1+\zeta)}, \quad (6.19)$$

where

$$\zeta = \frac{kT}{\Omega_0}\ln\frac{\omega_0 - \nu_p - \omega}{\Omega_0}.$$

It follows from (6.19) that when $\nu_1 \ll kT$, the first term ($p = 0$) plays the leading role in the sum. When $\omega_0 - \omega > \nu_1$ and $kT > \nu_1$, the main role in absorption when $\langle f, 0|0, \nu_1\rangle \neq 0$ is taken over by the second term ($p = 1$). This rule reflects qualitatively the experimentally observed regularity [34].

Thus, according to these elementary calculations, the Urbach rule must be satisfied in condensed media in which there is a large number of phonon states (interacting with an electronic excitation) that are thermally excited at a given temperature in accordance with Boltzmann's law. Because of this, the nature of the electronic excitation is not of great importance. All that is necessary is coupling of the electronic excitation with phonons that are associated with the quasi-continuous spectrum of lattice vibrations.

Appendix

1. Unitary Transformation of the Operators

Let F(M) be a function of the operator M. Let us consider the transformation

$$F' = SF(M)S^{+} \qquad (1.1)$$

by means of the unitary operator

$$S = \exp(iL), \qquad SS^{+} = 1, \qquad (1.2)$$

where L is a Hermitian operator, i.e., an operator that satisfies the operator equation

$$L = L^{+}.$$

If the function F(M) can be expanded into a series in powers of M, i.e.,

$$F(M) = \sum_{n=0}^{\infty} a_n \boldsymbol{M}^n, \qquad (1.3)$$

then

$$SF(M)S^{+} = F(SMS^{+}). \qquad (1.4)$$

Proof. According to (1.3), we have

$$SF(M)S^{+} = \sum_{n} a_n (SMS^{+})^n = F(SMS^{+}).$$

It follows from (1.4) that to determine the transformed function F' it is sufficient to find the transformation SMS^+ of the operator M. To calculate the transformation SMS^+, we introduce the auxiliary operator

$$S(x) = \exp(ixL), \tag{1.5}$$

where x is a real variable that varies from 0 to 1. From definitions (1.5) and (1.2) it follows that

$$S(0) = 1, \quad S(1) = S. \tag{1.6}$$

If we differentiate both sides of

$$M'(x) = S(x)\, MS^+(x)$$

with respect to x, we obtain, taking definition (1.2) into account, the differential equation

$$\frac{dM'(x)}{dx} = iS(x)\,[L, M]\,S^+(x).$$

Solving this equation under boundary conditions (1.6), we find

$$SMS^+ = M + i\int_0^1 S(x)\,[L, M]\,S^+(x)\,dx. \tag{1.7}$$

In particular, if

$$[L, M] = \lambda, \qquad [L, \lambda] = 0,$$

then expression (1.7) is considerably simplified:

$$SMS^+ = M + i\,[L, M]. \tag{1.8}$$

2. The Weyl Identity Operator[1]

If the commutator of the two operators L and M commutes with each of them, then

$$\exp(\alpha L)\exp(\gamma M) = \exp\{\gamma M + \alpha\gamma\,[L, M]\}\exp(\alpha L), \tag{2.1}$$

where α and γ are any operators that commute with the operators L and M.

To prove identity (2.1), we differentiate the equation

$$\Lambda(x) = \exp(\alpha L x)\, M \exp(-\alpha L x). \tag{2.2}$$

[1]H. Weyl, Z. Phys., 46:1 (1928); V. S. Nanda, Indian J. Phys., 24:181 (1950).

with respect to the real variable x. Integrating the obtained equation

$$\frac{d\Lambda(x)}{dx} = \alpha\, e^{\alpha L x}\,[L, M]\, e^{-\alpha L x}$$

with respect to x from 0 to 1 and taking (2.2) into account, we find

$$e^{\alpha L} M e^{-\alpha L} = M + \alpha\,[L, M]. \tag{2.3}$$

Having multiplied both sides of (2.3) by γ, we transform it to

$$e^{\alpha L}\,\gamma M = (\gamma M + \alpha\gamma\,[L, M])\, e^{\alpha L}. \tag{2.4}$$

Then, by induction, we find

$$e^{\alpha L}\,(\gamma M)^n = (\gamma M + \alpha\gamma\,[L,\ M])^n e^{\alpha L},$$

from which follows the desired identity (2.1).

In the particular case when the operators L and M coincide with the creation b_s^+ and annihilation b_s operators of particles in the state s, which satisfy the commutation relations

$$[b_s, b_{s'}^+] = \delta_{ss'}, \qquad [b_s, b_{s'}] = 0,$$

it follows from identity (2.1) that

$$e^{\alpha b_s} e^{\gamma b_{s'}^+} = \begin{cases} e^{\gamma b_{s'}^+} e^{\alpha b_s}, & \text{if } s \neq s'; \\ e^{\gamma b_s^+} e^{\alpha\gamma} e^{\alpha b_s}, & \text{if } s = s'. \end{cases} \tag{2.5}$$

If we apply identity operator (2.1) to the operator product $e^{(\alpha L+\beta M)} e^{-\alpha L}$, we find

$$e^{(\alpha L+\beta M)}\, e^{-\alpha L}\, e^{\frac{1}{2}\alpha\beta\,[M,\, L]} = e^{-\alpha L}\, e^{(\alpha L+\beta M)}\, e^{-\frac{1}{2}\alpha\beta\,[M,\, L]}.$$

After multiplying both sides of this equation by the operator we obtain a new identity operator, which, taking identity (2.1) into account, can be written as

$$e^{-\beta M}\, e^{(\alpha L+\beta M)}\, e^{-\alpha L}\, e^{\frac{1}{2}\alpha\beta\,[M,\, L]} = e^{-\alpha L}\, e^{(\alpha L+\beta M)}\, e^{-\beta M}\, e^{-\frac{1}{2}\alpha\beta\,[M,\, L]}. \tag{2.6}$$

From this equation and identity (2.1) it follows that the operator

$$\Lambda(\alpha, \beta) \equiv e^{-\beta M}\, e^{(\alpha L+\beta M)}\, e^{-\alpha L}\, e^{\frac{1}{2}\alpha\beta\,[M,\, L]} \tag{2.7}$$

commutes with the operators L and M and is independent of α and β. When $\alpha = \beta = 0$, we find directly from (2.7) that the operator Λ is equal to 1, and from the independence of Λ upon α and β it follows that the equation

$$\Lambda(\alpha, \beta) = 1 \tag{2.8}$$

is satisfied identically.

Identity (2.8) can be rewritten as

$$e^{(\alpha L+\beta M)} = e^{\beta M} e^{\alpha L} e^{-\frac{1}{2}\alpha\beta [L, M]}, \tag{2.9}$$

which is called the **Weyl identity**.

In the particular case when the operators L and M coincide with the creation b_s^+ and annihilation b_s operators for particles in state s, the Weyl identity takes the form

$$\exp(\alpha b_s^+ + \beta b_s) = \exp(\beta b_s) \exp(\alpha b_s^+) \exp\left(-\frac{1}{2}\alpha\beta\right). \tag{2.10}$$

Using (2.1), we can also write

$$\exp(\alpha b_s^+ + \beta b_s) = \exp(\alpha b_s^+) \exp(\beta b_s) \exp\left(\frac{1}{2}\alpha\beta\right). \tag{2.11}$$

3. Calculation of the Mean Values of the Bose Operators

The wave function $|n_s\rangle$ of the state of a system in which there are n_s quanta in the state s is expressed in terms of the creation operators b_s^+ and the ground-state function $|0\rangle$ by the formula

$$|n_s\rangle = \frac{1}{\sqrt{n_s!}} (b_s^+)^{n_s} |0\rangle. \tag{3.1}$$

The matrix elements of the operators $(b_s^+)^{m_s}$ are given by

$$\langle n'_{s'} | (b_s^+)^{m_s} | n_s\rangle = \delta_{ss'} \delta_{n'_s, m_s+n_s} \sqrt{\frac{(n_s+m_s)!}{n_s!}}. \tag{3.2}$$

Using Eq. (3.2), we find the mean values in the state $|n_s\rangle$ of the operator products

$$\langle n_s | (b_s)^{m_s} (b_s^+)^{m'_s} | n_s\rangle = \delta_{m_s m'_s} \frac{(n_s+m_s)!}{n_s!}, \tag{3.3}$$

$$\langle n_s|\,(b_s^+)^{m_s}\,(b_s)^{m_s'}\,|n_s\rangle = \delta_{m_s m_s'}\frac{n_s!}{(n_s-m_s)!}\,. \tag{3.4}$$

Expanding the exponential functions into series and using (3.3) and (3.4), we obtain

$$\langle n_s|\,e^{\alpha b_s}e^{\beta b_s^+}|\,n_s\rangle = \sum_{m_s=0}^{\infty}\frac{(\alpha\beta)^{m_s}(n_s+m_s)!}{(m_s!)^2\,n_s!}, \tag{3.5}$$

$$\langle n_s|\,e^{\beta b_s^+}\,e^{\alpha b_s}|\,n_s\rangle = \sum_{m_s=0}^{\infty}\frac{(\alpha\beta)^{m_s}\,n_s!}{(m_s!)^2\,(n_s-m_s)!}\,. \tag{3.6}$$

4. The Statistical Averages of the Phonon Operators

The phonon states in a crystal that is at temperature T are determined by the density matrix

$$\rho = \frac{\exp(-\beta H)}{\mathrm{Sp}\,\{\exp(-\beta H)\}}, \tag{4.1}$$

where $\beta = (kT)^{-1}$, and

$$H = \sum_s \left(b_s^+ b_s + \frac{1}{2}\right)\Omega_s \tag{4.2}$$

is the Hamiltonian (at $\hbar = 1$) of the system. The statistical average of any operator A_s that acts on phonons that are in the state s (for example, the operators b_s^+ or b_s) is given by the equation

$$\langle\!\langle A_s\rangle\!\rangle = \mathrm{Sp}\,(\rho A_s). \tag{4.3}$$

If the complete set of eigenfunctions $|n_s\rangle$ of operator (4.2) is used in calculating the spurs in Eq. (4.3), then (4.3) takes the form

$$\langle\!\langle A_s\rangle\!\rangle = (1-e^{-\beta\Omega_s})\sum_{n_s=0}^{\infty}\langle n_s|\,A_s\,|n_s\rangle\, e^{-\beta n_s\Omega_s}. \tag{4.4}$$

Let us apply Eq. (4.4) to particular cases.

a) Let the operator A_s coincide with the phonon-number operator $\hat{n} = b_s^+ b_s$. Then $\langle n_s|\,\hat{n}_s\,|n_s\rangle = n_s$ and (4.4) reduces to

$$\bar{n}_s \equiv \langle\!\langle \hat{n}_s\rangle\!\rangle = (1-e^{-\beta\Omega_s})\sum_{n_s=0}^{\infty} n_s e^{-\beta n_s\Omega_s} = (e^{\beta\Omega_s}-1)^{-1}. \tag{4.5}$$

It follows from (4.5) that

$$\bar{n}_s + 1 = \langle\langle b_s b_s^+ \rangle\rangle = (1 - e^{-\beta\Omega_s})^{-1}. \tag{4.6}$$

b) Let

$$A_s = e^{\alpha b_s} e^{\gamma b_s^+}.$$

In this case, substituting (3.5) into (4.4), we find

$$\langle\langle e^{\alpha b_s} e^{\gamma b_s^+} \rangle\rangle = (1 - e^{-\beta\Omega_s}) \sum_{n_s, m_s} \frac{(\alpha\gamma)^{m_s} (n + m_s)! e^{-\beta n_s \Omega_s}}{(m_s!)^2 n_s!}. \tag{4.7}$$

Considering the equation

$$\sum_{n=0}^{\infty} \frac{(n+m)!}{n!\, m!} e^{-\beta n} = \left(\sum_{n=0}^{\infty} e^{-\beta n} \right)^{m+1} = (1 - e^{-\beta})^{-m-1}$$

and Eq. (4.6), we can write

$$(1 - e^{-\beta\Omega_s}) \sum_{n_s=0}^{\infty} \frac{(n_s + m_s)!}{n_s!\, m_s!} e^{-\beta n_s \Omega_s} = (\bar{n}_s + 1)^{m_s}.$$

If we substitute this value into (4.7), we find the final expression

$$\langle\langle e^{\alpha b_s} e^{\gamma b_s^+} \rangle\rangle = \sum_{m_s=0}^{\infty} \frac{(\alpha\gamma)^{m_s}}{m_s!} (\bar{n}_s + 1)^{m_s} = \exp\{\alpha\gamma(\bar{n}_s + 1)\}. \tag{4.8}$$

Taking (4.6) into account, this expression can also be written as

$$\langle\langle e^{\alpha b_s} e^{\gamma b_s^+} \rangle\rangle = \exp\{\alpha\gamma \langle\langle b_s b_s^+ \rangle\rangle\}. \tag{4.9}$$

Thus, we can prove the equation

$$\langle\langle e^{\gamma b_s^+} e^{\alpha b_s} \rangle\rangle = \exp\{\alpha\gamma \langle\langle b_s^+ b_s \rangle\rangle\} = \exp\{\alpha\gamma \bar{n}_s\}. \tag{4.10}$$

c) Let

$$A_s = \exp[\alpha^* b_s^+ - \alpha b_s] \exp[\gamma b_s - \gamma^* b_s^+]. \tag{4.11}$$

Applying Weyl identity (2.10), we obtain

$$A_s = e^{-\alpha b_s} e^{\alpha^* b_s^+} e^{\gamma b_s} e^{-\gamma^* b_s^+} \exp\left\{\frac{1}{2}|\alpha|^2 + \frac{1}{2}|\gamma|^2\right\}.$$

If, with the aid of identity (2.1), we interchange the second and third terms of the cofactor, we obtain

$$A_s = e^{(\gamma-\alpha)\, b_s}\, e^{(\alpha^*-\gamma^*)\, b_s^+} \exp\left\{\frac{1}{2}|\alpha|^2 + \frac{1}{2}|\gamma|^2 - \alpha^*\gamma\right\}.$$

Considering that, according to (4.8),

$$\langle\!\langle e^{(\gamma-\alpha)\, b_s}\, e^{(\alpha^*-\gamma^*)\, b_s^+}\rangle\!\rangle = \exp\{(\bar{n}_s + 1)\,[\alpha\gamma^* + \gamma\alpha^* - |\alpha|^2 - |\gamma|^2]\},$$

we obtain the final expression

$$\langle\!\langle \exp\,[\alpha^* b_s^+ - \alpha b_s]\exp\,[\gamma b_s - \gamma^* b_s^+]\rangle\!\rangle =$$

$$= \exp\left\{\alpha\gamma^*(\bar{n}_s + 1) + \gamma\alpha^*\bar{n}_s - (|\alpha|^2 + |\gamma|^2)\left(\bar{n}_s + \frac{1}{2}\right)\right\}. \qquad (4.12)$$

References

Chapter I

1. V. M. Agranovich and V. L. Ginzburg, Crystal Optics with Allowance for Spatial Dispersion and the Theory of Excitons [in Russian], Nauka, Moscow (1965).
2. L. D. Landau and E. M. Lifshits, Electrodynamics of Continuous Media [in Russian], Gostekhizdat, Moscow (1957).
3. R. S. Knox, Theory of Excitons, Academic Press, New York (1963).
4. S. I. Pekar, ZhÉTF, 36:451 (1959); 33:1022 (1957).
5. V. M. Agranovich and V. L. Ginzburg, UFN, 76:643 (1962); 77:663 (1962).
6. M. Born and M. Heppert-Meier, Theory of Solids [Russian translation], ONTI, Moscow (1938).
7. M. Born and K. Huang, Dynamical Theory of Crystal Lattices, Oxford (1954).
8. V. L. Ginzburg, A. A. Rukhadze, and V. P. Silin, FTT, 3:1835, 2890 (1961).
9. W. Heitler, Quantum Theory of Radiation, Oxford (1954).
10. Ya. I. Frenkel', Phys. Rev., 37:17 (1931).

Chapter II

1. Ya.I.Frenkel', Phys. Rev., 37(17):1276 (1931); Phys. Z. Sowjetunion, 9:158 (1936).
2. Ya. I. Frenkel', Sow. Phys., 9:158 (1936).
3. R. E. Peierls, Ann. Phys., 13(5):905 (1932).
4. G. H. Wannier, Phys. Rev., 52:191 (1937).
5. C. Slater and W. Shockley, Phys. Rev., 50:705 (1936).
6. W. R. Heller and A. Marcus, Phys. Rev., 84:809 (1951).
7. A. S. Davydov, ZhÉTF, 18:210 (1948).
8. A. S. Davydov, Theory of Molecular Excitons, McGraw-Hill, New York (1962).
9. A. S. Davydov, UFN, 82:393 (1964).
10. A. F. Prikhot'ko, ZhÉTF, 19:383 (1949).
11. V. L. Broude, V. S. Medvedev, and A. F. Prikhot'ko, ZhÉTF, 21:673 (1951).
12. D. S. McClure and O. Schnepp, J. Chem. Phys., 23:1375 (1955).

13. D. P. Craig and P. C. Hobbins, J. Chem. Soc., 539:2309 (1955).
14. D. S. McClure, Solid State Phys., 8:1 (1959).
15. H. C. Wolf, Solid State Phys., 9:1 (1959).
16. N. F. Mott, Trans. Faraday Soc., 34:500 (1938).
17. G. Dresselhaus, Phys. Chem. Solids, 1:14 (1956).
18. R. I. Elliott, Phys. Rev., 108:1384 (1957).
19. S. A. Moskalenko and K. B. Tolpygo, ZhÉTF, 36:149 (1959).
20. R. S. Knox, Phys. Chem. Solids, 9:238, 265 (1959).
21. E. F. Gross and N. A. Karryev, DAN SSSR, 84:471 (1952).
22. V. P. Zhuze and S. M. Ryvkin, Izv. AN SSSR, Ser. Fiz., 16:93 (1952).
23. L. Apker and E. Taft, Phys. Rev., 81:698 (1951); 82:814 (1951).
24. D. G. Thomas and J. J. Hopfield, Phys. Rev. Letters, 5:505 (1960); Phys. Rev., 124:657 (1961).
25. G. Baldini, Phys. Rev., 128:1562 (1962).
26. R. S. Knox, Radiation Research, 20:77 (1963).
27. W. R. Heller and A. Marcus, Phys. Rev., 84:809 (1951).
28. U. Fano, Phys. Rev., 103:1202 (1956); 118:451 (1960).
29. S. I. Pekar, ZhÉTF, 35:522 (1958).
30. J. J. Hopfield and D. G. Thomas, Phys. Chem. Solids, 12:276 (1960).
31. V. M. Agranovich and V. L. Ginzburg, Crystal Optics with Allowance for Spatial Dispersion and the Theory of Excitons, Nauka, Moscow (1965).
32. D. Fox and S. Yatsiv, Phys. Rev., 108:938 (1957).
33. A. S. Davydov and E. F. Sheka, Phys. Stat. Solidi, 11:877 (1965).
34. A. S. Davydov and É. N. Myasnikov, DAN SSSR, 173:1040 (1967).
35. A. S. Davydov, ZhÉTF, 21:673 (1951).
36. A. S. Davydov, in: Recollections of S. I. Vavilov, Izd. AN SSSR, Moscow (1952), p. 210.
37. H. Winston, J. Chem. Phys., 19:156 (1951).
38. D. S. McClure, J. Chem. Phys., 22:1256 (1954).
39. L. E. Lyons, J. Chem. Phys., 23:1973 (1955).
40. D. P. Craig, J. Chem. Soc., 2302 (1955).
41. J. Tanaka, Progr. Theor. Phys., Suppl. 12:183 (1959).
42. H. Bethe, Ann. Phys., 3:133 (1929).
43. A. S. Davydov, Quantum Mechanics, Fizmatgiz, Moscow (1963).
44. A. I. Kitaigorodskii, Organic Crystal Chemistry, Consultants Bureau, New York (1961).
45. D. Fox and O. Schnepp, J. Chem. Phys., 23:767 (1955).
46. D. P. Craig and I. R. Walsh, J. Chem. Phys., 24:471 (1956); 25:588 (1956); J. Chem. Soc., 1913 (1958).
47. L. C. Brand and T. N. Goodwin, Trans. Faraday Soc., 53:295 (1957).
48. D. L. Peterson and W. T. Simpson, J. Am. Chem. Soc., 78:2375 (1957).
49. J. Tanaka, J. Chem. Soc. Japan, 79:1373 (1958).
50. D. P. Craig, L. E. Lyons, S. H. Walmley, and I. R. Walsh, Proc. Chem. Soc., 389 (1959).
51. D. P. Craig and I. R. Walsh, J. Chem. Soc., 1613 (1958).
52. R. Silbey, J. Jortner, and S. A. Rice, J. Chem. Phys., 42:1515 (1965).

53. P. P. Ewald, Ann. Phys., 64: 253 (1921).
54. M. H. Cohen and F. Keffer, Phys. Rev., 99: 1128 (1955).
55. M. S. Brodin and S. V. Marisova, Optika i Spektroskopiya, 10: 473 (1961).
56. A. F. Lubchenko and É. I. Rashba, Optika i Spektroskopiya, 4: 580 (1958).
57. G. Ya. Lyubarskii, Group Theory and Its Applications in Physics, Moscow (1957).
58. M. Hamermesh, Group Theory and Its Application to Physical Problems, Addison-Wesley, Reading, Mass. (1962).
59. E. Wigner, Group Theory and Its Application to Quantum Mechanics of Atomic Spectra, Academic Press, New York (1959).
60. G. F. Koster, Space Groups and Their Representations, Academic Press, New York (1964).
61. H. Eyring, D. Walter, and D. Kimball, Quantum Chemistry, Wiley, New York (1944).
62. L. P. Bouckaert, R. Smoluchowski, and E. Wigner, Phys. Rev., 50: 58 (1936).
63. J. W. Sidman, Phys. Rev., 102: 96 (1956).
64. V. L. Broude, UFN, 74: 577 (1961).
65. M. D. Borisov, Trudy IF AN USSR, 7:102 (1953).
66. A. Bree and L. E. Lyons, J. Chem. Soc., 2662 (1956).
67. M. S. Brodin and S. V. Marisova, Optika i Spektroskopiya, 19: 235 (1965).
68. I. Ferguson and W. G. Schneider, J. Chem. Phys., 28: 761 (1958).
69. M. S. Brodin and A. F. Prikhot'ko, Optika i Spektroskopiya, 7: 132 (1959).
70. T. A. Claxton, D. P. Craig, and T. Thirunamachahdran, J. Chem. Phys., 35: 1525 (1961).
71. A. Yu. Éichis, Trudy IF AN USSR, 5:137 (1954).
72. A. F. Prkhot'ko and A. F. Skorobogat'ko, Optika i Spektroskopiya, 20: 65 (1956).
73. A. F. Prikhot'ko and A. F. Skorobogat'ko, Ukr. Fiz. Zh., 10: 350 (1965).
74. S. V. Marisova, Optika i Spektroskopiya, 22: 566 (1967).
75. A. S. Davydov, Trudy IF AN SSSR, 3:36 (1952).
76. A. S. Davydov and É. I. Rashba, Ukr. Fiz. Zh., 3: 226 (1957).
77. A. S. Davydov and A. F. Lubchenko, ZhÉTF, 35: 1499 (1958).
78. Y. Toyozawa, Progr. Theor. Phys., 20: 53 (1958); 27: 89 (1962).
79. I. M. Robertson, Proc. Roy. Soc., A142: 684 (1933); D. M. Crushnack, Acta Cryst., 10: 504 (1957).
80. D. S. McClure, J. Chem. Phys., 22: 1668 (1954).
81. A. F. Prikhot'ko, Sow. Phys., 9: 34 (1936); ZhÉTF, 19: 383 (1949).
82. M. S. Soskin, Ukr. Fiz. Zh., 6: 806 (1961); 7: 27 (1962).
83. A. F. Prikhot'ko and M. S. Soskin, Optika i Spektroskopiya, 12: 101 (1962).
84. E. F. Sheka, Optika i Spektroskopiya, 10:684 (1961).
85. A. S. Davydov, ZhÉTF, 21: 673 (1951).
86. V. L. Broude and A. F. Prikhot'ko, ZhÉTF, 22: 605 (1952).
87. V. L. Broude, V. S. Medvedev, and A. F. Prikhot'ko, ZhÉTF, 2: 317 (1957).
88. E. G. Cox and J. Smith, Nature, 173: 75 (1954).
89. V. L. Broude and M. N. Onoprienko, Optika i Spektroskopiya, 10: 634 (1961).
90. I. W. Sidman and D. S. McClure, J. Chem. Phys., 24: 757 (1956).
91. H. C. Wolf, Z. Phys., 145: 166 (1956).
92. V. M. Agranovich, Izv. AN SSSR, Ser. Fiz., 23: 40 (1959).

93. V. M. Agranovich and A. N. Faidysh, Optika i Spektroskopiya, 1:885 (1956).
94. A. N. Faidysh, DAN SSSR, 6:215 (1955).
95. M. Trlifay, Czech. J. Phys., 6:6 (1956).
96. V. M. Agranovich, Optika i Spektroskopiya, 3:29 (1957).
97. V. M. Agranovich, Optika i Spektroskopiya, 3:84 (1957).
98. V. M. Agranovich, Optika i Spektroskopiya, 4:586 (1958); Izv. AN SSSR, Ser. Fiz., 23:40 (1959).
99. V. M. Agranovich and Yu. V. Konobeev, Optika i Spektroskopiya, 6:242, 648 (1959); 11:369 (1961).
100. Yu. V. Konobeev, Luminescence. Collection I of Optika i Spektroskopiya (1963), p. 135.
101. E. Lipsett and A. Dekker, Canad. J. Phys., 30:165 (1951).
102. V. N. Vishnevskii and M. D. Borisov, Ukr. Fiz. Zh., 1:17, 30 (1958).
103. A. F. Prikhot'ko and M. T. Shpak, Optika i Spektroskopiya, 4:17 (1958).
104. M. T. Shpak and E. F. Sheka, Izv. AN SSSR, Ser. Fiz., 9:57 (1960).
105. M. T. Shpak and E. F. Sheka, Optika i Spektroskopiya, 8:66 (1960).
106. M. T. Shpak and E. F. Sheka, Optika i Spektroskopiya, 9:57 (1960).
107. V. L. Broude, E. F. Sheka, and M. T. Shpak, Luminescence. Collection I of Optika i Spektroskopiya (1963), p. 102.
108. A. F. Prikhot'ko and N. Yu. Fugol', Optika i Spektroskopiya, 4:335 (1958).
109. P. W. Alexander, A. R. Lacey, and L. E. Lyons, J. Chem. Phys., 34:2200 (1961).
110. M. T. Shpak and N. I. Sheremet, Optika i Spektroskopiya, 14:816 (1963).

Chapter III

1. V. M. Agranovich, ZhÉTF, 37:430 (1959).
2. J. J. Hopfield, Phys. Rev., 112:1555 (1958).
3. A. A. Demidenko, FTT, 3:1164 (1961).
4. U. Fano, Phys. Rev., 103:1202 (1956).
5. V. M. Agranovich, FTT, 3:811 (1961).
6. D. P. Craig, J. Chem. Soc., 2302 (1955).
7. D. P. Craig and P. O. Hobbins, J. Chem. Soc., 539, 2309 (1955).
8. W. Heitler, Quantum Theory of Radiation, Oxford (1954).
9. M. Born and K. Huang, Dynamical Theory of Crystal Lattices, Oxford University Press, New York and London (1954).
10. K. B. Tolpygo, ZhÉTF, 20:497 (1950).
11. Kun Huang, Proc. Roy. Soc., A208:352 (1951).
12. V. M. Agranovich and A. A. Rukhadze, ZhÉTF, 35:982 (1958).
13. V. M. Agranovich and M. N. Kaganov, FTT, 4:1681 (1962).
14. J. J. Hopfield and D. G. Thomas, J. Phys. Chem. Solids, 12:276 (1960).
15. S. M. Neamtan, Phys. Rev., 92:1362 (1953); 94:327 (1954).
16. J. J. Hopfield, Phys. Rev., 112:1555 (1958).
17. V. M. Agranovich and Yu. V. Konobeev, FTT, 3:360 (1961).
18. A. S. Davydov, Quantum Mechanics [in Russian], Fizmatgiz, Moscow (1963).
19. M. M. Bogolyubov, Lectures in Quantum Statistics [in Ukrainian], Rad. Shkola, Kiev (1949).
20. R. S. Knox, Theory of Excitons, Academic Press, New York (1963).
21. S. I. Pekar, ZhÉTF, 33:1022 (1957).

Chapter IV

1. A. S. Davydov, Trudy IF AN SSSR, 3:36 (1952).
2. A. S. Davydov, Phys. Stat. Solidi, 20:143 (1967).
3. V. L. Bonch-Bruevich and S. V. Tyablikov, The Green Function Method in Statistical Mechanics, Fizmatgiz, Moscow (1961).
4. V. M. Agranovich and Yu. V. Konobeev, FTT, 6:831 (1964).
5. A. I. Ansel'm and Yu. D. Firsov, ZhÉTF, 28:151 (1955); 30:719 (1956).
6. Y. Toyozawa, Progr. Theor. Phys., 20:53 (1958).
7. A. S. Davydov, Quantum Mechanics [in Russian], Fizmatgiz, Moscow (1963).
8. D. N. Zubarev, UFN, 71:71 (1960).
9. H. L. Lehmann, Nuovo Cimento, 11:342 (1954).
10. L. D. Landau and E. M. Lifshits, Electrodynamics of Continuous Media [in Russian], Gostekhizdat, Moscow (1957).
11. M. M. Bogolyubov, Lectures in Quantum Statistics [in Ukrainian], Kiev (1949).
12. L. D. Landau and E. M. Lifshits, Statistical Physics [in Russian], Gostekhizdat, Moscow (1951).
13. S. A. Moskalenko, P. I. Khadzhi, A. I. Bobrysheva, and A. V. Lelyakov, ZhÉTF, 45:1189 (1963).
14. A. A. Abrikosov, L. P. Gor'kov, and I. E. Dzyaloshinskii, Quantum Field Theory Methods in Statistical Physics [in Russian], Fizmatgiz, Moscow (1962).
15. V. M. Galitskii and A. B. Migdal, ZhÉTF, 7:96 (1958).
16. P. C. Martin and J. Schwinger, Phys. Rev., 115:1342 (1959).
17. T. Matsubara, Progr. Theor. Phys., 14:351 (1955).
18. A. A. Abrikosov, L. P. Gor'kov, and I. E. Dzyaloshinskii, ZhÉTF, 36:900 (1959).
19. E. S. Fradkin, ZhÉTF, 36:1286 (1959).
20. N. N. Bogolyubov and S. V. Tyablikov, DAN SSSR, 126:53 (1959).
21. V. A. Onishchuk, Ukr. Fiz. Zh., 7:10 (1968).
22. A. S. Davydov and V. A. Onishchuk, Phys. Stat. Solidi, 24:373 (1967).
23. U. Fano, Phys. Rev., 103:1202 (1956).
24. J. J. Hopfield, Phys. Rev., 112:1555 (1958).
25. V. M. Agranovich, ZhÉTF, 37:430 (1959).
26. V. M. Agranovich, UFN, 71:141 (1960).
27. V. M. Agranovich and Yu. V. Konobeev, FTT, 3:360 (1961).
28. M. Born and K. Huang, Dynamical Theory of Crystal Lattices, Oxford (1954).
29. V. M. Agranovich and V. L. Ginzburg, UFN, 76:643 (1962), 77:663 (1962); Fortschr. Phys., 11:163 (1963).
30. V. L. Ginzburg, A. A. Rukhadze, and V. P. Silin, FTT, 3:1835, 2890 (1961); Phys. Chem. Solids, 23:85 (1962).
31. V. M. Agranovich and V. L. Ginzburg, Crystal Optics with Allowance for Spatial Dispersion and the Theory of Excitons, Nauka, Moscow (1965).
32. S. I. Pekar, ZhÉTF, 33:1022 (1957); UFN, 77:309 (1962).
33. I. E. Dzyaloshinskii and L. P. Pitaevskii, ZhÉTF, 36:1797 (1959).

Chapter V

1. A. S. Davydov, Trudy IF AN SSSR, 3:36 (1952).
2. A. S. Davydov and É. I. Rashba, Ukr. Fiz. Zh., 2:226 (1957).

3. A. S. Davydov and A. F. Lubchenko, ZhÉTF, 35:1499 (1958).
4. Y. Toyozawa, Progr. Theor. Phys., 20:53 (1958).
5. R. E. Merifield, J. Chem. Phys., 28:647 (1958); 36:2519 (1962); 40:445 (1964).
6. A. Suna, Phys. Rev., 135A:111 (1964).
7. A. S. Davydov and B. M. Nitsovich, FTT, 9:2230 (1967).
8. É. I. Rashba, Optika i Spektroskopiya, 2:88 (1957).
9. A. B. Migdal, ZhÉTF, 34:1438 (1958).
10. V. M. Agranovich and Yu. V. Konobeev, FTT, 6:831 (1964).
11. A. S. Davydov and E. F. Sheka, Phys. Stat. Solidi, 11:877 (1965).
12. A. S. Davydov and É. N. Myasnikov, DAN SSSR, 171:1069 (1966); Phys. Stat. Solidi, 20:153 (1967).
13. Y. Toyozawa, Progr. Theor. Phys., 27:89 (1962).
14. K. Huang and R. Rhys, Proc. Roy. Soc., 204A:406 (1950).
15. A. S. Davydov, ZhÉTF, 24:197 (1953).
16. A. S. Davydov, Fiz. Sb. Kievskogo Universiteta, 7:5 (1955).
17. A. S. Davydov and A. F. Lubchenko, Ukr. Fiz. Zh., 1:5, 15, 111 (1956).
18. A. F. Lubchenko, Izv. AN SSSR, Ser. Fiz., 18:718 (1954).
19. A. F. Lubchenko, Ukr. Fiz. Zh., 1:120 (1956).
20. A. F. Lubchenko, Ukr. Fiz. Zh., 1:256 (1956).
21. T. A. Ratner and G. E. Zil'berman, FTT, 1:1697 (1957); 4:687 (1961).
22. Yu. E. Perlin, UFN, 53:553 (1963).
23. Ya. I. Frenkel', Phys. Rev., 37:17 (1931).
24. S. I. Pekar, Studies in the Electronic Theory of Crystals [in Russian], Gostekhizdat, Moscow (1951).
25. M. S. Brodin and A. F. Prikhot'ko, Optika i Spektroskopiya, 7:132 (1959).
26. M. S. Brodin and S. I. Pekar, ZhÉTF, 38:74 (1960).
27. A. S. Davydov, ZhÉTF, 45:723 (1964).
28. R. Kubo and K. Tomita, J. Phys. Soc. Japan, 9:888 (1954).
29. R. Kubo, J. Phys. Soc. Japan, 12:570 (1957).
30. A. Haug, Phys. Rev., 147:612 (1966).
31. F. Urbach, Phys. Rev., 92:1324 (1953).
32. A. Matsui, J. Phys. Soc. Japan, 21:2212 (1966).
33. H. Mahr, Phys. Rev., 125:1510 (1962).
34. R. S. Knox, Theory of Excitons, Academic Press, New York (1963).
35. D. L. Dexter, Nuovo Cimento, Suppl. 7:245 (1958).
36. Y. Toyozawa, Progr. Theor. Phys., 20:53 (1958); 22:455 (1959); Suppl. 12:111 (1959).
37. G. D. Maham, Phys. Rev., 145:602 (1966).
38. K. Thomas, Phys. Rev., 144:582 (1966).
39. V. M. Agranovich and Yu. V. Konobeev, FTT, 3:360 (1961).
40. A. S. Davydov, Ukr. Fiz. Zh., 8:13 (1968); Phys. Stat. Solidi, 27:51 (1968).
41. A. S. Davydov and A. F. Lubchenko, DAN SSSR, 179:1301 (1968); "Theory of the form of the absorption band when localized excitations are produced in a crystal," Preprint ITF 68-17, Kiev (1968).

Index